AutoCAD Plant 3D 2014 for Designers

CADCIM Technologies
525 St. Andrews Drive
Schererville, IN 46375, USA
(www.cadcim.com)

Contributing Author

Sham Tickoo
Professor
Department of Mechanical Engineering Technology
Purdue University Calumet
Hammond, Indiana
USA

CADCIM Technologies

AutoCAD Plant 3D 2014 for Designers
Sham Tickoo

Published by CADCIM Technologies, 525 St Andrews Drive, Schererville, IN 46375 USA.

ISBN 978-1-936646-15-9

www.cadcim.com

DEDICATION

*To teachers, who make it possible to disseminate knowledge
to enlighten the young and curious minds
of our future generations*

*To students, who are dedicated to learning new technologies
and making the world a better place to live in*

THANKS

*To the faculty and students of the MET department of
Purdue University Calumet for their cooperation*

To employees at CADCIM Technologies for their valuable help

Online Training Program Offered by CADCIM Technologies

CADCIM Technologies provides effective and affordable virtual online training on various software packages including Computer Aided Design and Manufacturing (CAD/CAM), computer programming languages, animation, architecture, and GIS. The training is delivered 'live' via Internet at any time, any place, and at any pace to individuals, as well as the students of colleges, universities, and CAD/CAM training centers. The main features of this program are:

Training for Students and Companies in a Classroom Setting

Highly experienced instructors and qualified Engineers at CADCIM Technologies conduct the classes under the guidance of Prof. Sham Tickoo of Purdue University Calumet, USA. This team has authored several textbooks that are rated "one of the best" in their categories and are used in various colleges, universities, and training centers in North America, Europe, and in other parts of the world.

Training for Individuals

CADCIM Technologies with its cost effective and time initiative strives to deliver the training in the comfort of your home or work place, thereby relieving you from the hassles of traveling to training centers.

Training Offered on Software Packages

CADCIM provides basic and advanced training on the following software packages:

CAD/CAM/CAE*: CATIA, Pro/ENGINEER Wildfire, Creo Parametric, SolidWorks, Autodesk Inventor, Solid Edge, NX, AutoCAD, AutoCAD LT, Customizing AutoCAD, AutoCAD Electrical, EdgeCAM, and ANSYS*

Computer Programming*: C++, VB.NET, Oracle, AJAX, and Java*

Animation and Styling*: Autodesk 3ds Max, 3ds Max Design, Maya, and Alias Design*

Architecture and GIS*: Autodesk Revit Architecture, AutoCAD Civil 3D, Autodesk Revit Structures, and AutoCAD Map 3D*

For more information, please visit the following link:
http://www.cadcim.com

Note
If you are a faculty member, you can register by clicking on the following link to access the teaching resources: ***http://www.cadcim.com/Registration.aspx***. The student resources are available at ***http://www.cadcim.com***. We also provide **Live Virtual Online Training** on various software packages. For more information, write us at ***sales@cadcim.com***.

Table of Contents

Chapter 2: Creating Projects and P&IDs

Chapter 3: Creating Structures

Chapter 4: Creating Equipment

Chapter 5: Editing Specifications and Catalogs

Chapter 6: Routing Pipes

Chapter 7: Adding Valves, Fittings, and Pipe Supports

Chapter 8: Creating Isometric Drawings

Chapter 9: Creating Orthographic Drawings

Chapter 10: Managing Data and Generating reports

Preface

AutoCAD PLANT 3D

AutoCAD Plant 3D 2014, a product of Autodesk Corp., is one of the world's leading applications that is designed specifically to create and modify P&IDs and Plant 3D models. This software is designed built to carry out the design process in a project oriented manner. After designing the plant model, you can generate and share isometric drawings, orthographic drawing, and materials reports.

The **AutoCAD Plant 3D 2014 for Designers** textbook explains the readers to effectively use the designing tools in AutoCAD Plant 3D. Also, the chapters are structured in a pedagogical sequence that makes this textbook very effective in learning the features and capabilities of the software. The accompanying tutorials and exercises, which relate to the real world projects, help you understand the usage and abilities of the tools available in AutoCAD Plant 3D 2014. Each of these tutorials and exercises, though complete in themselves will be step toward accomplishing the larger projects.

The salient features of this textbook are as follows:

- **Tutorial Approach**

  The author has adopted the tutorial point-of-view and the learn-by-doing theme throughout the textbook. This approach guides the users understand the concepts and processes easily.

- **Tips and Notes**

  The additional information related to various topics is provided to the users in the form of tips and notes.

- **Learning Objectives**

  The first page of every chapter summarizes the topics that are covered in that chapter.

- **Self-Evaluation Test, Review Questions, and Exercises**

  Every chapter ends with Self-Evaluation Test so that the users can assess their knowledge of the chapter. The answers to the Self-Evaluation Test are given at the end of the chapter. Also, the Review Questions and Exercises are given at the end of each chapter and they can be used by the Instructors as test questions and exercises.

• **Heavily Illustrated Text**

  The text in this book is heavily illustrated with about 600 line diagrams and screen capture images.

Naming Conventions Used in the Text
Tool

If you click on an item in a panel of the **Ribbon** and a command is invoked to create/edit an object or perform some action, then that item is termed as **tool**.

For example:
To Create: **Route Pipe** tool, **Create Supports** tool, **Create Ortho View** tool
To Edit: **Modify Equipment** tool, **Attach Supports** tool, **Detach Supports** tool
Action: **Zoom** tool, **Move** tool, **Copy** tool

If you click on an item in a panel of the **Ribbon** and a dialog box is invoked wherein you can set the properties to create/edit an object, then that item is also termed as **tool**, refer to Figure 1.

For example:
To Create: **Create Equipment** tool, **Create Supports** too
To Edit: **Edit Attributes** tool, **Block Editor** tool

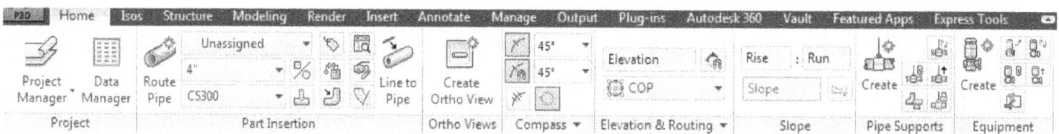

Figure 1 *Various tools in the* **Ribbon**

Button

If you click on an item in a toolbar or a panel of the **Ribbon** and the display of the corresponding object is toggled on/off, then that item is termed as **Button**. For example, **Grid** button, **Snap** button, **Ortho** button, **Properties** button, **Tool Palettes** button, and so on; refer to Figure 2.

Figure 2 *Various buttons displayed in the* **Status Bar** *and* **Ribbon**

The item in a dialog box that has a 3d shape like a button is also termed as **Button**. For example, **OK** button, **Cancel** button, **Apply** button, and so on. Refer to Figure 3 for the terminology used for the components in a dialog box.

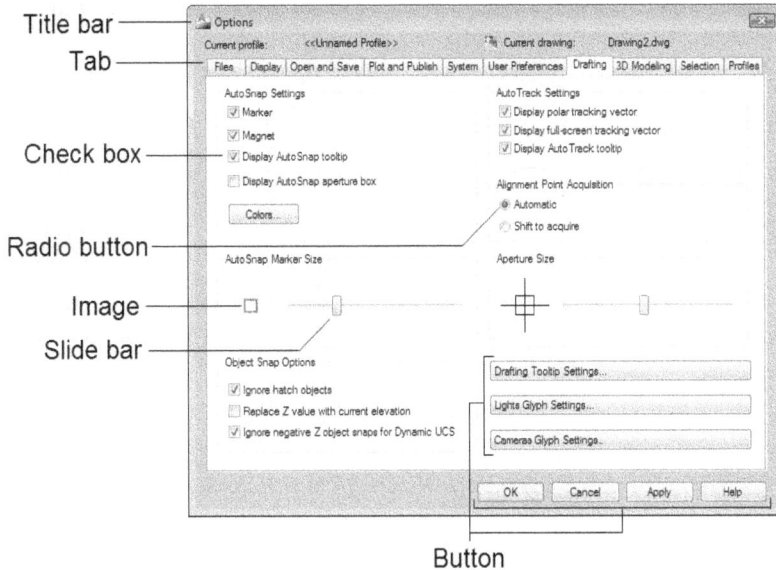

Figure 3 *The components in a dialog box*

Drop-down

A drop-down is one in which a set of common tools are grouped together for creating an object. You can identify a drop-down with a down arrow on it. These drop-downs are given a name based on the tools grouped in them. For example, **Project** drop-down, **Settings** drop-down, **Create Light** drop-down, and so on; refer to Figure 4.

Figure 4 *The **Project**, **Settings**, and **Create Light** drop-downs*

Drop-down List

A drop-down list is one in which a set of options are grouped together. You can set various parameters using these options. You can identify a drop-down list with a down arrow on it.

To know the name of a drop-down list, move the cursor over it; its name will be displayed as a tool tip. For example, **Lineweight** drop-down list, **Linetype** drop-down list, **Object Color** drop-down list, **Visual Styles** drop-down list, and so on; refer to Figure 5.

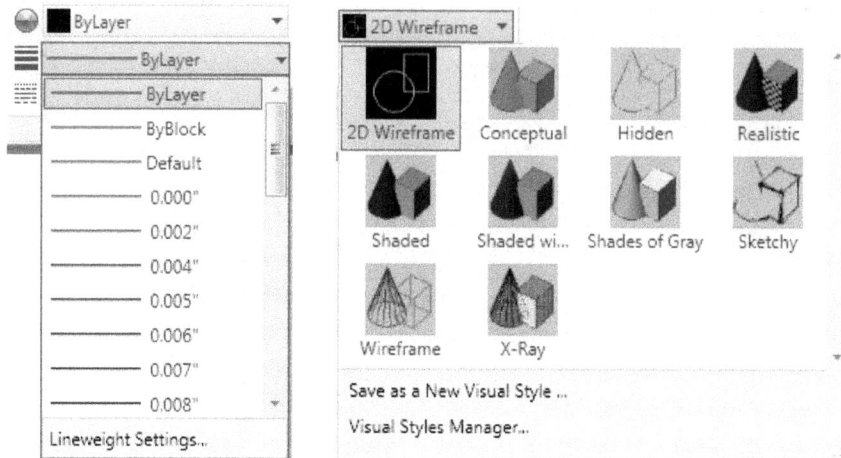

Figure 5 The LineWeight and Visual Styles drop-down lists

Options

Options are the items that are available in shortcut menu, drop-down list, Command prompt, **Properties** panel, and so on. For example, choose the **Properties** option from the shortcut menu displayed on right-clicking in the drawing area, refer to Figure 6.

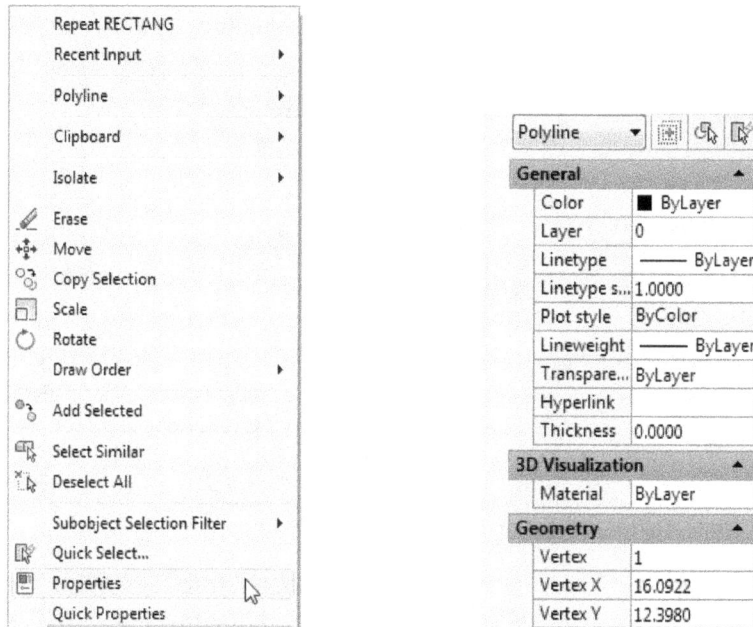

Figure 6 Options in the shortcut menu and the Properties palette

Symbols Used in the Text

Note
The author has provided additional information related to various topics in the form of notes.

Tip
The author has provided a lot of useful information to the users about the topic being discussed in the form of tips.

Free Companion Website

It has been our constant endeavor to provide you the best textbooks and services at affordable price. In this endeavor, we have come out with a Free companion website that will facilitate the process of teaching and learning of AutoCAD Plant 3D. If you purchase this textbook, you will get access to the companion website.

To access the files, you need to register by visiting the **Resources** section at *www.cadcim.com*. The following resources are available for the faculty and students in this website:

Faculty Resources

• **Technical Support**
    The faculty can get online technical support by contacting *techsupport@cadcim.com*.

• **Instructor Guide**
    Solutions to all review questions and exercises in the textbook are provided in this link to help faculty members test the skills of the students.

• **PowerPoint Presentations**
    The contents of the book are arranged in customizable powerpoint slides that can be used by the faculty for their lectures.

• **Part Files**
    The part files used in illustrations, tutorials, and exercises are available for free download.

Student Resources

• **Technical Support**
    The students can get online technical support by contacting *techsupport@cadcim.com*.

• **Part Files**
    The part files used in illustrations and tutorials are available for free download.

- **Additional Student Projects**
  Various projects are provided for the students to practice.

Stay Connected

To get the latest information about our textbooks, videos, teaching/learning resources, and other such updates, follow us on Facebook (*www.facebook.com/cadcim*) and Twitter (*@cadcimtech*). For more information about our latest video tutorials, you can also subscribe to our YouTube channel (*www.youtube.com/cadcimtech*).

You can access additional learning resources by visiting *http://allaboutcadcam.blogspot.com*.

If you face any problem in accessing these files, please contact the publisher at *sales@cadcim.com* or the author at *stickoo@purduecal.edu* or *tickoo525@gmail.com*.

Chapter 1

Introduction to AutoCAD Plant 3D

Learning Objectives

After completing this chapter, you will be able to:

- *Start AutoCAD Plant 3D*
- *Understand the components of the AutoCAD Plant 3D interface*
- *Invoke AutoCAD Plant 3D commands from the keyboard, menus, shortcut menus, Tool Palettes, and Ribbon*
- *Learn about the components of dialog boxes in AutoCAD Plant 3D*

INTRODUCTION

AutoCAD Plant 3D is a purpose-built plant design software. This software is used to design and document process plants. AutoCAD Plant 3D contains various predefined shapes of plant components. These predefined shapes, which are solid models, carry the intelligence of AutoCAD Plant 3D drawings. While installing AutoCAD Plant 3D, the standard specifications and part catalogs related to piping are also installed.

AutoCAD Plant 3D comes along with AutoCAD P&ID which is used to create piping and instrumentation drawings. AutoCAD P&ID contains various piping and instrumentation symbols. These symbols which are mostly the AutoCAD blocks with attributes carry the intelligence of AutoCAD P&ID drawings.

STARTING AutoCAD Plant 3D

When you install AutoCAD Plant 3D 2014, its icon is created and displayed on the desktop. You can launch the application by double-clicking on that icon. You can also start AutoCAD Plant 3D by using the Taskbar. To do so, choose **Start > All Programs > Autodesk > AutoCAD Plant 3D 2014 > AutoCAD Plant 3D 2014-English** from the Taskbar, refer to Figure 1-1; the interface screen of AutoCAD Plant 3D along with the **AutoCAD Plant 3D** window will be displayed, as shown in Figure 1-2

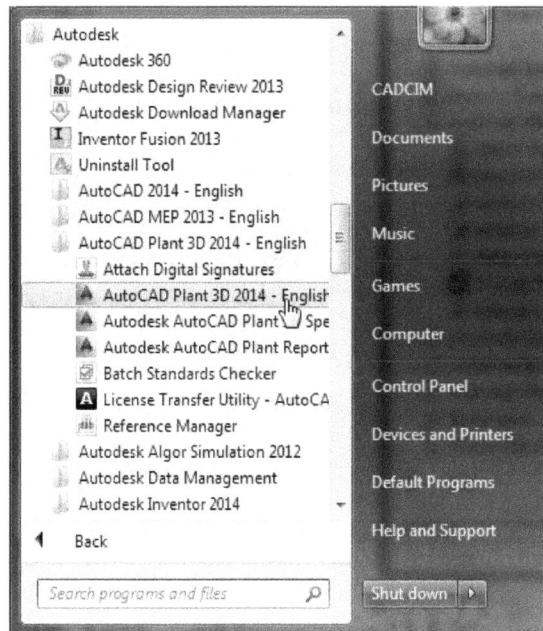

Figure 1-1 *Starting AutoCAD Plant 3D 2014 using the Taskbar*

In the **AutoCAD Plant 3D** window, various links are available that provide information about the recently opened files, projects, and other resources such as AutoCAD Plant Exchange.

The information available in this window is in the form of text and image. You can close this window by clicking on the **Close** button located on the top right corner.

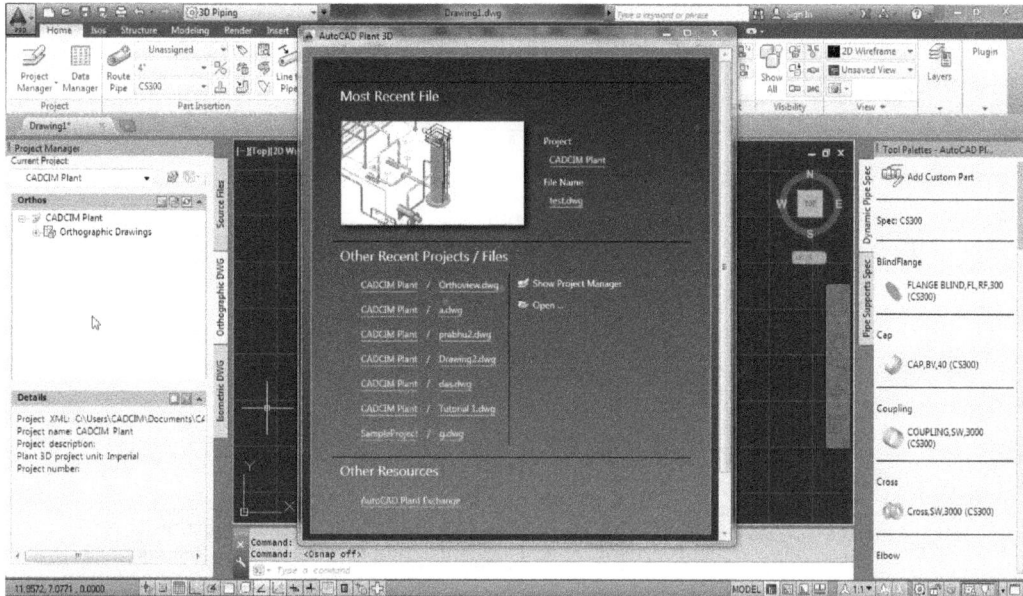

Figure 1-2 The AutoCAD Plant 3D 2014 window

WORKING IN A PROJECT

A piping design project includes the drawing and the other forms of data. These data sources are inter-related to each other. The drawing data includes P&IDs, 3D models, Isometric drawings, and Orthographic drawings. The other data forms include catalogs and specifications for piping, process data, and so on. Figure 1-3 shows a project flow diagram.

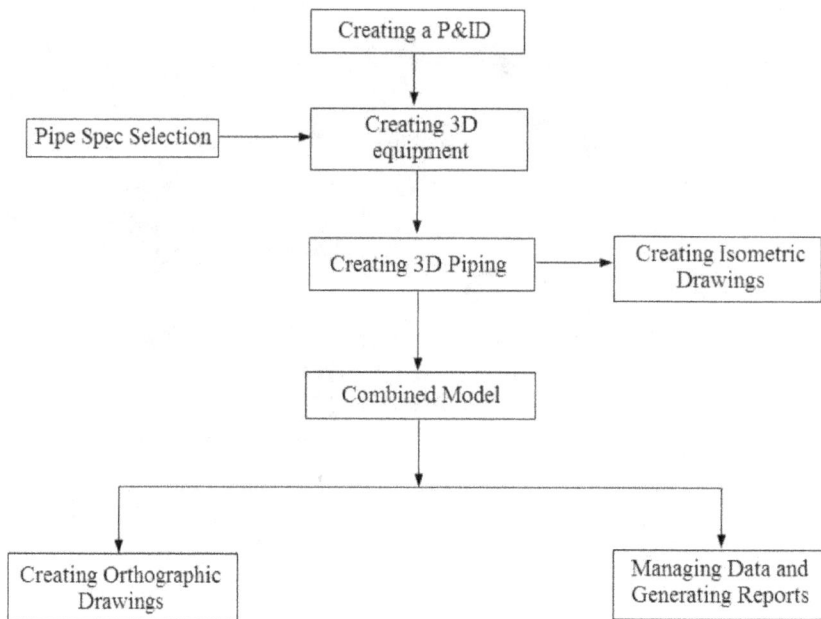

Figure 1-3 A project flow diagram

AutoCAD Plant 3D USER INTERFACE

The AutoCAD Plant 3D interface screen consists of drawing area, command window, menu bar, toolbars, status bar, and so on, refer to Figure 1-4. These components are discussed next.

Drawing Area

The drawing area covers the major portion of the screen. In this area, you can draw the objects and use the commands. To draw the objects, you need to define the coordinate points, which can be done by using the pointing device. The position of the pointing device is represented on the screen by the cursor. The window also has the standard Windows buttons such as close, minimize, and so on, at the top right corner. These buttons have the same functions as for any other standard window.

Command Window

The command window at the bottom of the drawing area has the Command prompt where you can enter the commands. It also displays the subsequent prompt sequences and the messages. You can change the size of the window by placing the cursor on the top edge (double line bar known as the grab bar) and then dragging it. This way you can increase its size to see all the previous commands you have used.

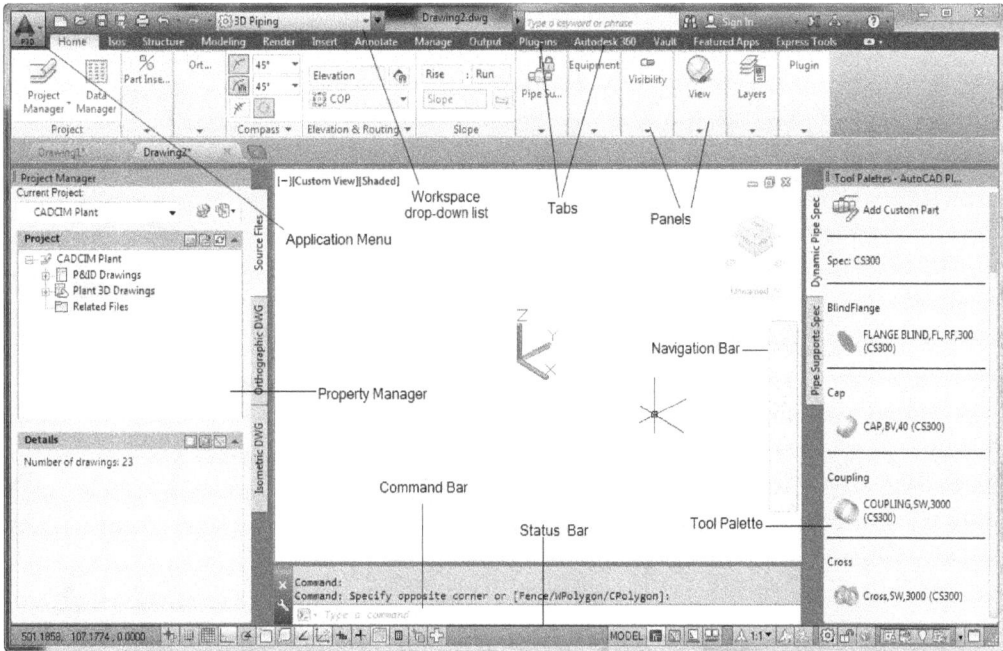

Figure 1-4 *Interface screen of AutoCAD Plant 3D*

Project Manager

Project Manager is used to access and manage all the drawings of a project. In addition to that, you can configure project settings, and export and import data from the **Project Manager**. Figure 1-5 shows the **Project Manager**. There are three tabs in the **Project Manager**: **Source Files**, **Orthographic DWG**, and **Isometric DWG**. You can create and access the P&ID and Plant 3D drawing files using the **Source Files** tab. The other two tabs can be used to create and access the orthographic and isometric drawings.

The data used in a project is arranged in their respective folders. For example, if you create a project with the name CADCIM, a folder named CADCIM will be created as the project folder. The data related to the project will then be stored in that folder in a systematic manner, as shown in Figure 1-6.

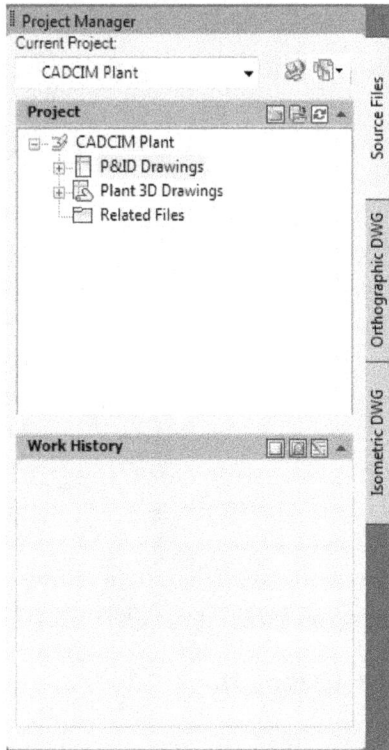

Figure 1-5 The **Project Manager**

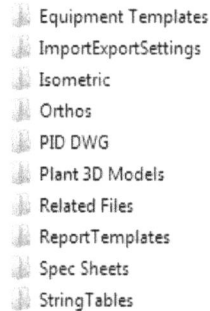

Figure 1-6 Project data arranged in folders

Data Manager

When you create a P&ID or a Plant 3D model, each and every part of the drawing is assigned with some properties. These properties can be accessed through Data Manager. Using the Data Manager, you can view, import, export, and create reports of the project data. Choose the **Data Manager** button from the **Project** panel to display the Data Manager. Figure 1-7 shows a partial view of the Data Manager.

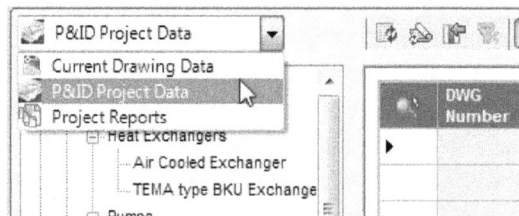

Figure 1-7 Partial view of the Data Manager

Navigation Bar

The Navigation Bar is displayed in the drawing area and contains navigation tools. These tools are grouped together, refer to Figure 1-8, and are discussed next.

SteeringWheels

The SteeringWheels has a set of navigation tools such as Pan, Zoom, Orbit, and Showmotion.

Pan

This tool allows you to view the portion of the drawing that is outside the current display area. To do so, choose this tool, press and hold the left mouse button, and then drag it in drawing area. You can exit this command by pressing ESC.

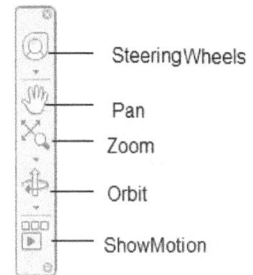

Figure 1-8 *The tools in the Navigation Bar*

Zoom

The zoom tools are used to enlarge the view of the drawing on the screen without affecting the actual size of the object.

Orbit

The rotate tools are used to rotate the view in the 3D space.

ShowMotion

This button is used to capture different views in a sequence and animate them when required.

ViewCube

ViewCube is available on the top right corner of the drawing area and is used to switch between the standard and isometric views or roll back to the current view.

In-Canvas Viewport Controls

In-canvas Viewport Controls is available on the top left corner of the drawing screen. It enables you to change the view, the visual style as well as the viewport.

Status Bar

There are two types of Status Bars in AutoCAD Plant 3D, Application Status Bar and Drawing Status Bar. The Status Bar is displayed at the bottom of the screen is called Application Status Bar. It contains some useful information and buttons, refer to Figure 1-9, that make it easy to change the status of some AutoCAD functions. You can toggle between the on and off states of most of these functions by selecting or deselecting them.

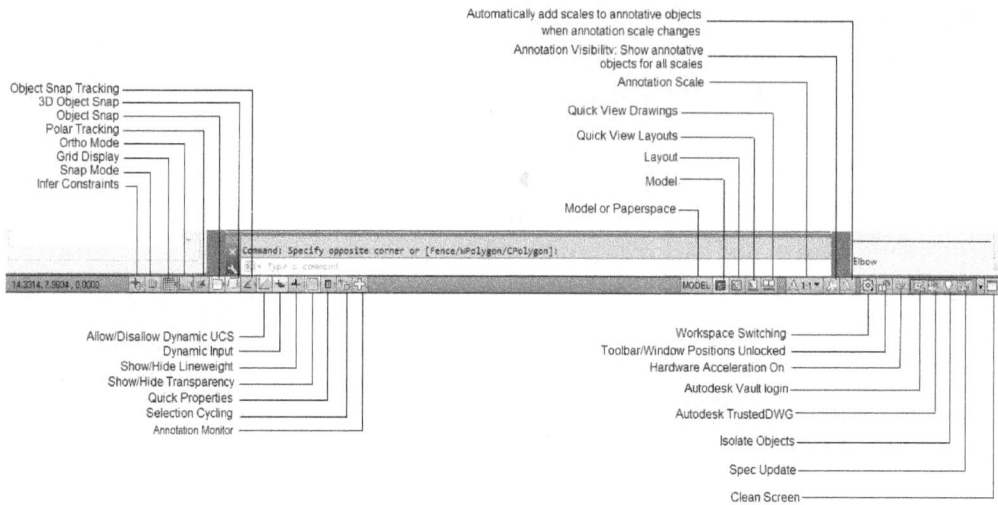

Figure 1-9 *The Application Status Bar*

Drawing Status Bar

The **Drawing Status Bar** is displayed in between the drawing area and the command window. If it is not displayed, choose the **Application Status Bar Menu** arrow and then choose the **Drawing Status Bar** option from the flyout; the **Drawing Status Bar** will be displayed, as shown in Figure 1-10. Turn on the **Drawing Status Bar**; the **Annotation Scale**, **Annotation Visibility**, and **Automatically Add Scale** buttons will move automatically to the **Drawing Status Bar**. If you turn off the **Drawing Status Bar**, these buttons will move back to the **Application Status Bar**.

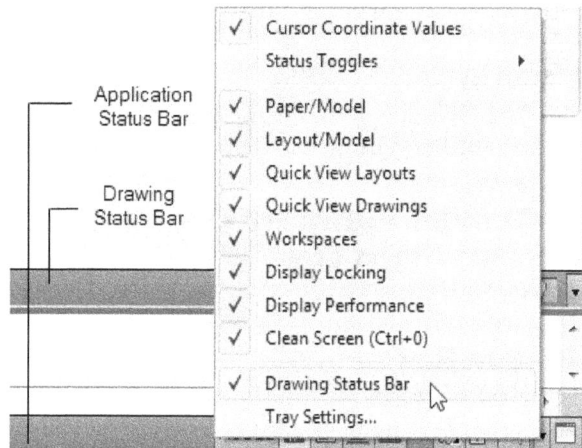

Figure 1-10 *The Drawing Status Bar*

Tray Settings

Choose the **Tray Settings** option from the flyout displayed on clicking the arrow in the **Application Status Bar**; the **Tray Settings** dialog box will be displayed, as shown in Figure 1-11. You can control the display of icons and notifications in the tray at the right end of the status bar by selecting appropriate options.

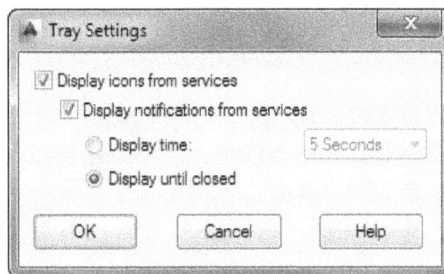

*Figure 1-11 The **Tray Settings** dialog box*

Clean Screen

The **Clean Screen** button is at the lower right corner of the screen. This button, when chosen, displays an expanded view of the drawing area by hiding all the toolbars except the command window, Status Bar, and menu bar. The expanded view of the drawing area can also be displayed by choosing **View > Clean Screen** from the menu bar or by using the CTRL+0 keys. You can choose the **Clean Screen** button again to restore the previous display state.

Status Toggles

You can hide the display of some of the buttons in the Status Bar. To do so, right-click on the **Application Status Bar**; a shortcut menu will be displayed. Move the cursor on the **Status Toggles** option in the shortcut menu; a cascading menu will be displayed. In this menu, clear the check mark next to the names of the buttons you need to hide.

Plot/Publish Details Report Available

This icon is displayed when some plotting or publishing activity is performed in the background. When you click on this icon, the **Plot and Publish Details** dialog box, which provides the details about the plotting and publishing activities, will be displayed. You can copy this report to the clipboard by choosing the **Copy to Clipboard** button from the dialog box.

Manage Xrefs

The **Manage Xrefs** icon is displayed whenever an external reference drawing is attached to the selected drawing. This icon displays a message and an alert whenever the Xreffed drawing needs to be reloaded. To find detailed information regarding the status of each Xref in the drawing and the relation between various Xrefs, click on the **Manage Xrefs** icon; the **External References Palette** will be displayed.

Properties Palette

The **Properties** palette is used to set the current properties and to change the general properties of the selected objects. The **Properties** palette is displayed on right-clicking on an object and then choosing the **Properties** option from the shortcut menu. Right-clicking in the **Properties** palette displays a shortcut menu from where you can choose **Allow Docking** or **Hide** to dock or hide the palette. When you double-click on an object, the **Properties** palette will display the properties of the selected object.

In AutoCAD Plant 3D, the **Properties** palette displays an additional section for the properties specific to the selected object, refer to Figure 1-12. For example, if you select a Plant 3D object, the Plant 3D section will be displayed in the **Properties** palette with the 3D properties of the object. The list of properties displayed varies depending on the object selected. For example, if you select a pipe support, the dimensions of the selected geometry will be displayed in the **Part Geometry** section, refer to Figure 1-13. In this section, you can edit the geometry of the selected pipe support. Similarly, if you select a valve from a Plant 3D model, properties of the valve will be displayed. You can change the valve actuator using the **Properties** palette.

Plant 3D	
Class	Tank
Tag	
Tag	TK-106
General	
Short Description	
Long Description (Size)	Tank
Long Description (Fa...	
Compatible Standard	
Manufacturer	
Item Code	
Design Std	
Design Pressure Factor	
Weight	
Weight Unit	
Flange Thickness	1.12
Content Iso Symbol D...	
Status	New
Number	106
Area	
Nozzles	
Tag	N-1
Size	8"x10"
Pressure Class	150
Short Description	Nozzle, flanged

Figure 1-12 The additional section displayed for a Plant 3D object

Part Properties	
Part Data	
Material	
Material Code	
Pressure Class	
Part Geometry	
Preview	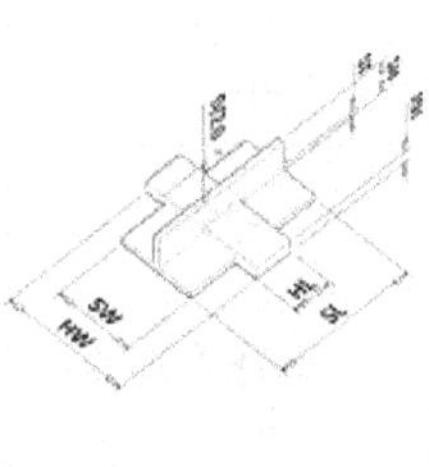

Dimensions	
D	10 15/16"
SL	1'-4 15/16"
SW	9 3/8"
ST	3/4"
SH	6 9/16"
HL	4 11/16"
HW	1'-3 1/16"
HH	0"
Length	

Figure 1-13 The dimensions of the selected geometry displayed in the **Properties** palette

DIFFERENT WORKSPACES IN AutoCAD Plant 3D

A workspace is defined as a customized arrangement of **Ribbon**, toolbars, menus, and window palettes in the AutoCAD environment. In AutoCAD Plant 3D, there are different workspaces to create a P&ID and Plant 3D model. These workspaces are given next.

3D Piping
P&ID PIP
P&ID ISO
P&ID ISA
P&ID DIN
P&ID JIS-ISO

You will notice that there are different workspaces for creating a P&ID. These workspaces are based on different standards. When you invoke a P&ID workspace of a particular standard, the tool palette will display the symbols related to that standard.

You can select any of the predefined workspaces from the **Workspace** drop-down list available in the title bar located next to the **Quick Access Toolbar**, see Figure 1-14. You can also set the workspace from the flyout that will be displayed on choosing the **Workspace Switching** button in the Status Bar, refer to Figure 1-15.

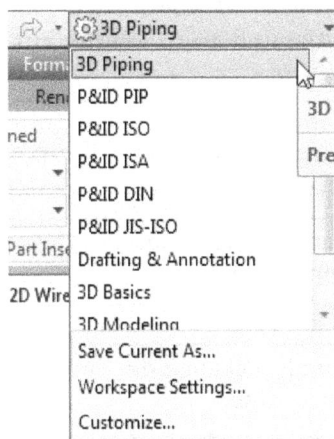

Figure 1-14 *The predefined workspaces*

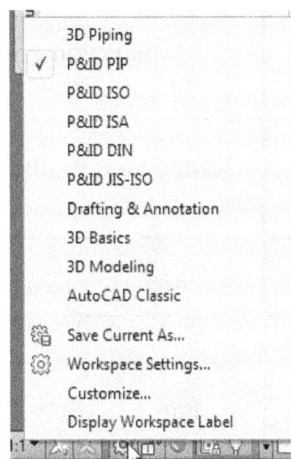

Figure 1-15 *The flyout displayed on choosing the **Workspace Switching** button*

GRIPS

Grips provide a convenient and quick means of editing objects. Grips are displayed on the key points of an object when the object is selected. There are different grips available in AutoCAD Plant 3D. The usage of these grips is explained in the following table.

Grip Symbol	Grip Name	Usage
✛	Continuation grip	It is used to start or continue routing a pipe.
▽	Substitution grip	It is used to substitute a component.
✛▯	Add Nozzle grip	You can add a nozzle to an equipment using this grip.
△	Elevation grip	You can change the elevation of a pipe using this grip.

	Edit nozzle grip	It is used to modify a nozzle.
	Rotate Part grip	It is used to rotate a valve or fitting.
	Flip grip	It is used to flip a valve or a fitting.
	Connection grip	It is displayed when a schematic line is connected to a component or an equipment.
	Move SLine Parallel grip	It is displayed at the midpoint of a schematic line. You can move the schematic line using this grip.
	Continue/Shorten SLine grip	You can increase or decrease the length of a schematic line using this grip.
	Flip grip (in P&ID)	It is used to change the direction of the schematic line.
	Gap grip	It is displayed when a gap is added on the schematic line. You can increase or decrease the gap using this grip.

INVOKING COMMANDS IN AutoCAD Plant 3D

When you are in the drawing area, you need to invoke AutoCAD Plant 3D commands to perform required operations. For example, to draw a line, first you need to invoke the **LINE** command and then define the start point and the endpoint of the line. Similarly, if you want to erase objects, you must invoke the **ERASE** command and then select the objects for erasing. In AutoCAD Plant 3D, you can invoke the commands using different methods which are discussed next.

Invoking Commands Using Command Prompt

You can invoke any AutoCAD Plant 3D command using the keyboard by typing the command name at the Command prompt and then pressing the ENTER key. As you type the first letter of command, AutoCAD Plant 3D displays all available commands starting with the letter typed. You can also use the **Dynamic Input** button to directly enter the command in the **Pointer Input** box. The **Pointer Input** box is a small box displayed on the right of the cursor, as shown in Figure 1-16. However, if the cursor is currently placed on any toolbar or menu bar, or if the **Dynamic Input** is turned off, the command will be entered through the Command prompt. The following example shows how to invoke the **LINE** command using the keyboard:

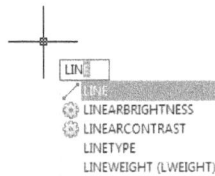

Figure 1-16 The *Pointer Input* box displayed
when the *Dynamic Input* is on

Command: **LINE** or **L** [Enter]   (L is command alias)

Invoking Commands Using Ribbon

In AutoCAD Plant 3D, you can also invoke a tool from the **Ribbon**. In the Ribbon, the tools
for creating pipes, equipment, supports, and the other Plant 3D components are available in
different panels instead of being spread out in the entire drawing area in different toolbars
and menus, see Figure 1-17.

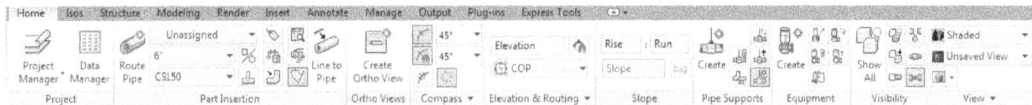

Figure 1-17 The **Ribbon** for the **3D Piping** workspace

In AutoCAD Plant 3D, there are different tabs and ribbons for performing different tasks
such as creating a P&ID, Isometric drawings, orthographic drawing, and so on. These are
discussed next.

Home tab of the P&ID workspace

The **Home** tab of the P&ID workspaces contains tools that are used to create a P&ID. This
tab is available in P&ID workspaces, refer to Figure 1-18.

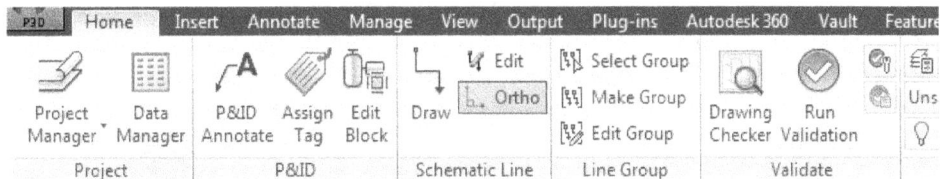

Figure 1-18 The **Home** tab of the P&ID workspace

Home tab of the 3D Piping workspace

This is one of the most important tabs provided in the 3D Piping workspace. This tab provides
all tools that are used to create 3D piping, equipment, and pipe supports. The **Home** tab of
the **3D Piping** workspace is shown in Figure 1-19.

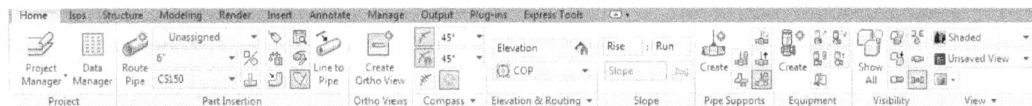

Figure 1-19 The **Home** tab of the **3D Piping** workspace

Isos Tab

The tools in the **Isos** tab are used to generate isometric drawings. The **Isos** tab is shown in Figure 1-20.

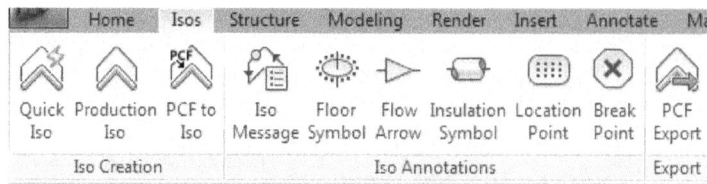

*Figure 1-20 The **Isos** tab*

Structure Tab

The tools in the **Structure** tab are used to create and modify structures. The **Structure** tab is shown in Figure 1-21.

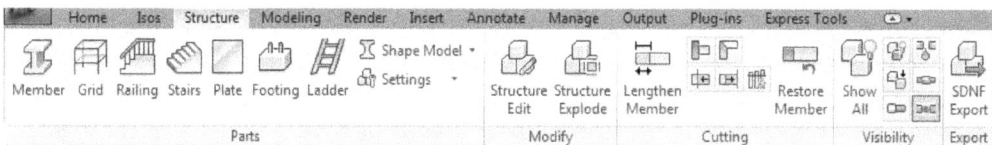

*Figure 1-21 The **Structure** tab*

Ortho Editor Tab

The tools in the **Ortho Editor** tab are used to create orthographic views. The **Ortho Editor** tab is shown in Figure 1-22.

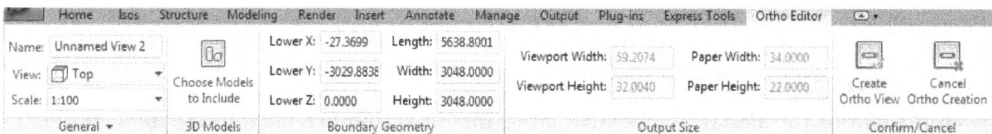

*Figure 1-22 The **Ortho Editor** tab*

Ortho View Tab

The tools in the **Ortho View** tab are used to annotate and dimension the view. Also, you can create adjacent views and locate the view components in a 3D Model. The **Ortho View** tab is shown in Figure 1-23.

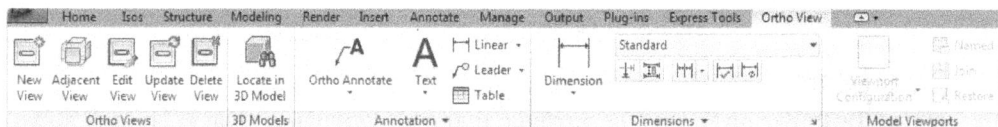

*Figure 1-23 The **Ortho View** tab*

Tool Palettes

AutoCAD Plant 3D has provided different Tool Palettes as an easy and convenient way of placing symbols and 3D parts in the current drawing. The Tool Palettes display items based on the workspace in which you are currently working. The Tool Palettes in different workspaces are discussed next.

Tool Palettes in P&ID workspace

The P&ID Tool Palettes contain various tabs such as **Lines**, **Equipment**, **Valves** and so on, refer to Figure 1-24. The symbols in each tab are grouped into different areas. You can create more custom symbols and add to the Tool Palettes. You can change the Tool Palette by choosing the **Properties** button and then selecting the required tool palette from the flyout displayed.

Tool Palettes in Plant 3D workspace

In the 3D Piping workspace, the Tool Palettes contain tabs such as **Dynamic Pipe Spec**, and **Pipe Supports Spec**, refer to Figure 1-25. The **Dynamic Pipe Spec** tab contains the piping components from the selected specification. You can add more components to the **Dynamic Pipe Spec** tab by invoking the **Spec Viewer**. You will learn more about the **Spec Viewer** in later chapters. In addition, you can add a custom part to the Tool Palette. Also, you can change the components displayed in this tab by selecting a different specification. The **Pipe Supports Spec** tab contains pipe supports. You can dynamically select a support and place it in the 3D model.

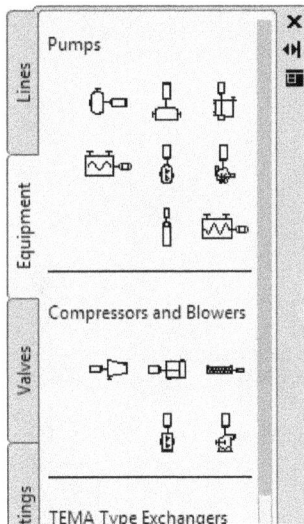

Figure 1-24 *Tool Palettes in P&ID workspace*

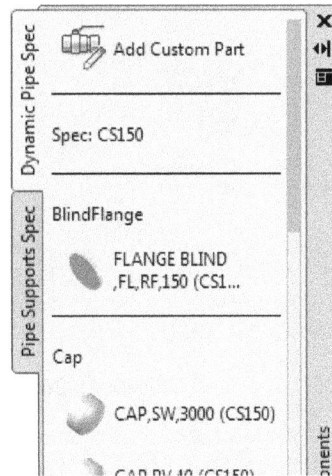

Figure 1-25 *Tool Palettes in 3D Piping workspace*

Application Menu

The **Application Menu** is available at the top left corner of the AutoCAD Plant 3D window. It contains some of the tools that are available in the **Standard** toolbar. Click on the down arrow on the **Application Menu** to display the tools, as shown in Figure 1-26. You can search a command using the search field on the top of the **Application Menu**. To search a tool, enter the complete or partial name of the command in the search field; the possible tool list will be displayed. If you click on a tool from the list, the corresponding command will get activated.

By default, the **Recent Document** button is chosen in the **Application Menu**. Therefore, the recently opened drawings will be listed. If you have opened multiple drawing files, choose the **Open Documents** button; the documents that are opened will be listed in the **Application Menu**. To set the preferences of the file, choose the **Options** button available at the bottom right of the **Application Menu**. To exit AutoCAD Plant 3D, choose the **Exit Autodesk AutoCAD Plant 3D 2014** button next to the **Options** button.

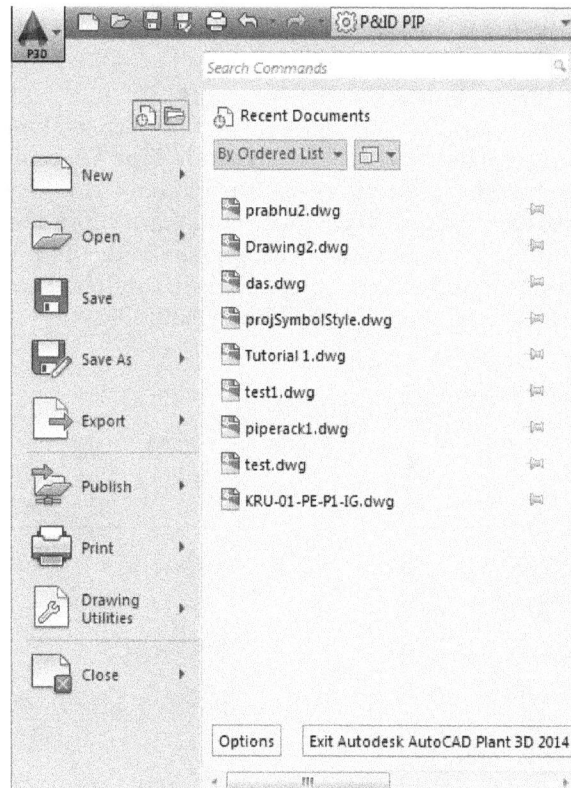

Figure 1-26 The Application Menu

Menu Bar

You can also invoke commands from the menu bar. Menu bar is not displayed by default in AutoCAD Plant 3D. To display the menu bar, click on the down arrow in the **Quick Access Toolbar**; a flyout will be displayed. Choose the **Show Menu Bar** option from it; the menu bar will be displayed. As you move the cursor over the menu bar, different titles get highlighted. You can click on the desired item to display a menu. You can invoke a command by left-clicking on it in the menu. Some of the menu items display an arrow on the right side, which indicates that they have a cascading menu. The cascading menu provides various options to execute the same AutoCAD Plant 3D command.

Shortcut Menu

AutoCAD Plant 3D provides shortcut menus to invoke the recently used tools easily. These shortcut menus are context-specific, which means that the tools available in them are dependent on the place/object for which they are displayed. A shortcut menu is invoked by right-clicking in the drawing area. It generally contains an option to select the previously invoked tool again, apart from the common tools for Windows, refer to Figure 1-27.

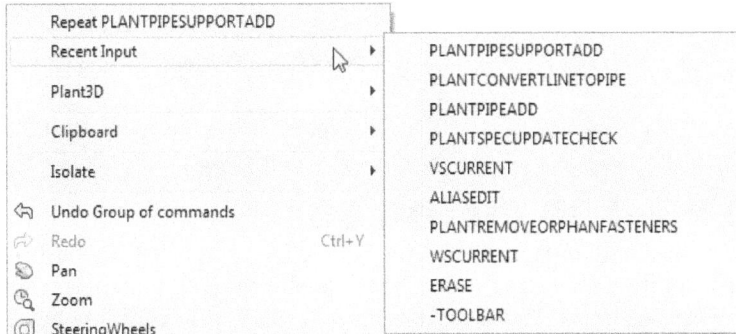

Figure 1-27 Partial view of the shortcut menu with the recently used commands

If you right-click in the drawing area while a command is active, a shortcut menu is displayed, containing the options corresponding to that particular command. Figure 1-28 shows the shortcut menu when the **Route Pipe** tool is active.

You can also right-click on the command window to display the shortcut menu. This menu displays the six recently used commands and some of the Windows options such as **Copy** and **Paste**, refer to Figure 1-29. The commands and their prompt entries are displayed in the **History** window (previous command lines not visible) and can be selected, copied, and pasted in the command line using the shortcut menu. As you press the up arrow key, the previously entered commands are displayed in the command window. Once the desired command is displayed at the Command prompt, you can execute it by simply pressing the ENTER key. You can also copy and edit any previously invoked command by locating it in the **History** window and then selecting the lines. After selecting the desired command lines from the **History** window, right-click to display a shortcut menu. Choose **Copy** from the menu and then paste the selected lines at the end of the command line.

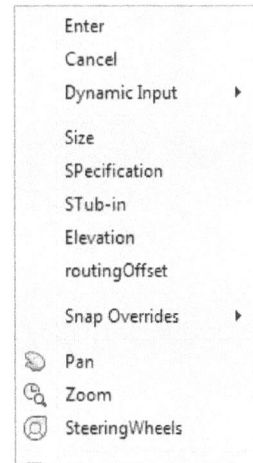

*Figure 1-28 Shortcut menu with the **Route Pipe** tool active*

You can right-click on the coordinate display area of the Application Status Bar to display the shortcut menu. This menu contains the options to modify the display of coordinates, as shown in Figure 1-30. You can also right-click on any of the toolbars to display the shortcut menu from where you can choose any toolbar to be displayed.

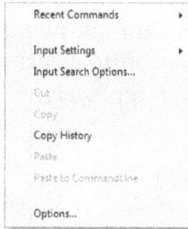

Figure 1-29 *Command line window shortcut menu*

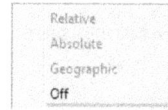

Figure 1-30 *The Application Status Bar shortcut menu*

AutoCAD Plant 3D DIALOG BOXES

On invoking certain commands in AutoCAD Plant 3D, the related dialog box is displayed. When you choose an item in the menu bar with the ellipses [...], it displays a dialog box. For example, when **Options** in the **Tools** menu is chosen, the **Options** dialog box is displayed. A dialog box contains a number of components like the dialog label, radio buttons, text or edit boxes, check boxes, slider bars, image boxes, and command buttons. Some of the components in a dialog box are shown in Figure 1-31.

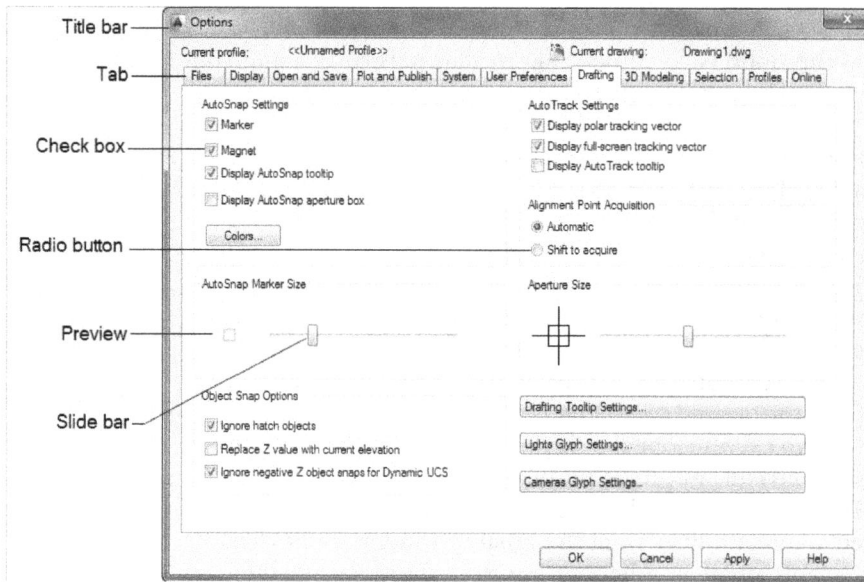

Figure 1-31 *Components of a dialog box*

The title bar displays the name of the dialog box. The tabs specify the various sections that contain various groups of related options under them. The check boxes are toggle options for making the particular option available or unavailable. When you click on an option and a list of options is displayed, then it is termed as drop-down list. You can select options using the radio buttons. Note that only one button can be selected at a time. The image displays the preview image of the item selected. The text box is an area where you can enter a text such as a file name. It is also called an edit box, because you can make any change to the text entered. In some dialog boxes, you will find the [...] button, which displays another related dialog box. There are certain buttons such as **OK, Cancel**, and **Help** that are also displayed at the bottom of the dialog box. The button with a dark border is the default button.

CREATING BACKUP FILES

If the drawing file already exists and you use **Save** or **Save As** tools to update the current drawing, AutoCAD Plant 3D creates a backup file. AutoCAD Plant 3D takes the previous copy of the drawing and changes it from a file type *.dwg* to *.bak*, and the updated drawing is saved as a drawing file with the *.dwg* extension. For example, if the name of the drawing is *myproj.dwg*, AutoCAD Plant 3D will change it to *myproj.bak* and save the current drawing as *myproj.dwg*.

Changing Automatic Timed Saved and Backup Files into AutoCAD Format

Sometimes, you may need to change the automatic timed saved and backup files into AutoCAD format. To change the backup file into an AutoCAD format, open the folder in which you have saved the backup or the automatic timed saved drawing using **Computer** or **Windows Explorer**. Choose **Organize > Folder and Search Options** from the menu bar to invoke the **Folder Option**s dialog box. Choose the **View** tab and under the **Advanced settings** area, clear the **Hide extensions for known file types** text box, if selected. Exit the dialog box. Rename the automatically saved drawing or the backup file with a different name and also change the extension of the drawing from *.sv$* or *.bak* to *.dwg*. After you rename the drawing, you will notice that the icon of the automatically saved drawing or the backup file is replaced by the AutoCAD icon. This indicates that the automatically saved drawing or the backup file is changed to an AutoCAD Plant 3D drawing.

Using the Drawing Recovery Manager to Recover Files

The files that are saved automatically can also be retrieved by using the **Drawing Recovery Manager**. You can open the **Drawing Recovery Manager** again by choosing **Drawing Utilities > Open the Drawing Recovery Manager** from the **Application Menu** or by entering **DRAWINGRECOVERY** at the Command prompt.

If the automatic save operation is performed in a drawing and the system crashes accidentally, the next time when you run AutoCAD Plant 3D, the **Drawing Recovery** message box will be displayed, as shown in Figure 1-32. The message box informs you that the program unexpectedly crashed and you can open the most relevant among the backup files created by AutoCAD Plant 3D. Choose the **Close** button from the **Drawing Recovery** message box; the **Drawing Recovery Manager** is displayed on the left of the drawing area, as shown in Figure 1-33.

The **Backup Files** rollout lists the original files, the backup files, and the automatically saved files. Select a file; its preview will be displayed in the **Preview** rollout. Also, the information corresponding to the selected file will be displayed in the **Details** rollout. To open a backup file, double-click on its name in the **Backup Files** rollout. Alternatively, right-click on the file name and then choose **Open** from the shortcut menu. It is recommended that you save the backup file at the desired location before you start working on it.

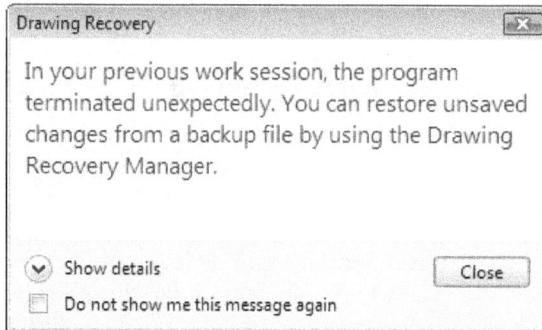

Figure 1-32 The **Drawing Recovery** message box

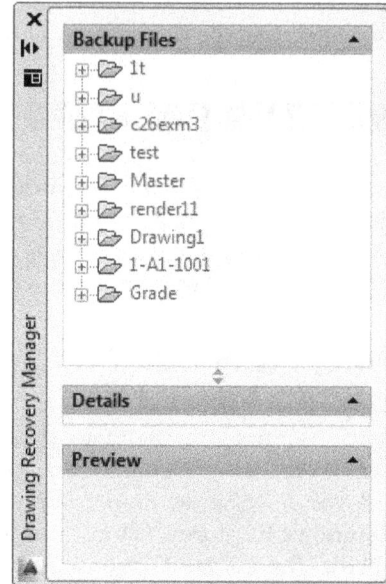

Figure 1-33 The **Drawing Recovery Manager**

CLOSING A DRAWING

You can use the **CLOSE** command to close the current drawing file without actually quitting AutoCAD Plant 3D. If you choose **Close > Current Drawing** from the **Application Menu** or enter **CLOSE** at the Command prompt, the current drawing file will be closed. If multiple drawing files are opened, choose **Close > All Drawings** from the **Application Menu**. If you have not saved the drawing after making the last changes to it and you invoke the **CLOSE** command, AutoCAD Plant 3D displays a dialog box that allows you to save the drawing before closing. This box gives you an option to discard the current drawing or the changes made to it. It also gives you an option to cancel the command. After closing the drawing, you are still in AutoCAD Plant 3D from where you can open a new or an already saved drawing file. You can also use the close button (**X**) of the drawing area to close the drawing.

Note
You can close a drawing even if a command is active.

OPENING A PROJECT DRAWING

You can open an existing drawing file that has been saved previously in a project. The drawings are located under the P&ID drawings and Plant 3D drawings nodes in the **Project Manager**. To open a drawing, expand the respective drawings node and right-click to display the shortcut menu. Next, choose the **Open** option from the shortcut menu, refer to Figure 1-34; the drawing will be opened. Alternatively, double-click on the drawing in the **Project Manager** to open it.

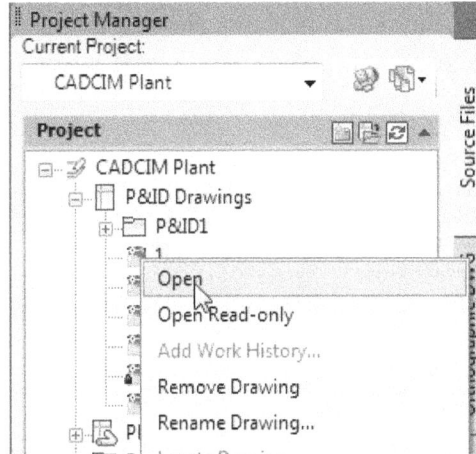

Figure 1-34 *Opening a drawing from the* **Project**
Manager

To view a drawing without altering it, you must select the **Open Read-Only** option from shortcut menu. In other words, opening read only protects the drawing file from changes. AutoCAD Plant 3D does not prevent you from editing the drawing. But if you try to save the opened drawing with the original file name, AutoCAD Plant 3D warns you that the drawing file is write-protected. However, you can save the edited drawing to a file with a different file name using the **SAVEAS** command. This way you can preserve your drawing.

OPENING A DRAWING THAT IS NOT IN THE PROJECT

Application Menu: Open > Drawing	**Quick Access Toolbar:** Open
Menu Bar: File > Open	**Command:** OPEN

You can open a drawing file that does not exist in the currently opened project using the **Select File** dialog box. The method of invoking drawing using **Select File** dialog box is discussed next.

Opening an Existing Drawing Using the Select File Dialog Box

If you are already in the drawing editor and you want to open a drawing file, choose the **Open** tool from the **Quick Access Toolbar**; the **Select File** dialog box will be displayed. Alternatively, invoke the **OPEN** command to display the **Select File** dialog box by using the Command prompt, as shown in Figure 1-35. You can select the drawing to be opened using this dialog box. This dialog box is similar to the standard dialog boxes. You can choose the file that you want to open, from the folder in which it is stored. You can access the required folder using the **Look in** drop-down list. You can then select the name of the drawing from the list box or you can enter the name of the drawing file you want to open in the **File name** edit box. After selecting the drawing file, you can select the **Open** button to open the file.

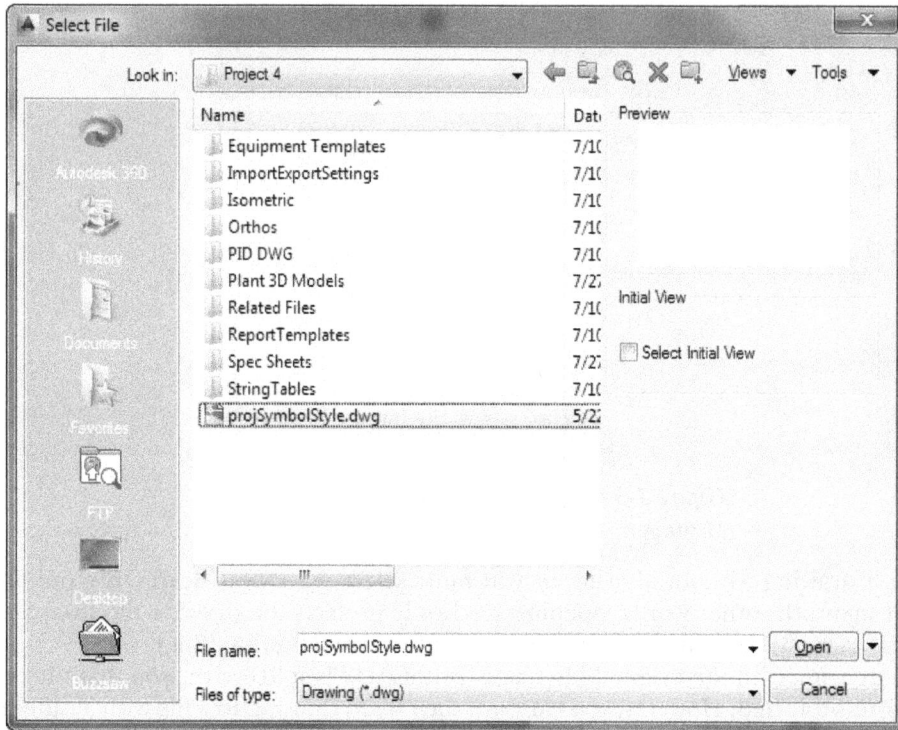

*Figure 1-35 The **Select File** dialog box*

When you select a file name, the preview of the selected drawing file will be displayed in the **Preview** box. If you are not sure about the file name of a particular drawing but know the contents, you can still select the file names and look for the that drawing in the **Preview** box. You can also change the file type by selecting it in the **Files of type** drop-down list. Apart from the *dwg* files, you can open the *dwt* (template) files or the *dxf* files. You have all the standard icons in the **Places** list that can be used to open drawing files from different locations. The **Open** button has a drop-down list, as shown in Figure 1-36. You can choose any of the methods given in this list for opening the file.

Note that you cannot make any changes to the drawing file that is not in the current project. If you do so, the **Alert** message box will be displayed, as shown in Figure 1-37. This message box warns you that the object can only be inserted into a project drawing, and if want to add this drawing to the current project. Choose the **Yes** button; the message box will be closed and the drawing will be added to the currently opened project.

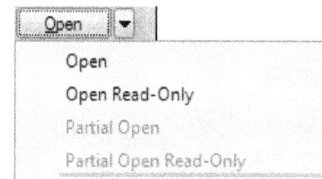

*Figure 1-36 The **Open** drop-down list*

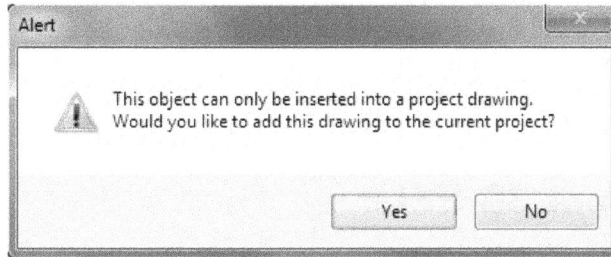

*Figure 1-37 The **Alert** message box*

Alternatively, to add files to the current project, right-click on the **P&ID Drawings** or **Plant 3D Drawings** node and choose the **Copy Drawing to Project** option from the shortcut menu displayed, refer to Figure 1-38; the **Select Drawings to Copy to Project** dialog box will be displayed. Browse to the file location and double-click on the drawing file to be added to the project; the drawing file will be added to the current project and the **Project Data Merged** message box will be displayed, as shown in Figure 1-39. Choose **OK** to close the message box.

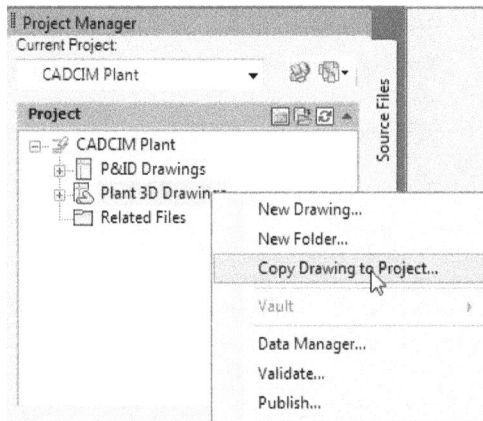

Figure 1-38 Copying a drawing to project

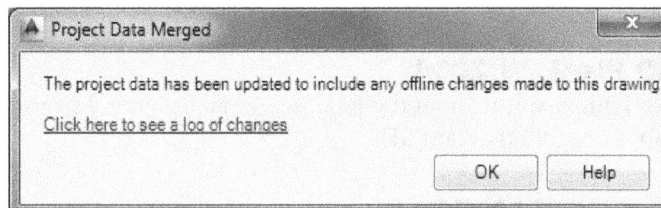

*Figure 1-39 The **Project Data Merged** message box*

QUITTING AutoCAD Plant 3D

You can exit the AutoCAD Plant 3D program by using the **EXIT** or **QUIT** command. Even if you have an active command, you can choose **Exit Autodesk AutoCAD Plant 3D 2014** from the **Application Menu** to close the AutoCAD Plant 3D program. In case the drawing has not been saved, it allows you to first save the work through a dialog box. Note that if you choose **No** in this dialog box, all the changes made in the current list till the last save will be lost. You can also use the **Close** button (**X**) of the main AutoCAD Plant 3D window (present in the title bar) to end the AutoCAD Plant 3D session.

AutoCAD Plant 3D HELP

Titlebar: ? > Help   **Shortcut Key:** F1   **Command:** HELP or ?

You can get the on-line help and documentation about the working of AutoCAD Plant 3D 2014 commands from the **Help** menu in the title bar, see Figure 1-40. You can also access the **Help** menu by pressing the F1 function key. Some important options in the **Help** menu are discussed next.

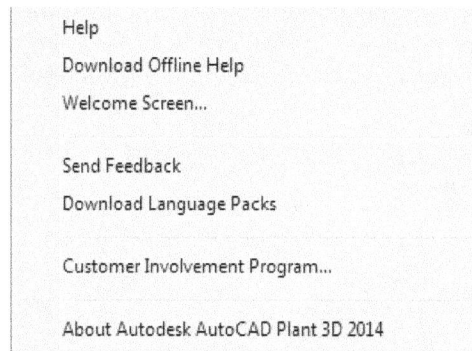

*Figure 1-40 The **Help** menu*

Figure 1-41 shows the **Autodesk AutoCAD Plant 3D 2014 - Help** window after choosing the **Help** button from the **Infocenter** bar.

Customer Involvement Program

This option is used to share the system configuration information and uses of Autodesk products with Autodesk. The collected information is used by Autodesk for the improvement of Autodesk software.

About AutoCAD Plant 3D 2014

This option gives you information about the Release, Serial number, Licensed to, and also the legal description about AutoCAD Plant 3D.

AUTODESK EXCHANGE Apps

Autodesk Exchange Apps helps you download various applications for AutoCAD, get connected to the AutoCAD network, share information and designs, and so on. On choosing the **AutoCAD**

Exchange Apps button from the title bar, the **AUTODESK EXCHANGE Apps** window will be opened in Internet Explorer window, as shown in Figure 1-42.

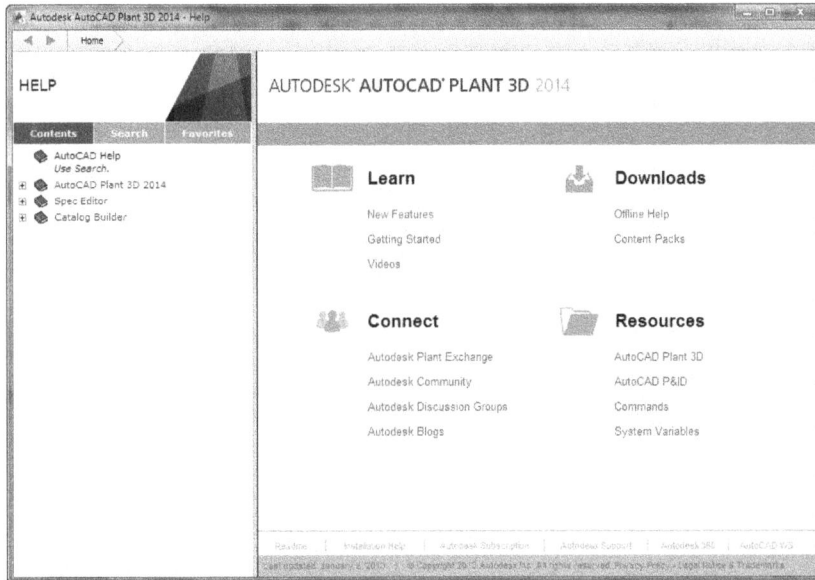

*Figure 1-41 The **Autodesk AutoCAD Plant 3D 2014 Help** window*

You can download various Autodesk apps such as Screenshot, DWF PDF Batch Publish, and so on from this page. Some of them are free of cost. You can also publish your own Autodesk products for other users of Autodesk.

Also, you can download applications for software other than AutoCAD such as Autodesk Alias, Revit, Simulation, and so on. You can also search for the applications by entering the name of the app in the **Search Exchange Apps** search box.

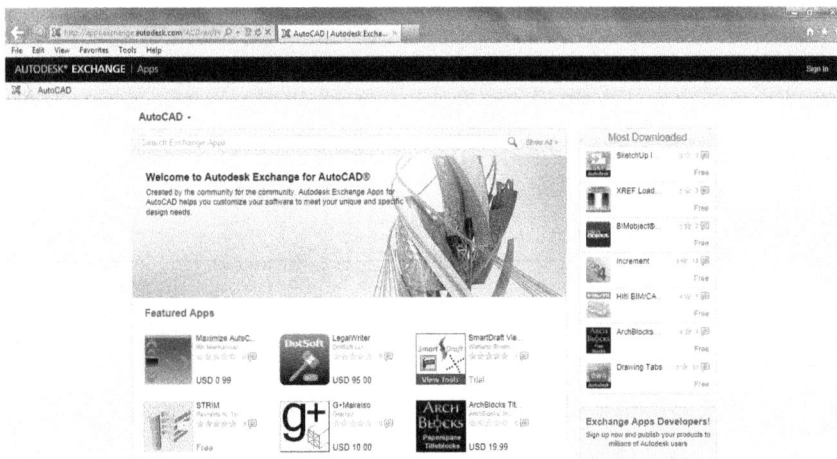

*Figure 1-42 The **AUTODESK EXCHANGE** Apps window*

ADDITIONAL HELP RESOURCES

1. You can get help for a command while working by pressing the F1 key. On doing so, the help html containing information about the command is displayed. You can exit the dialog box and continue with the command.

2. You can get help about a dialog box by choosing the **Help** button in that dialog box.

3. Autodesk has provided several resources that you can use to get assistance with your AutoCAD Plant 3D questions. The following is a list of some of the resources:

    a. Autodesk website *http://www.autodesk.com*
    b. AutoCAD Plant 3D Technical Assistance website
       http://autocad.autodesk.com/?nd=plant_home

4. You can also get help by contacting the author, Sham Tickoo, at *stickoo@purduecal.edu* and *tickoo525@gmail.com*.

5. You can download AutoCAD Plant 3D drawings, programs, and special topics by registering yourself at the faculty website by visiting: *http://cadcim.com/Registration.aspx*

Self-Evaluation Test

Answer the following questions and then compare them to those given at the end of this chapter:

1. You can press F3 key to invoke the **AutoCAD** text window. This window displays the previous commands and prompts. (T/F)

2. If you do not have an internet connection, you cannot access the Help files. (T/F)

3. You can close a drawing in AutoCAD Plant 3D 2014 even if a command is active. (T/F)

4. If the current drawing is unnamed and you save the drawing for the first time, you will be prompted to specify the file name in the **Save Drawing As** dialog box. (T/F)

5. The items in the _____ tab in the Tool Palette contain the piping components from the selected specification.

6. The _____ grip is used to modify a nozzle.

7. If you want to work on a drawing without altering the original drawing, you must select the _____ option.

8. The _____ is used to start or continue routing a pipe.

9. The_____ palette displays an additional section for the properties specific to the selected item.

10. You can use the _____ command to close the current drawing file without actually quitting AutoCAD Plant 3D.

Review Questions

Answer the following questions:

1. The shortcut menu invoked by right-clicking in the command window displays the most recently used commands and some of the window options such as **Copy**, **Paste**, and so on. (T/F)

2. The P&ID workspaces are based on different standards. (T/F)

3. The Tool Palettes display items based on the workspace currently in use. (T/F)

4. You cannot make any changes to the drawing file which is not in the current project. (T/F)

5. The file name that you enter to save a drawing in the **Save Drawing As** dialog box can be 255 characters long, but cannot contain spaces and punctuation marks. (T/F)

6. Which of the following combination of keys should be pressed to turn on or off the display of the **Tool Palettes** window?

    (a) CTRL+3   (b) CTRL+0
    (c) CTRL+5   (d) CTRL+2

7. Which of the following commands is used to exit the AutoCAD Plant 3D program?

    (a) **QUIT**   (b) **END**
    (c) **CLOSE**   (d) None of these

8. When you choose **Save** from the **File** menu or choose the **Save** tool in the **Quick Access Toolbar**, which of the following commands is invoked?

    (a) **SAVE**   (b) **LSAVE**
    (c) **QSAVE**   (d) **SAVEAS**

9. By default, the angles are positive if measured in the _____ direction.

Answers to Self-Evaluation Test
1. T, 2. F, 3. F, 4. T, 5. **Dynamic Pipe Spec**, 6. **Edit Nozzle**, 7. **Open Read Only**, 8. **Continuation grip**, 9. **Properties**, 10. **CLOSE**

Chapter 2

Creating Projects and P&IDs

Learning Objectives

After completing this chapter, you will be able to:

• *Create a new project*
• *Create a new drawing*
• *Create a P&ID drawing*
• *Add an equipment to a P&ID*
• *Assign a tag to a line*
• *Add valves*
• *Add instruments and instrumentation lines*
• *Add fittings*
• *Add off-page connectors*
• *Validate the P&ID drawing*
• *Correct the errors*
• *Edit the drawing*
• *Substitute components*
• *Convert AutoCAD components into P&ID Symbols*

INTRODUCTION

AutoCAD Plant 3D is a project-based software which is available with the P&ID module. In this chapter, you will learn to create a new project and a new drawing. Also, you will be briefly introduced to the P&ID module.

PROJECT MANAGER

The **Project Manager** is used to access existing projects, create new projects, add new drawings to a project, re-order drawing files, and modify the existing information in a project. It also allows you to export and import data, create project reports, include referenced drawings (xrefs), and link or copy files to the project folders. By default, the **Project Manager** is opened and docked on the left of your screen, as shown in Figure 2-1.

The **Project Manager** contains three tabs: **Source Files**, **Orthographic DWG**, and **Isometric DWG**. The **Source Files** tab contains a project tree that displays P&ID drawings, Plant 3D drawings and all related files of the project. The **Orthographic DWG** tab contains the list of all the orthographic drawing files. The **Isometric DWG** tab contains the isometric line diagrams of the plant layout.

CREATING A NEW PROJECT IN AutoCAD Plant 3D

To create a new project, choose the **New Project** button from the **Project Manager** drop-down of the **Project** panel in the **Home** tab; the **Project Setup Wizard** dialog box will be displayed, as shown in Figure 2-2. Alternatively, use the **NEWPROJECT** command to create a new project. Steps to create a new project are discussed next.

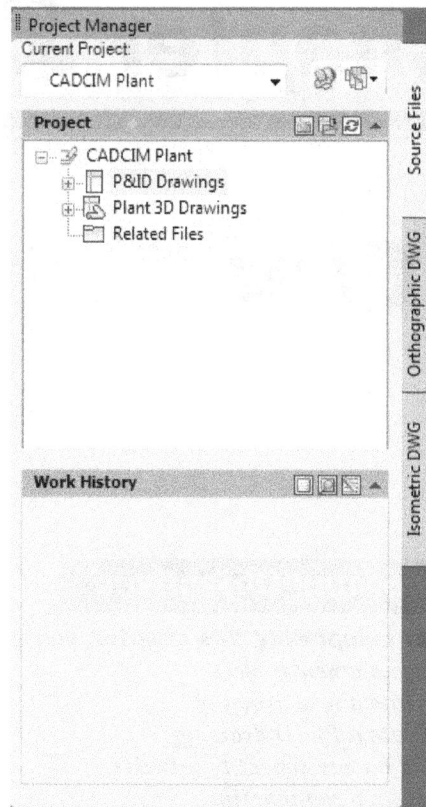

Figure 2-1 The Project Manager

1.   Enter a project name in the **Enter a name for this project** edit box. Next, enter a description of the project in the **Enter an optional description** edit box.

2.   Choose the **Browse** button adjacent to the **Specify the directory where program-generated files are stored** edit box; the **Select Project Directory** dialog box is displayed. Browse to the directory *C:\Users\user_name\Documents* and choose the **Open** button; the location of the project is automatically displayed in the edit box.

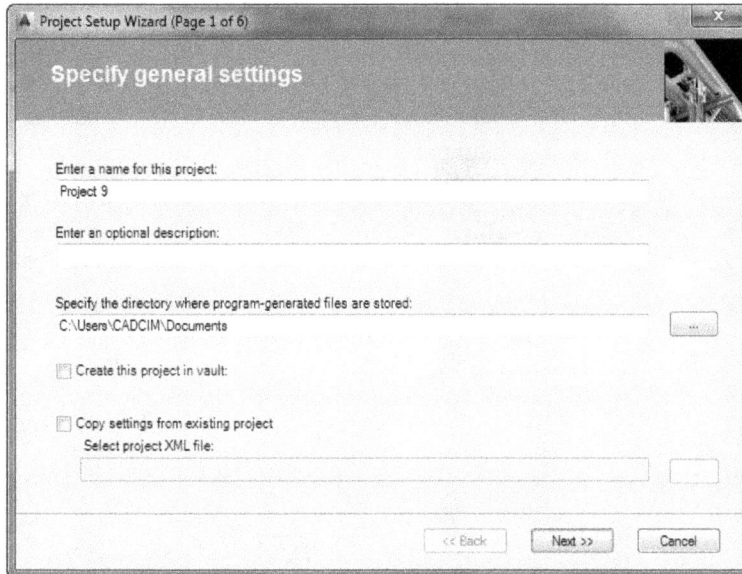

Figure 2-2 The **Project Setup Wizard** *dialog box*

3. Clear the **Copy settings from existing project** check box and choose the **Next** button; the **Specify unit settings** page (Page 2 of 6) will be displayed, as shown in Figure 2-3.

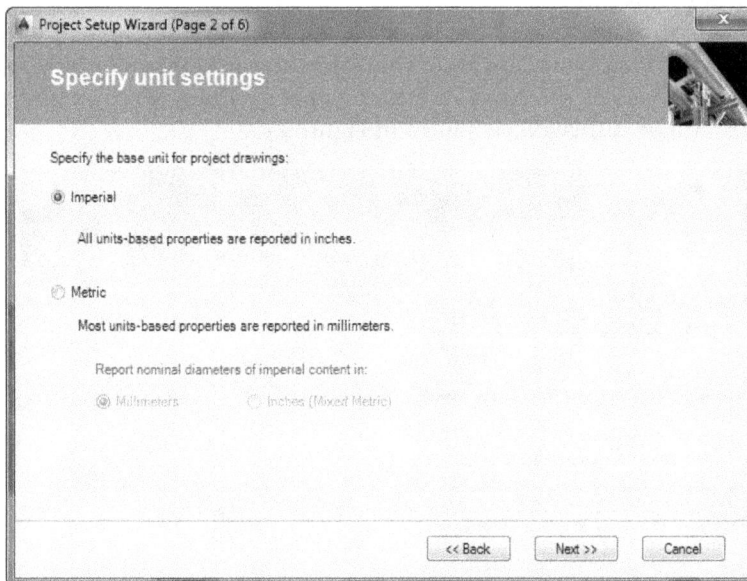

Figure 2-3 The **Specify unit settings** *page*

4. Choose the **Imperial** radio button and the **Next** button; the **Specify P&ID settings** page (Page 3 of 6) will be displayed, as shown in Figure 2-4.

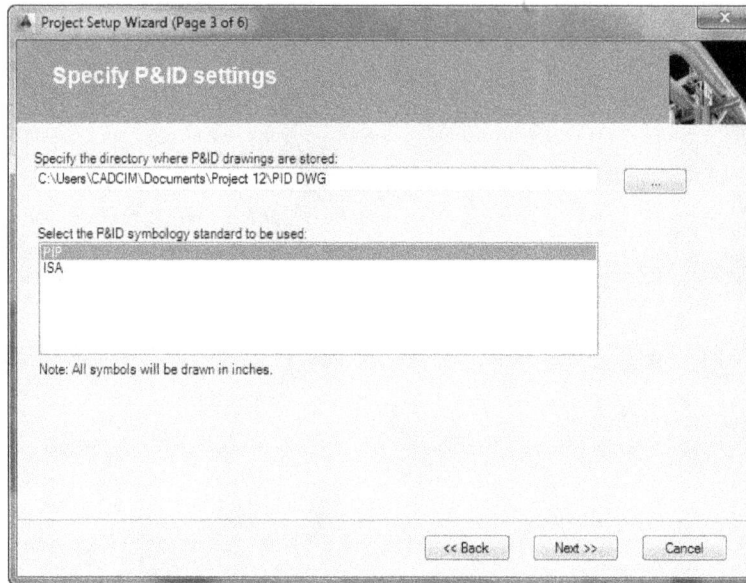

Figure 2-4 The Specify P&ID settings page

5. Specify the directory to *C:\Users\user_name\Documents\My Project\PID DWG* in the **Specify the directory where P&ID drawings are stored** edit box by using the **Browser** button available next to this exit box.

6. Select **PIP** as the P&ID standard from the **Select the P&ID symbology standard to be used** list box and choose the **Next** button; the **Specify Plant 3D directory settings** page (Page 4 of 6) will be displayed, as shown in Figure 2-5.

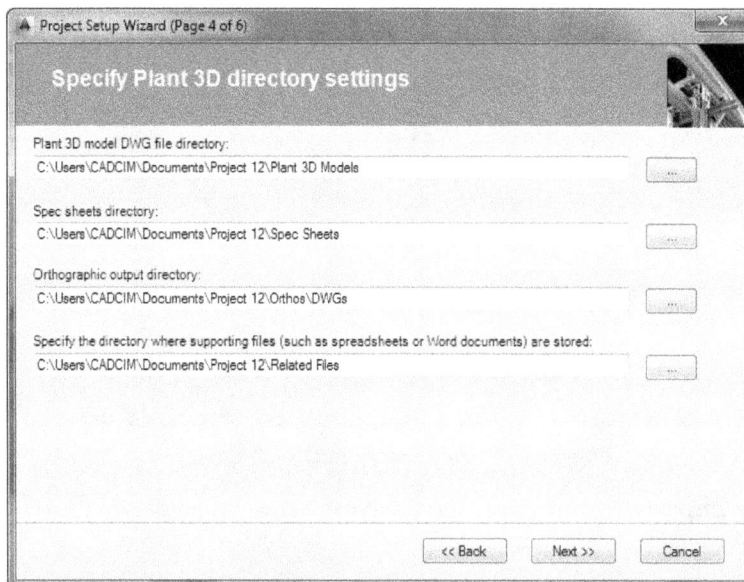

Figure 2-5 The Specify Plant 3D directory settings page

Note

In this book, Imperial standard has been used throughout.

7. Accept the default settings in this page and choose the **Next** button; the **Specify database settings** page (Page 5 of 6) will be displayed, as shown in Figure 2-6.

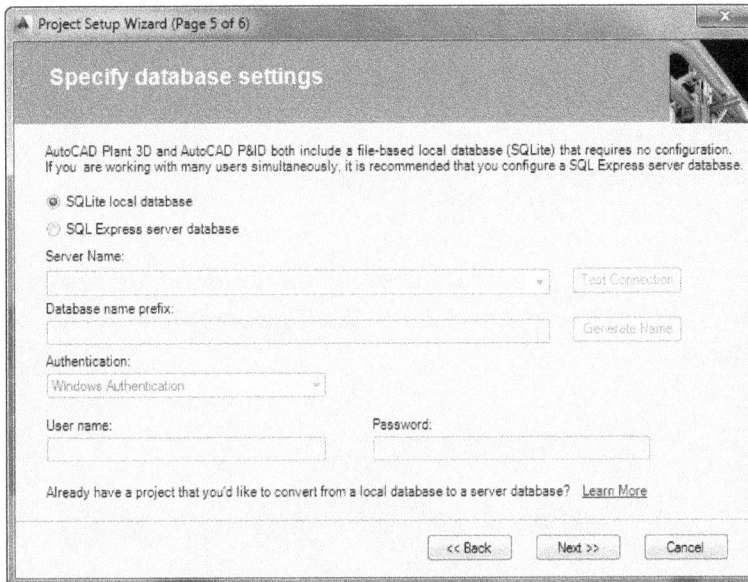

*Figure 2-6 The **Specify database settings** page*

8. In this page, select the option to specify the database settings. Select the **SQLite Local database** radio button, if you are working on a stand-alone system.

    Note that you need to skip Steps 10 through 12, if you have selected the **SQLite Local database** radio button.

9. If your system is connected to a server, select the **SQL Express server database** radio button. Next, you need to specify the server name in the **Server Name** edit box and then choose the **Test Connection** button next to it.

10. After connecting to the server, you need to enter a database prefix in the **Database name Prefix** edit box. You can also choose the **Generate Name** button to automatically generate a prefix.

11. Next, specify the authentication type by using the **Authentication** drop-down list. If you select **SQL Server Authentication**, you need to specify the **Username** and **Password** in the respective edit boxes.

12. After specifying the database settings, choose the **Next** button; the **Finish** page will be displayed.

13. Choose the **Finish** button; a new project will be created and listed in the **Project Manager**.

Creating a New Drawing

To create a new drawing, first select the required node (For example: **P&ID Drawings**) from the **Project Manager**. Next, choose the **New Drawing** button from the **Project** toolbar, refer to Figure 2-7; the **New DWG** dialog box will be displayed, as shown in Figure 2-8. In this dialog box, enter the file name, author, and then select a **dwg** template by choosing the **Browse** button next to the **DWG template** edit box. Choose the **OK** button; a drawing file with the specified name will be created.

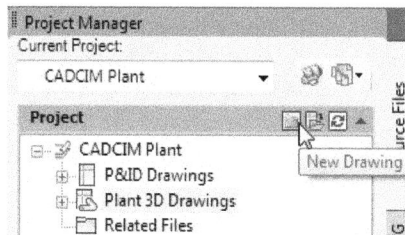

*Figure 2-7 Choosing the **New Drawing** button*

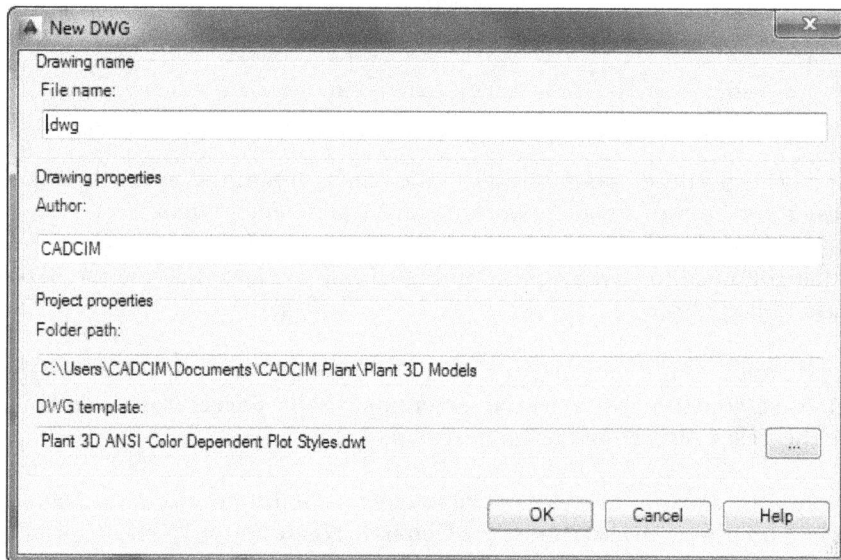

*Figure 2-8 The **New DWG** dialog box*

Grouping Project Files

You can group drawings into a folder. This ensures that the folder path retains the same folder hierarchy. even if the project files are moved to another computer. To group drawings, right-click on the required node (For example **P&ID Drawings** node) in the **Project Manager** tree and choose the **New Folder** option from the shortcut menu displayed; the **Project Folder Properties** dialog box will be displayed, as shown in Figure 2-9. Next, enter a name in the

Folder name edit box. Now, choose the **OK** button; the new folder will be added to the **P&ID Drawings** node in the project tree. Click and drag the drawing file to the newly created folder to move the file, refer to Figure 2-10.

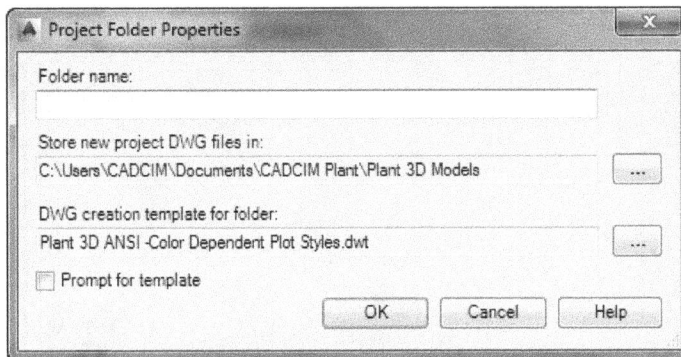

*Figure 2-9 The **Project Folder Properties** dialog box*

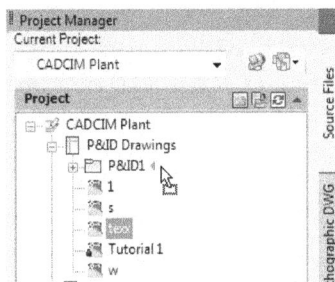

Figure 2-10 Dragging the drawing file into the folder

DESIGNING A P&ID

A P&ID is a graphical representation of a plant process and is created by combining various types of components, piping and instrumentation required to design, construct and operate the plant. To design a P&ID, you need to create a new P&ID drawing file. The procedure to create a new drawing file is explained earlier. After creating a P&ID drawing file, the P&ID environment will be displayed, as shown in Figure 2-11.

The **Tool Palettes** which is at the right side of the window is loaded with Plant 3D components. In order to display P&ID components in it, choose the **Workspace Switching** button available at the right side of the status bar; a flyout will be displayed. Select the **P&ID PIP** option from the flyout, refer to Figure 2-12; the **Tool Palettes** will be loaded with P&ID components, as shown in Figure 2-13.

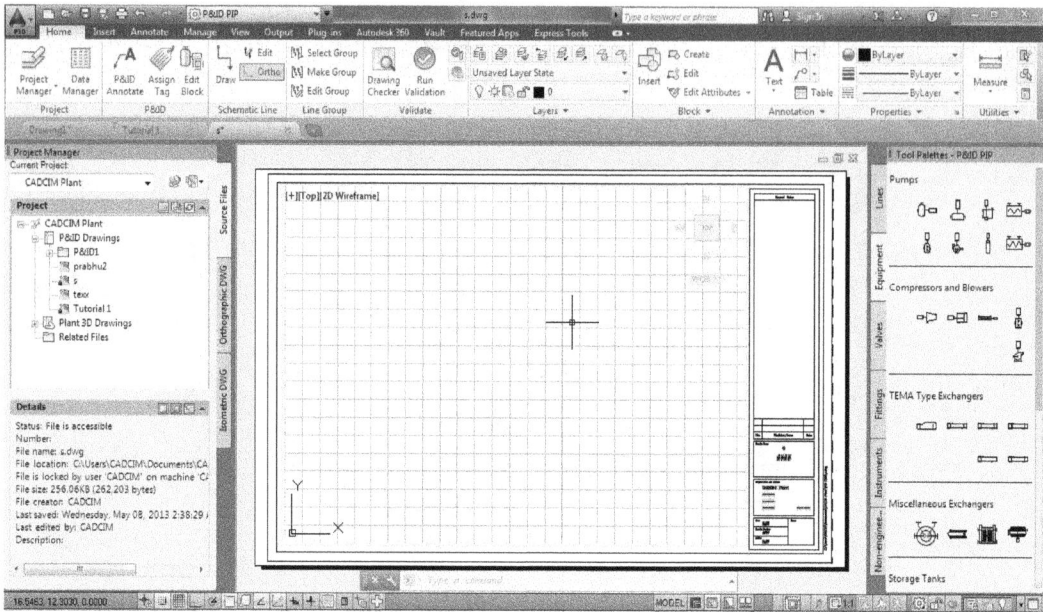

*Figure 2-11 The **P&ID** environment*

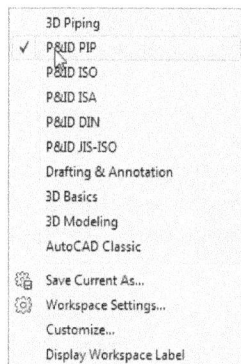

*Figure 2-12 Selecting the **P&ID PIP** option from the flyout*

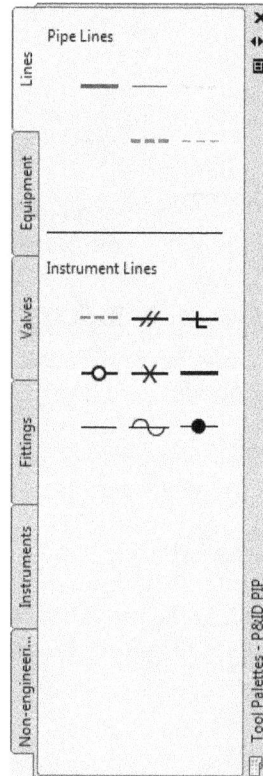

*Figure 2-13 The **Tool Palettes** loaded with P&ID components*

Adding Equipment to a P&ID

Equipment includes all the components that are associated with the process plant such as pumps, towers, heat exchangers, compressors, blowers, tanks, vessels and so on. To place equipment in a P&ID, first you need to see whether the **Tool Palettes - P&ID PIP** is displayed or not. If it is not displayed, choose the **Tool Palettes** button from the **Palettes** panel in the **View** tab; the **Tool Palettes - P&ID PIP** will be displayed on the right side of the window. In the **Tool Palettes - P&ID PIP**, select the **Equipment** tab; the symbols of various equipments will be displayed in it. Choose the symbol of the desired equipment from the **Tool Palette**; it will be attached to the cursor and you will be prompted to specify the insertion point. Click in the drawing area to place the symbol; you will be prompted to enter XY scale factor. You can scale the equipment horizontally or vertically by dragging the cursor in the respective direction. Alternatively, you can enter a scale factor in the command prompt. Next, press ENTER; the **Assign Tag** dialog box will be displayed, as shown in Figure 2-14.

In the **Assign Tag** dialog box, enter a numeric value in the **Number** edit box and select the **Place annotation after assigning tag** check box. Next, choose the **Assign** button; the **Assign Tag** dialog box will be closed and you will be prompted to select the annotation position. Place the annotation below the equipment at an appropriate location.

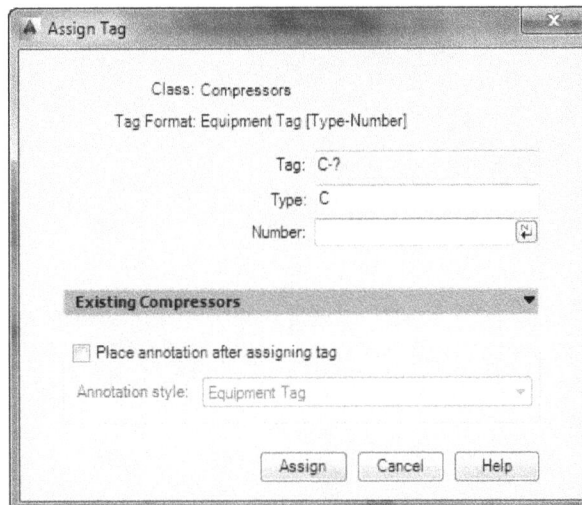

*Figure 2-14 The **Assign Tag** dialog box*

Adding Pipe Lines

Pipe Lines represent pipe lines which are connected to an equipment. You can select a pipe line from the **Pipe Lines** area in the **Line** tab of the **Tool Palettes - P&ID PIP**. For example, to connect a vessel and a condenser with a primary line, choose the **Primary Line Segment** tool from the **Pipe Lines** area in the **Lines** tab of the **Tool Palettes - P&ID PIP**; you will be prompted to select the start point. Select a point on the top of the vessel. Next, connect the line to the top nozzle of the condenser, as shown in Figure 2-15; a nozzle will be created automatically on the vessel and an arrow will show the direction of flow. Note that the direction of flow is determined by the direction in which you move the cursor while creating the line.

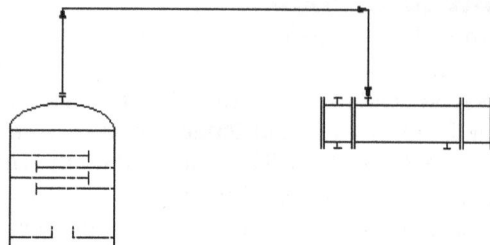

Figure 2-15 *Pipe line connecting the vessel and the condenser*

You can easily distinguish between a line connected to a component, and a line not connected. On selecting the line that is connected to a component, a connection grip is displayed at the end of the pipe line, as shown in Figure 2-16. But if the line is not connected to a component, an end grip is displayed at the end of the pipe line, as shown in Figure 2-17.

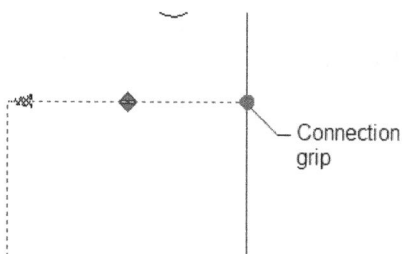

Figure 2-16 *A connected line showing the connection grip*

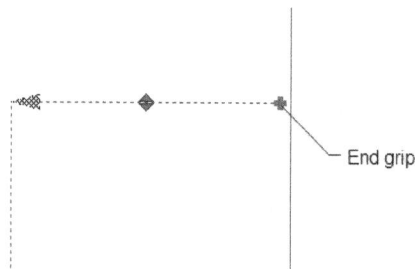

Figure 2-17 *A line not connected to any line or a component*

Note
After connecting two components with a pipe line, the 'From' and 'To' information is automatically added to the line. This information is displayed when you hover the cursor over the line. The 'From' field shows the component from which the line originates and the 'To' field indicates the destination component.

Assigning Tags to a Line

After creating a line, you need to assign a tag to it. The tag data consists of information such as pipe size, specification, service, and line number, refer to Figure 2-18, for the representation of tag data on the line. In order to assign a tag to a line, choose the **Assign Tag** tool from the **P&ID** panel and select a line. Next, press ENTER; the **Assign Tag** dialog box will be displayed. Enter the data in the respective edit boxes and choose the **Assign** button; the tag data will be assigned to the line. Also, you can select the **Place annotation after assigning tag** check box from the **Assign Tag** dialog box in order to place the annotation near the line.

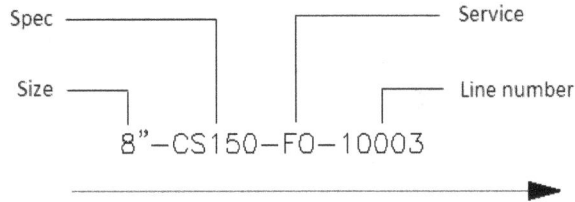

Figure 2-18 *Annotation of a pipeline Tag*

Adding Valves

The **Valve** tab in the **Tool Palettes - P&ID PIP** contains a group of valve symbols. You can add any type of valve to a pipe line. For example, to place a gate valve on the pipe line, you need to select the **Valves** tab from **Tool Palettes - P&ID PIP** and choose the **Gate Valve** tool from the **Valves** area; the valve symbol will be attached to the cursor and you will be prompted to select the insertion point. Select an insertion point on the line; the valve will be placed, as shown in Figure 2-19.

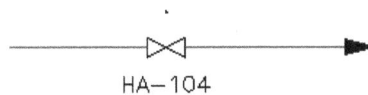

Figure 2-19 *A gate valve placed on the line*

Adding Instruments and Instrumentation Lines

Instruments play a major role in maintaining the safety and streamlining the process in the plant. The **Instruments** tab in the **Tool Palettes - P&ID PIP** contains the **Control Valve, Relief Valves, Primary Element Symbols (Flow),** and **General Instruments** areas. Figure 2-20 shows various instrument symbols available in the **Instruments** tab.

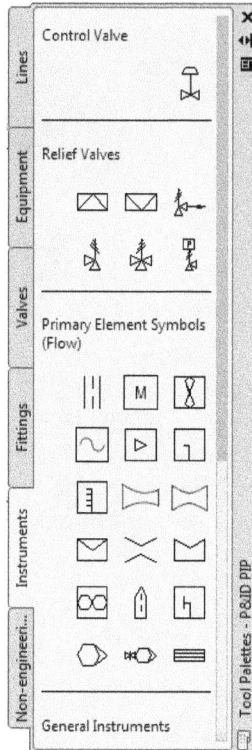

Figure 2-20 *Instrument symbols*
*available in the **Instruments** tab*

For example, to add a control valve, choose the **Control Valve** tool from the **Instrument** tab; the **Control Valve Browser** dialog box will be displayed, if you are adding a control valve for the first time. Select the required valve body from the **Select Control Valve Body** tree view. Next, select the actuator type from the **Select Control Valve Actuator** tree view, as shown in Figure 2-21 and choose **OK**; the dialog box will be closed and you will be prompted to select an insertion point. Select an insertion point on the line; the control valve will be added and will be prompted to select annotation position.

Note
*If the **Control Valve Browser** dialog box is not displayed, choose the **Change body or actuator** option from the Command Bar to change the actuator of valve body.*

Move the cursor and click to position the annotation; the **Assign Tag** dialog box will be displayed. Enter values in the **Area**, **Type,** and **Loop Number** edit boxes, and clear the **Place annotation after assigning tag** check box. Next, choose the **Assign** button from the **Assign Tag** dialog box; an annotation bubble will be placed at the specified location.

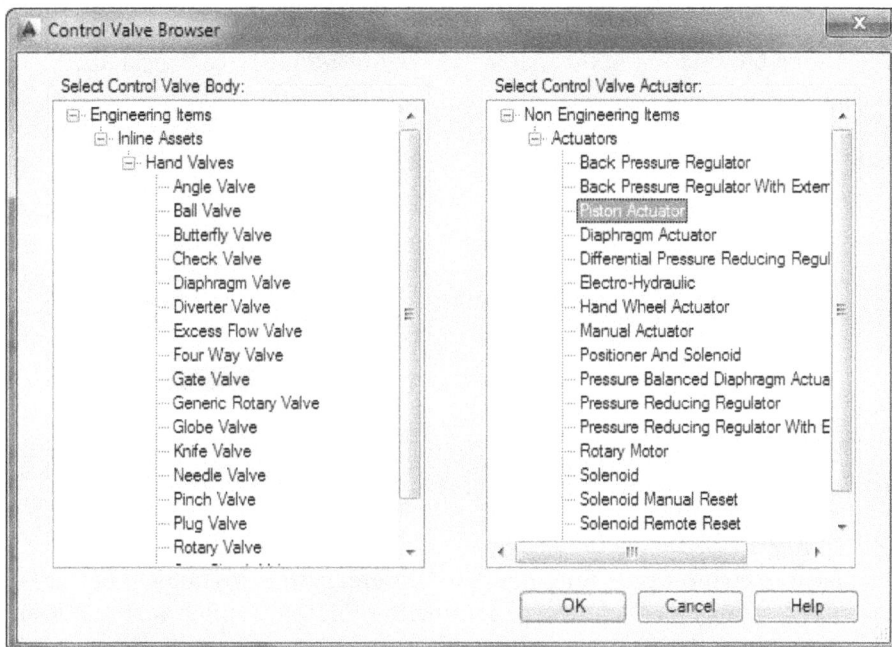

Figure 2-21 *The **Control Valve Browser** dialog box*

After adding the control valve, you need to add the signal lines connecting the instruments and the process line. Figure 2-22 shows various instrument lines used in a P&ID. For example, to add an electrical signal, choose the **Electrical Signal** line from the **Instrument Lines** area in the **Lines** tab; you will be prompted to select the start point. Select the point on the process line; you will be prompted to select the next point. Move the cursor toward left and connect the line to the control valve.

Next, you need to place control instruments on the signal lines. For example, to add a temperature controller, choose the **Field Descrete Instrument** tool from the **General Instruments** area in the **Tool Palette**; you will be prompted to specify the insertion point. Select a point on the signal line connecting the control valve; the instrument will be placed and the **Assign Tag** dialog box will be displayed.

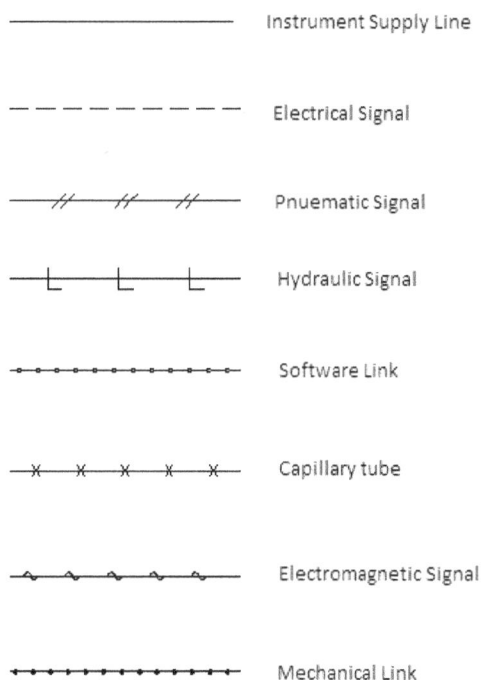

Figure 2-22 *The instrument line symbols*

In the **Assign Tag** dialog box, enter appropriate values in the **Area** and the **Loop Number** edit boxes. Next, select the **Temperature Controller** option from the **Type** drop-down list and choose the **Assign** button to close the dialog box. Figure 2-23 shows a typical instrumentation loop.

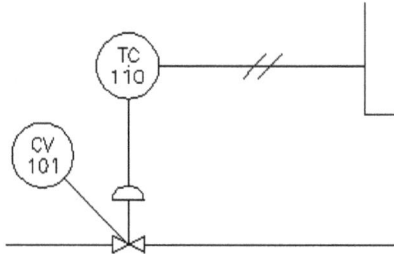

Figure 2-23 *A typical instrumentation loop*

Adding Fittings

Fittings are used to connect pipe lines that have different sizes or shapes. They are used for regulating or measuring flow. You can add a fitting to a P&ID drawing from the **Fittings** tab in the Tool Palette. It contains symbols of **Piping Fittings**, **Piping Speciality Items** and **Nozzles**. For example, to add a concentric reducer, choose the **Concentric Reducer** tool from the **Piping Fittings** area in the **Fittings** tab; you will be prompted to specify the insertion point. Select a point on the line; the reducer will be placed at the specified position.

Adding the Off page Connectors

The off page connectors are used to indicate the continuation of the process line from one drawing sheet to another. To add an off-page connector, choose the **Off Page Connector** tool from the **Off Page Connectors and Tie-In Symbol** area in the **Non-engineering** tab of the **Tool Palettes - P&ID PIP**; you will be prompted to specify the insertion point. Select the endpoint of the pipe line, as shown in Figure 2-24; the off page connector will be placed at the specified location.

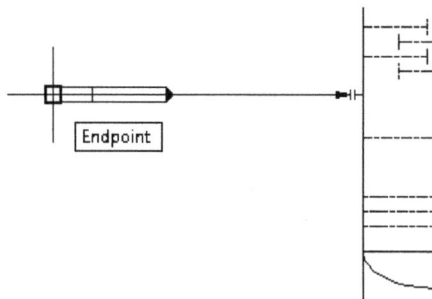

Figure 2-24 *Placing the off-page connector*

Connecting the Off page Connectors

To connect two off page connectors, select any one of the off page connectors; a plus symbol will be displayed on it, refer to Figure 2-25. Select the plus symbol and choose the **Connect To** option, refer to Figure 2-26; the **Create Connection** dialog box will be displayed, as shown in Figure 2-27.

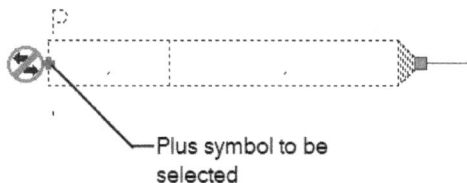

Plus symbol to be selected

Figure 2-25 The plus symbol diplayed on the off page connector

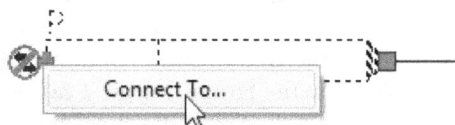

*Figure 2-26 Choosing the **Connect To** option*

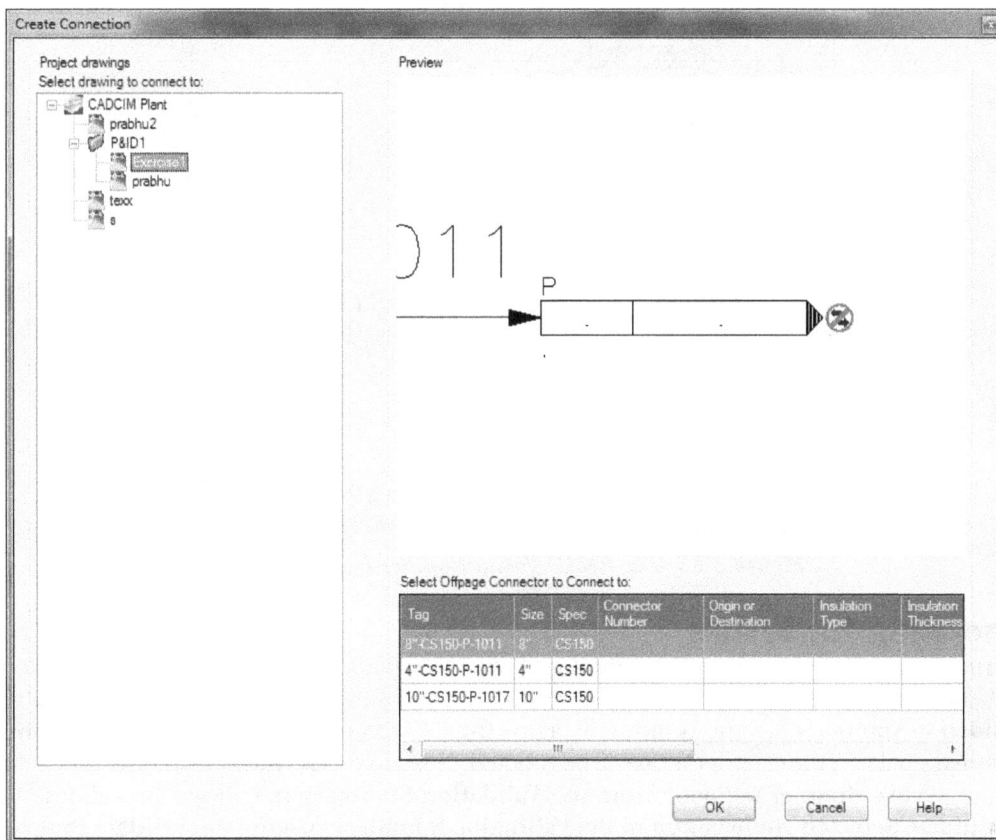

*Figure 2-27 The **Create Connection** dialog box*

In this dialog box, select the drawing to which the off page connector is to be connected from the **Project drawings** area. Next, select the off page connector to be connected from the **Select Offpage Connector to Connect to** table. Choose the **OK** button from the **Create Connection** dialog box; the off page connectors will be connected.

VALIDATING THE DRAWING

You need to validate the drawing to detect the errors and to correct them. To do so, first you need to select the conditions to validate a P&ID. Choose the **Validate Config** tool from the **Validate** panel in the **Home** tab; the **P&ID Validation Settings** dialog box will be displayed, as shown in Figure 2-28. Select the check boxes under the **P&ID objects** and the **Base AutoCAD objects** nodes and choose the **OK** button; the **P&ID Validation Settings** dialog box will be closed.

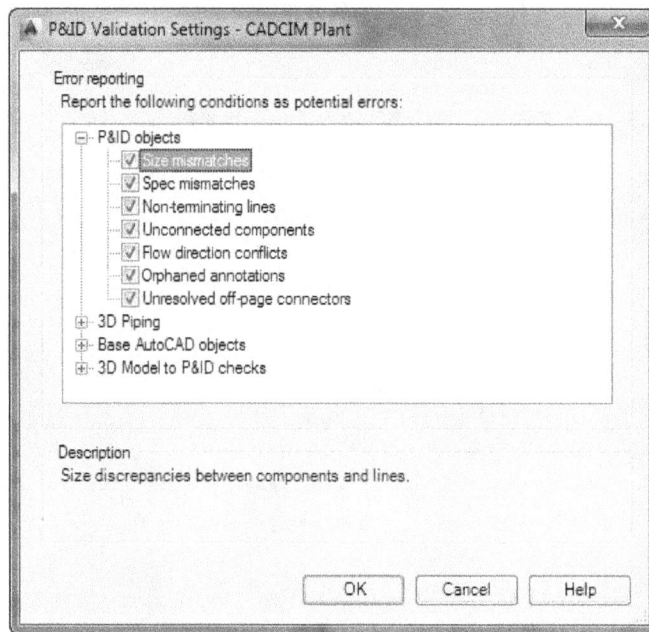

*Figure 2-28 The **P&ID Validation Settings** dialog box*

Checking for Errors

To run check, choose the **Run Validation** tool from the **Validate** panel; the validation process will start and the errors are checked. The list of errors detected will be displayed in the **Validation Summary** palette, as shown in Figure 2-29. Click on the error type in the **Validation Summary** palette; the error location will be zoomed. Next, to correct the individual errors, select the respective drawing file node from the **Validation Summary** tree view. Now, choose the **Revalidate Selected Node** button in the **Validation Summary** palette; the **Validate Progress** message box will be displayed and errors are checked again. You can also check for errors that are not corrected and close the **Validation Summary** palette.

Correcting the Errors

After validating a P&ID, you need to correct the errors displayed in the **Validation Summary** palette. The methods to correct the errors are discussed next.

Correcting the Base AutoCAD Object Errors

You can correct a base AutoCAD object error by converting it into a P&ID component. To do so, first you need to expand the **Base AutoCAD Object errors** node in the **P&ID Validation Settings** dialog box. Next, click on the error; the error will zoom in the drawing. Next, right-click on it and choose the **Convert to P&ID object** option from the shortcut menu displayed; the **Convert to P&ID** dialog box will be displayed. In this dialog box, select the required component type from the **Classes** tree and choose the **OK** button; the base object will be converted into a P&ID component.

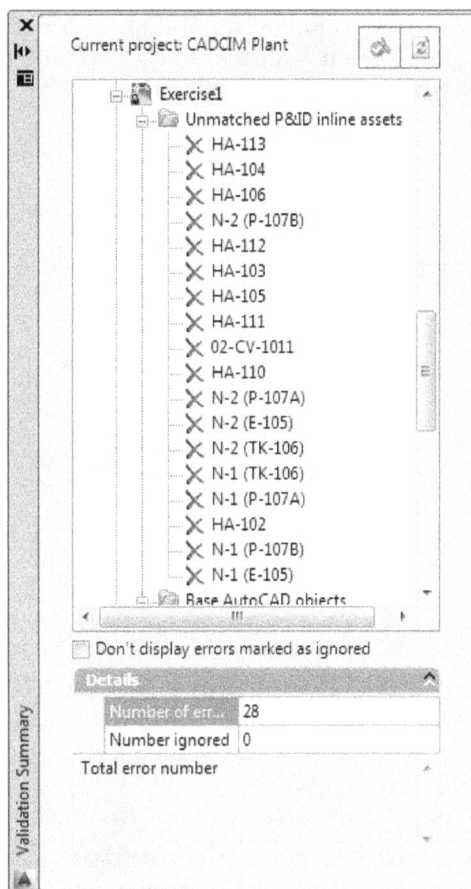

Figure 2-29 The Validation Summary palette

You can also ignore the base AutoCAD object error. To do so, right-click on it in the **Validation Summary** palette; a shortcut menu will be displayed. Next, choose the **Ignore** option to ignore the error. You can also delete the object by choosing the **Erase** option.

Correcting the Size Mismatch Errors

To correct the size mismatch error, expand the **Size mismatches** node in the **Validation Summary** palette and then click on the error; the error will zoom in the drawing area. Next, manually change the size of the component or the size of the pipe line connected to it.

Correcting the Spec mismatch Errors

To correct the spec mismatch errors, zoom in the error location by clicking on the error in the **Spec mismatches** node. Next, manually change the spec of the pipe line by invoking the **Assign Tag** dialog box.

Correcting the Orphaned Annotation Errors

To correct the orphaned annotation error, expand the **Orphaned annotations** node in the **Validation Summary** palette. Next, select the error to zoom it in the drawing area. Next, drag the annotation and place it near the parent object. You can also right-click on the annotation and then choose the **Place from Parent** option from the shortcut menu to place it near the parent object. Figure 2-30 shows an orphaned annotation and Figure 2-31 shows the annotation after correcting the error.

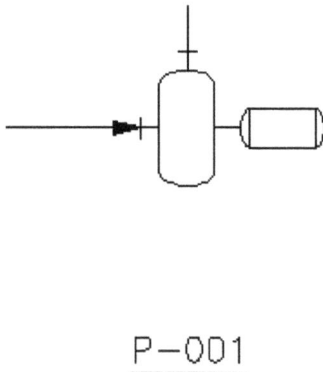

P−001

Figure 2-30 Orphaned annotation

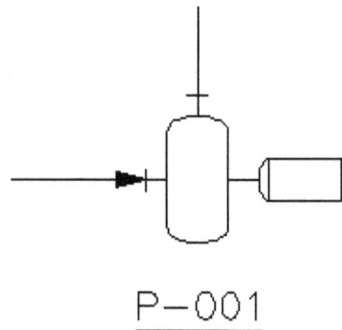

P−001

Figure 2-31 Annotation after placing it close to the component

Correcting the Unconnected Component Errors

To correct the unconnected component error, select the error from the **Unconnected Components** node; the component in the drawing area will be zoomed in. Next, manually connect the component to the line near it.

Correcting the Unresolved Off Page Connectors Errors

Select the error from the **Unresolved Off-Page Connectors** node in the **Validation Summary** palette. Next, right-click on the off-page connector and choose **Off-page connector > Connect** from the shortcut menu displayed; the **Create Connection** dialog box will be displayed, refer to Figure 2-27. Next, select the drawing to which you want to connect the off page connector and then choose the **OK** button.

Correcting the Non-Terminating Line Errors

Select the error from the **Non-terminating lines** nodes to zoom it in the drawing area. Next, manually reconnect the line to the corresponding object by moving its endpoint. You can also recreate the connection.

Correcting the Flow Direction errors

To correct a flow direction error, click on the error under the **Flow Direction Conflicts** node; the error location will zoom in the drawing area. Next, click on the direction arrow; the **Flip Component** grip will be displayed. Click on this grip to flip the flow direction.

EDITING THE DRAWING

You can edit or modify a drawing while creating it or after it is created. The components of the drawing such as equipment, valves, instruments, lines, and so on can be modified as per the requirement. The various editing tasks are discussed next.

Moving an Equipment

You can move or change the location of an equipment. To do so, first you need to select the equipment by clicking on its border; the **Move Component** grip (square grip) will be displayed on it. Select it and drag the equipment and click to place it at a new location, as shown in Figure 2-32.

Moving a Valve

To move a valve, first select it to display the **Move Component** grip. Next, click on the grip, drag the valve along the horizontal or vertical line, and then click to place it at the new location. You can move the valve which is on the horizontal line to the vertical line or vice-versa, refer to Figure 2-33.

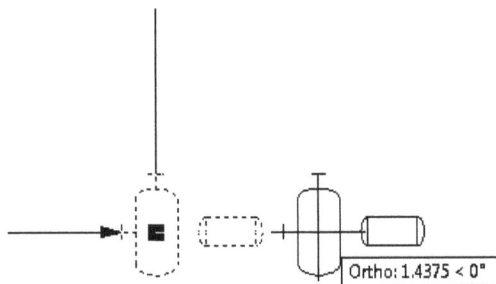

Figure 2-32 *Moving an equipment*

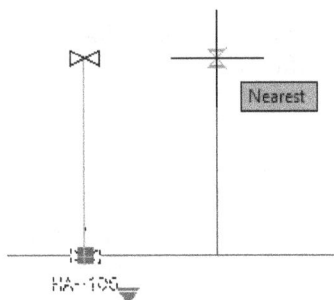

Figure 2-33 *Moving a valve from horizontal line to the vertical line*

Moving a Line

To move a line, first you need to select it; the stretch grip will be displayed on the line and the components connected to the line will be automatically selected. Select it from the middle of the line and drag it to the new location, as shown in Figure 2-34; the valves and the bubbles on the line will be moved. Note that the equipment connected to the line will not move.

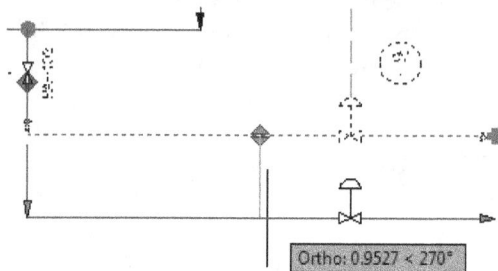

Figure 2-34 *Moving a line using the stretch grips*

Editing a Line

To edit a line, choose the **Edit** tool from the **Schematic line** panel in the **Home** tab; you will be prompted to select a line. Select a pipe line or a signal line from the drawing area; you will be prompted to enter an option at the Command prompt. The options available at the Command prompt are discussed next.

Attach

This option is used to attach a line to a component without physically connecting them. On invoking this option, you will be prompted to select a component to attach to the line. Select an equipment, valve, or any other P&ID component; you will be prompted to select an endpoint on the line. Select an endpoint on the line and press ENTER; the component will be attached to the line.

Detach

This option is used to detach the attached line from the component. On invoking this option, you will be prompted to select the endpoint on the line which has been attached to the source line. Select the attached endpoint on the line and press ENTER; the component will be detached from the line.

Gap

This option is used to create a gap on the line without breaking it. On invoking this option, you will be prompted to specify first gap point on the line. Specify the first gap point and move the cursor upto the required distance, and select the second gap point; a gap will be created on the line between two specified points.

uNgap

This option removes the gap created on a line. Invoke this option and select the line segment with a gap or gap symbol; the gap will be removed from the line.

Straighten

This option is used to straighten an inclined or a non-orthogonal line. On invoking this option, you will be prompted to select the line segment to be straightened. Select the line segment; you will be prompted to select an endpoint on the line to which the source line will be aligned. Select the endpoint on the line; the line will be straightened in alignment with the endpoint.

Corner

This option is used to create a corner by dividing the line into two sides. On invoking this option, you will be prompted to select a point on the line. Select a point on the line; the line will be divided into two sides and you will be prompted to specify the second point of the corner. Move the cursor in a direction perpendicular to the line and select a point in the drawing area; you will be prompted to specify a point on the side of the pipe line to move it. Select a point on one of the two sides; the selected side will be moved and aligned with the second point in the prompt sequence. As a result, a corner will be created.

Reverseflow

This option is used to reverse the direction of a pipe line. Invoke this option and press ENTER; the direction of flow of the pipe line will be reversed.

Break

This option is used to break the line into two segments. On invoking this option, you will be prompted to select a break point. Select a point on the line; the line will be divided into two segments and you will be prompted to select an additional break point. Press ENTER to exit.

Join

This option is used to join two line segments. On invoking this option, you will be prompted to select one or more lines to join the source line. Select a line segment that lies inline with the source line; the line will be joined with the source line.

Link

This option is used to link data of two line segments. On invoking this option, the following prompt will be displayed:

Select a sline segment to link to: *Select a line*

Sline segment data "?-?-?-?" will be deleted and segment will be linked to "?-?-?-?". Continue? [Yes/No]: *Enter 'Yes' to continue*

Unlink

This option unlinks the data link created between the two line segments. On invoking this option, the following prompt will be displayed:

Sline segment will be unlinked from "?-?-?-?". Continue? [Yes/No]: *Enter 'Yes' to continue*

eXit

This option is used to exit the command.

Grouping Lines

[⁞⁞] Make Group   You can group pipe lines together and assign tag to the group. To do so, choose the **Make Group** tool from the **Line Group** panel; you are prompted to select the source line to group additional lines to it. Select the source line; you will be prompted to select pipe lines to group. Select a single line or multiple lines and press ENTER; the lines will be grouped together.

Editing a P&ID Symbol

You can edit a P&ID symbol and also retain the original symbol by using the **Edit Block** tool. To do so, choose the **Edit Block** tool from the **P&ID** panel in the **Home** tab; you will be prompted to select a P&ID component to edit its block. Select an equipment, valve or any other P&ID component; the **Block Editor** will be invoked. The drawing area of the **Block Editor** has a dull background and is called Authoring area. In this area, you can edit the existing entities or add new ones to the block of a P&ID component. In addition to the authoring area, the **Block Editor** tab, the **Block Authoring Palettes** and the **Edit P&ID Object's Block** dialog box will also be displayed. After editing the block, choose the **Save Changes and Exit Block Editor** button from the **Edit P&ID Object's Block** dialog box; the **Block Editor** will be closed and the edited P&ID component will be displayed in the drawing area.

Substituting Components

You can substitute a valve or any other component of a drawing with another similar component. For example, you can substitute a gate valve with a ball valve. To do so, first you need to select the component to display the grips on it. Next, select the substitution grip (down arrow); the **Substitution** palette containing valves will be displayed, as shown in Figure 2-35. Select the ball valve symbol from the palette; it will replace the existing gate valve.

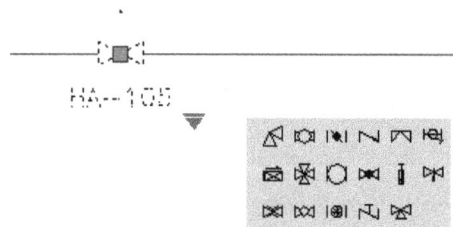

*Figure 2-35 **Substitution** palette showing valves*

Converting AutoCAD Components into P&ID Symbols

You can add new symbols which are not available in the **Tool Palettes - P&ID PIP** by creating a block and converting it into a P&ID symbol. To convert a block into a P&ID symbol, first you need to create a block by using the AutoCAD drawing tools. Next, select the component, right-click on it and choose **Convert to P&ID Object** from the shortcut menu displayed; the **Convert to P&ID Object** dialog box will be displayed, as shown in Figure 2-36. The tree view in the dialog box shows a list of components arranged in groups. Expand the tree view by clicking on the + sign adjacent to the required group and select the object from it. Next, choose the **OK** button; you will be prompted to select the insertion point. Select an insertion point on the component; the component will be converted into the selected P&ID object.

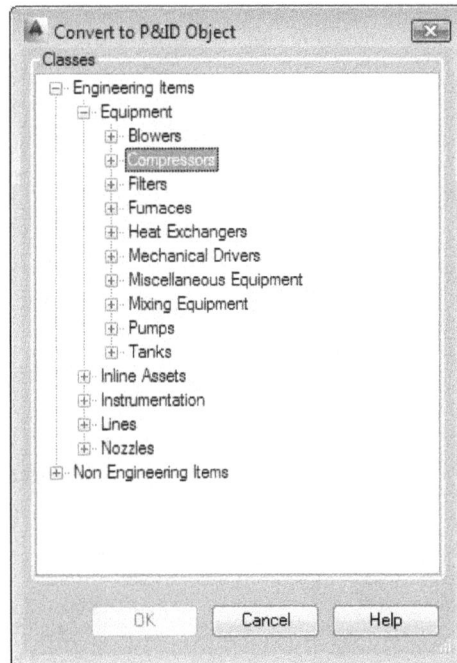

*Figure 2-36 The **Convert to P&ID Object** dialog box*

To add the converted component to the **Tool Palettes - P&ID PIP**, first you need to save the current file. Next, open the tab in the **Tool Palette** to which you want to add the symbol. Then, click and drag the object and place it in the desired group in the **Tool Palette**; the symbol will be displayed in the Tool Palette.

TUTORIAL

Tutorial 1

In this tutorial, you will create a P&ID, as shown in Figure 2-37. You will create it by placing equipments, instruments, and inline components and then connecting them.

(Expected Time: 2hr)

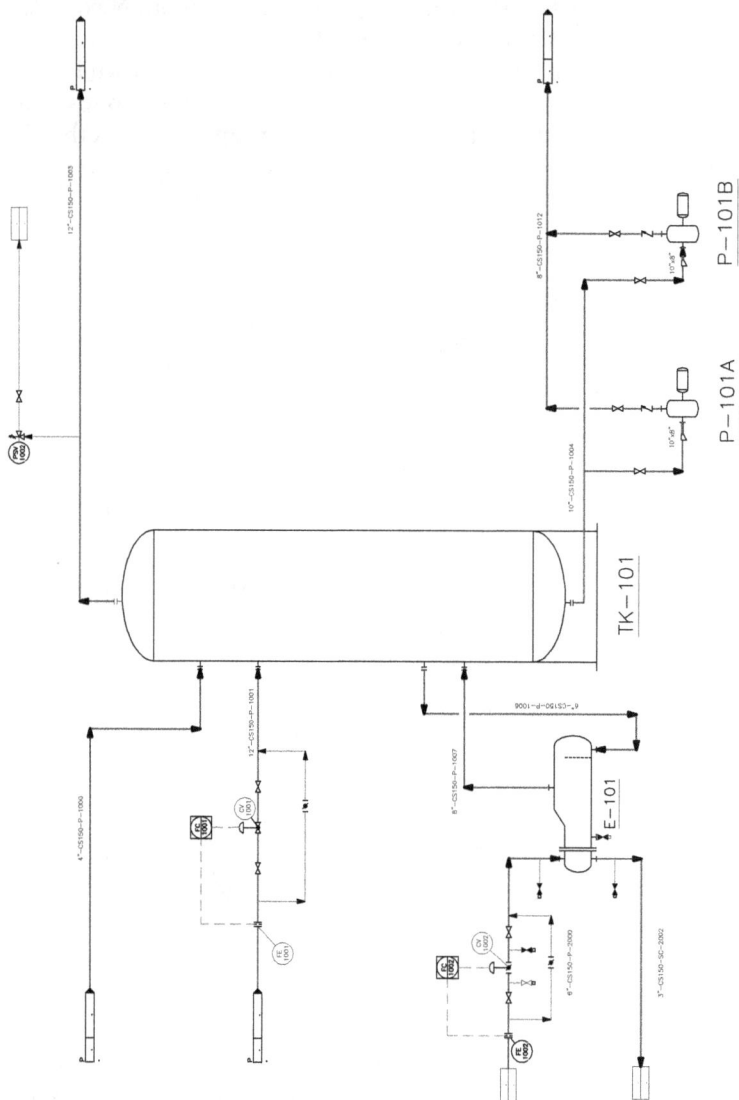

Figure 2-37 *P&ID for Tutorial 1*

The following steps are required to complete this tutorial:

a. Start AutoCAD Plant 3D and then create a new project.
b. Create a new drawing.

c.   Place equipments in the drawing.
d.   Connect equipments using pipe lines.
e.   Place valves on the lines connecting the components.
f.   Add off-page connectors to the drawing.
g.   Add tags to pipe lines.
h.   Validate the drawing.
i.   Save and close the drawing file.

Starting AutoCAD Plant 3D and Creating a New Project

1.   Choose **Start > All Programs (or Programs) > Autodesk > AutoCAD Plant 3D 2014 > AutoCAD Plant 3D 2014**; AutoCAD Plant 3D 2014 is started. Alternatively, double-click on the shortcut icon of **AutoCAD Plant 3D 2014** on the desktop of your computer to start **AutoCAD Plant 3D 2014**.

2.   The **Project Manager** is displayed by default on the left of the screen. In the **Project Manager**, choose the **New Project** option from the drop-down list; the **Project Setup Wizard** dialog box is displayed.

3.   Enter **CADCIM** in the **Enter a name for this project** edit box and choose the **Next** button; the **Specify unit settings** page is displayed.

4.   Select the **Imperial** radio button and choose the **Next** button; the **Specify P&ID settings** page is displayed.

5.   Select the **PIP** option from the **Select the P&ID symbology standard to be used** list box and choose the **Next** button; the **Specify Plant 3D directory settings** page is displayed.

6.   Accept the default directory settings and choose the **Next** button; the **Specify database settings** page is displayed.

7.   Select the **SQLite local database** radio button and choose the **Next** button; the **Finish** page is displayed.

8.   Choose the **Finish** button; you will notice that the CADCIM project is displayed in the **Project** drop-down list displayed in the **Project Manager**.

Creating a New Drawing

1.   In the **Project Manager**, right-click on the **P&ID Drawings** node in the **Project** tree and then choose the **New Drawing** option; the **New DWG** dialog box is displayed.

2.   Enter **P&ID1.dwg** in the **File name** edit box and then specify the **PID ANSI D - Color Dependent Plot Styles.dwt** template in the **DWG template** edit box.

3.   Choose the **OK** button from the **New DWG** dialog box; the new P&ID drawing file is created with the **Tool Palette** displayed on the right side of the screen.

4.   Choose the **Workspace Switching** button on the right-side of the status bar; a flyout is

displayed. Choose the **P&ID PIP** option from the flyout; the **P&ID PIP** workspace is invoked and the **Tool Palette** is loaded with the P&ID symbols.

Placing Equipment in the Drawing

1. Choose the **Vessel** tool from the **Vessel and Miscellaneous Vessel Details** area in the **Equipment** tab of the **Tool Palettes - P&ID PIP**; you are prompted to specify an insertion point. Next, specify **(13,5)** as the insertion point of the vessel; you are prompted to specify the scale factor.

2. Enter **1.5** as the scale factor; the **Assign Tag** dialog box is displayed. Enter **101** in the **Number** edit box and make sure that the **Place annotation after assigning tag** check box is selected. Next, choose the **Assign** button and specify the position of the annotation tag near the vessel, as shown in Figure 2-38.

3. Choose **TEMA type BKU Exchanger** from the **TEMA Type Exchangers** area in the **Tool Palettes - P&ID PIP** and place it at the point **(7,5)**. Next, assign tag **E-101** to the heat exchanger. Figure 2-38 shows the P&ID after placing the vessel and the heat exchanger.

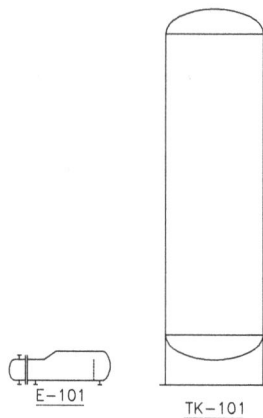

Figure 2-38 *P&ID after placing the vessel and the heat exchanger*

4. Choose **Horizontal Centrifugal pump** from the **Pump** area in the **Equipment** tab of the **Tool Palettes - P&ID PIP** and place at the point **(17,3)**. Next, assign tag **P-101A** to the pump.

5. Similarly, place another horizontal centrifugal pump at the point **(21,3)**. Next, assign the tag **P-101B** to the pump. Figure 2-39 shows the P&ID after placing the centrifugal pumps.

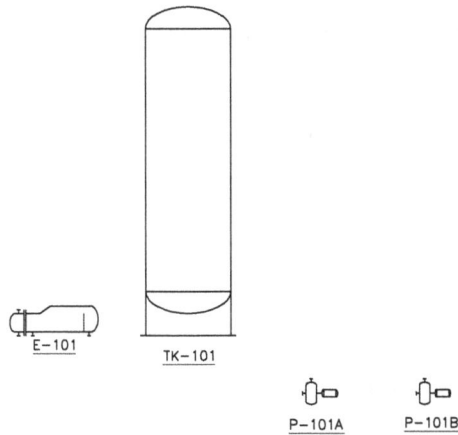

Figure 2-39 *P&ID after placing two centrifugal pumps*

Connecting the Heat Exchanger and the Vessel

Now, you need to create pipe lines connecting the equipment. Before doing that, you need to place a nozzle at the required location on the equipment.

1. Choose **Single Line Nozzle** from the **Nozzle** area in the **Fittings** tab of the **Tool Palettes - P&ID PIP**; you are prompted to select the asset to place nozzle.

2. Select the heat exchanger from the drawing sheet; the nozzle is attached to the cursor and you are prompted to specify the insertion point.

3. Specify the insertion point, as shown in Figure 2-40, and then specify the rotational angle as 90-degrees.

   Next, you need to connect the heat exchanger and the vessel by drawing pipe lines. To do so, you need to first make sure that the **Ortho Mode** is turned on to draw straight lines. Also, the **Object Snap** should be turned on for easy selection of the points.

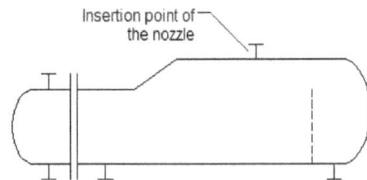

Figure 2-40 *Insertion point of the nozzle*

4. Select the **Lines** tab from the **Tool Palette** and choose the **Primary Line Segment** line type from the **Pipe Lines** area; you are prompted to select the start point.

5. Specify the start point on the newly created nozzle of the heat exchanger. Next, move the cursor upward and enter **2** at the Command prompt. Next, press ENTER.

6. Move the cursor horizontally toward right and specify the endpoint on the vessel; the line connecting the vessel and heat exchanger is created, as shown in Figure 2-41.

Next, you need to assign a tag to the line.

7. Choose the **Assign Tag** tool from the **P&ID** panel in the **Home** tab and select the previously created line.

8. Press the ENTER key; the **Assign Tag** dialog box is displayed. Specify the following values in the dialog box:

Size:	**8"**
Spec:	**CS150 - 150# Carbon Steel**
Pipe Line Group.Service:	**P-GENERAL PROCESS**
Pipe Line Group.Line Number:	**1007**

9. Clear the **Place annotation after assigning tag** check box choose the **Assign** button; the tag is assigned to the line.

> **Note**
> *You can also annotate the line with the assigned tag. To do so, you need to select the **Place annotation after assigning tag** check box and then select the **Pipeline Tag** option from the **Annotation style** drop-down list in the **Assign Tag** dialog box.*

10. Invoke the **Primary Line Segment** tool from the **Lines** tab of the Tool Palette and connect the vessel to the nozzle located at the bottom of the heat exchanger. The line created should be similar to the one shown in Figure 2-42.

11. Assign the tag **6"-CS150-P-1006** to the line.

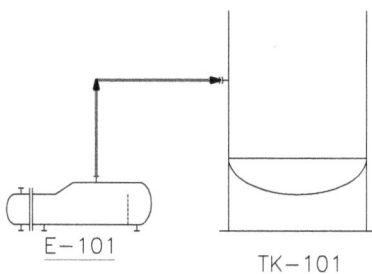

Figure 2-41 Line connecting the vessel and the heat exchanger

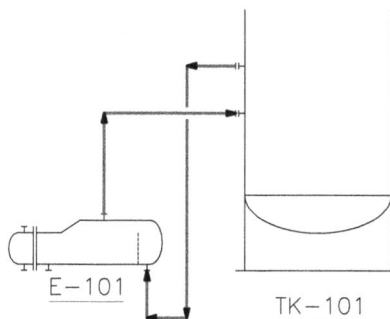

Figure 2-42 Line connecting the vessel and the bottom nozzle of the heat exchanger

12. Similarly, create lines connecting the other two nozzles (steam inlet and outlet nozzles) of the heat exchanger, as shown in Figure 2-43. Next, assign the tags **6"-CS150-P-2000** and **3"-CS150-SC-2002** to steam inlet line and condensate lines, respectively.

13. Choose the **Secondary Line Segment** tool from the **Lines** tab in the **Tool Palettes - P&ID PIP** and create a loop on the line connecting the steam inlet nozzle of the heat exchanger, as shown in Figure 2-44.

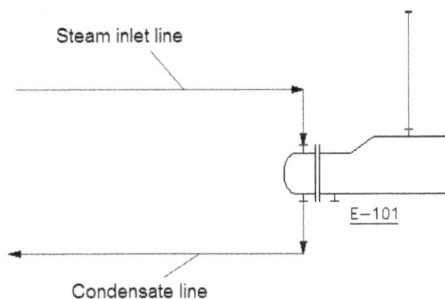

Figure 2-43 *The steam inlet line and condensate line of the heat exchanger*

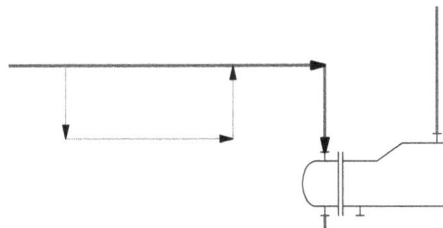

Figure 2-44 *Loop created on line connecting the steam inlet nozzle*

Connecting the Vessel to Pumps

Next, you need to create pipe lines connecting the vessel and pumps.

1. Invoke the **Primary Line Segment** tool from the **Lines** tab in the **Tool Palettes - P&ID PIP** and specify the start point at the bottom of the vessel, as shown in Figure 2-45.

2. Connect the line to the horizontal nozzle of the pump tagged **P-101B**. The line created should be similar to the one shown in Figure 2-46. Note that in this tutorial, this line is referred as inlet line of the pump.

3. Assign the tag **10"-CS150-P-1004** to the line.

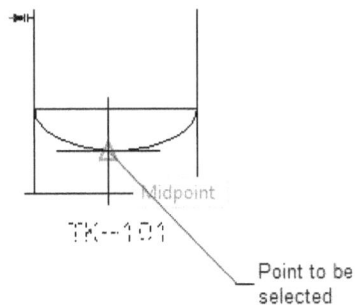

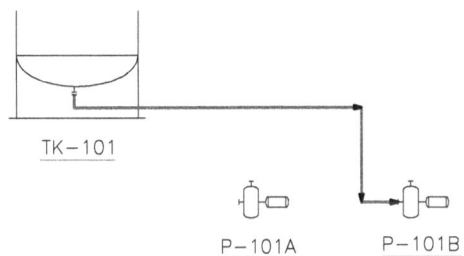

Figure 2-45 *Point to be selected on the vessel*

Figure 2-46 *Line connecting the vessel and the pump*

4. Create a line connecting the previous line and the horizontal nozzle of the pump tagged **P-101A**, refer to Figure 2-47. Assign the tag **10"-CS150-P-1004** to the line.

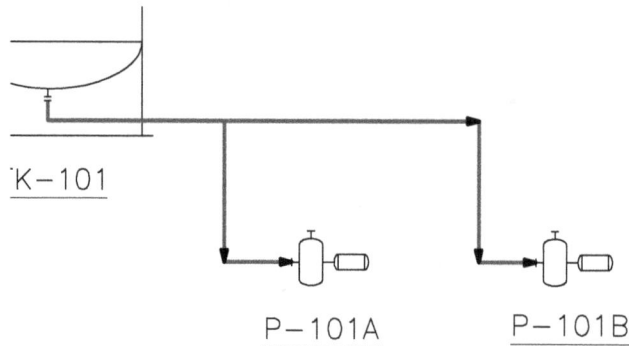

Figure 2-47 *Line connecting the horizontal nozzle of the pump tagged* **P-101A**

5. Invoke the **Draw** tool from the **Schematic Lines** panel in the **Home** tab; you are prompted to specify the start point of the pipe line.

6. Specify the start point on the vertical nozzle of the pump tagged **P-101A**, and then create a line, as shown in Figure 2-48.

7. Assign the tag **8"-CS150-P-1012** to the line.

8. Connect the vertical nozzle of the other pump to the previously created line, refer to Figure 2-49. Assign the tag **8"-CS150-P-1012** to the line.

   Note that in this tutorial, this line is referred as outlet line of the pump.

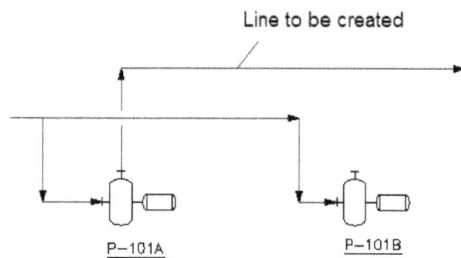

Figure 2-48 *Line connecting the vertical nozzle of the pump*

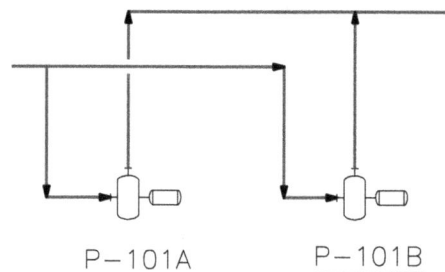

Figure 2-49 *Line connecting another pump and previously created line*

Creating the Remaining Lines Connecting the Vessel

1. Invoke the **Primary Line Segment** tool and specify the start point as (4,13). Next, move the cursor horizontally toward right and specify the endpoint on the vessel; the line is created, as shown in Figure 2-50. Note that this line is referred as feed line in this tutorial.

2. Assign tag **12"-CS150-P-1001** to the line.

3. Choose **Secondary Line Segment** from the **Lines** tab in the **Tool Palettes - P&ID PIP** and create the loop, as shown in Figure 2-51.

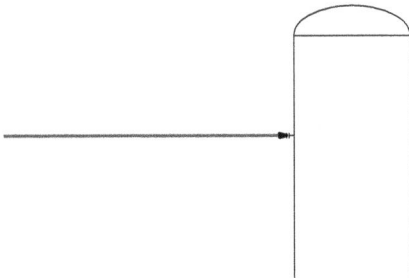

Figure 2-50 *Line connected to the vessel*   *Figure 2-51* *Loop created on the previous line*

4. Create another line by specifying the start point as (4,17). It should be similar to the line shown in Figure 2-52. Assign the tag **4"-CS150-P-1000** to the line.

   This line is referred as reflux line.

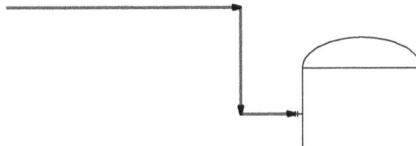

Figure 2-52 *Reflux line connected to the vessel*

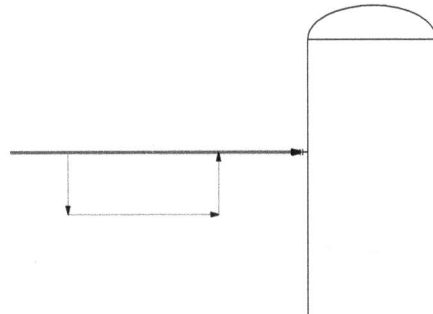

Next, you need to create a line originating from the top of the vessel.

5. Invoke the **Primary Line Segment** tool and specify the start point on the top of the vessel.

6. Move the cursor upward and enter **1** at the Command prompt. Next, press ENTER

7. Move the cursor horizontally toward right and create a line, as shown in Figure 2-53. Assign the tag **12"-CS150-P-1006** to the line.

8. Invoke the **Secondary Line Segment** tool and create a line, as shown in Figure 2-54.

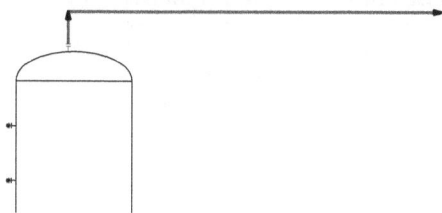

Figure 2-53 Line drawn from the top of the vessel

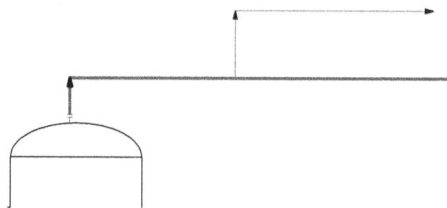

Figure 2-54 Drawing the secondary line connecting the previous line

Grouping lines

Next, you need to group the lines connecting the pumps.

1. Choose the **Make Group** tool from the **Line Group** panel in the **Home** tab; you are prompted to select the source line to group additional lines.

2. Select the lines connecting the vertical nozzles of the pumps, refer to Figure 2-55. Next, press ENTER; the selected lines are grouped.

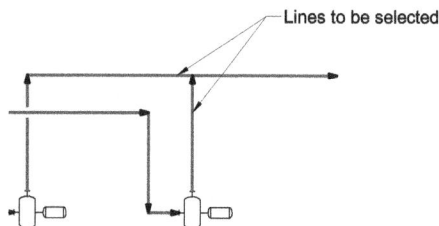

Figure 2-55 Lines to be selected to create a group

Adding Valves to the P&ID

1. Choose the **Control Valve** tool from the **Control Valve** area of the **Valves** tab of the Tool Palette; you are prompted to pick the insertion point.

2. Choose the **Change body or actuator** option in the Command prompt to invoke the **Control Valve Browser** dialog box.

3. Select the **Globe Valve** option from the **Select Control Valve Body** tree view and **Diaphragm Actuator** from the **Select Control Valve Actuator** tree view and choose the **OK** button.

4. Insert the control valve on the feed line, as shown in Figure 2-56.

5. Specify the position of the annotation balloon near the control valve; the **Assign Tag** dialog box is displayed.

6. Enter the values given next in this dialog box and then choose the **Assign** button.

    Area: 01
    Type: CV
    Loop Number: 1001

7. Similarly, place a control valve on the steam inlet line of the heat exchanger, as shown in Figure 2-57. Note that you need to specify **Butterfly Valve** and **Diaphragm Actuator** as the valve body and actuator, respectively.

8. Enter the following values in the **Assign Tag** dialog box:

    Area: 01
    Type: CV
    Loop Number: 1002

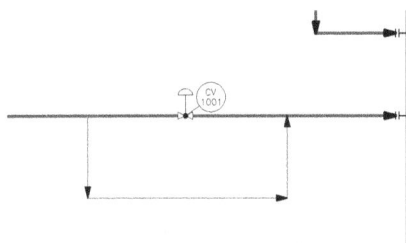

Figure 2-56 *Location of the control valve*

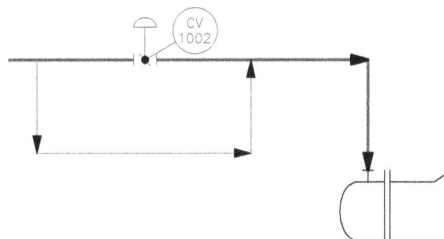

Figure 2-57 *Location of the control valve on the steam inlet line*

9. Choose **Gate Valve** from the **Valves** area in the **Valve** tab of **Tool Palettes - P&ID PIP**, and place at two locations, as shown in Figure 2-58.

10. Choose **Butterfly Valve** and place it at the location, shown in Figure 2-58.

11. Similarly, place two gate valves and a butterfly valve on the pipe which is connected to the heat exchanger, as shown in Figure 2-59.

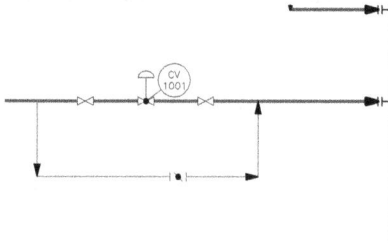

Figure 2-58 *Location of the gate valve and butterfly valve on the line connected to the vessel*

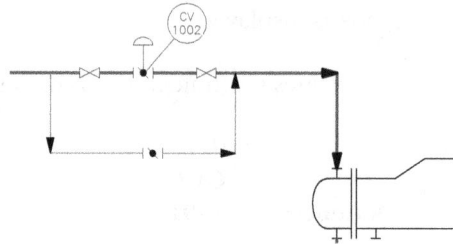

Figure 2-59 *Location of gate valves and the butterfly valve*

12. Create two secondary lines connecting the steam inlet pipe, as shown in Figure 2-60.

13. Choose **Plug** from the **Pipe Fittings** area in the **Fittings** tab of the **Tool Palettes - P&ID PIP**, and then place it at the endpoints of the secondary lines, refer to Figure 2-60.

14. Next, delete the arrows on the secondary lines and place gate valves, as shown in Figure 2-61.

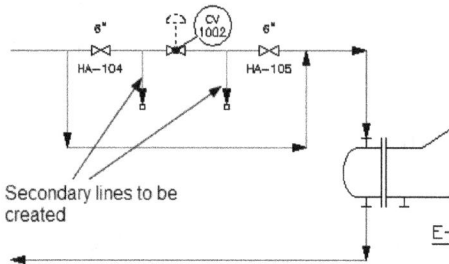

Figure 2-60 *Secondary lines with plugs placed at their endpoints*

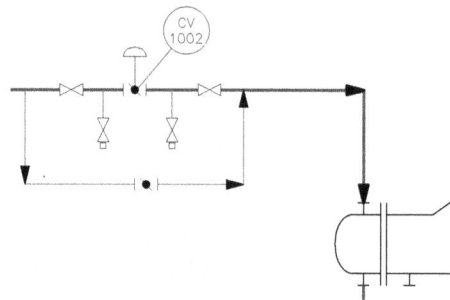

Figure 2-61 *Gate valves placed on the secondary lines*

15. Invoke the **Properties** palette of the gate valves by double-clicking on them.

16. Select the **Gate Valve Closed Style** option from the **Graphical style** drop-down list, refer to Figure 2-62; the gate valve is closed, as shown in Figure 2-63.

Figure 2-62 *Selecting the **Gate Valve Closed Style** option from the **Graphical style** drop-down list*

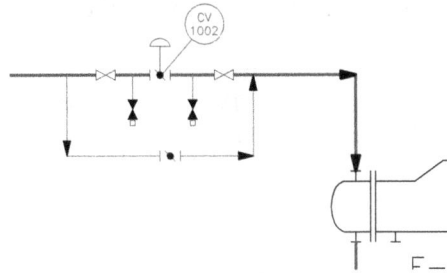

Figure 2-63 *The graphical style of gate valves changed to the closed type*

17. Similarly, place gate valves at the locations shown in Figure 2-64.

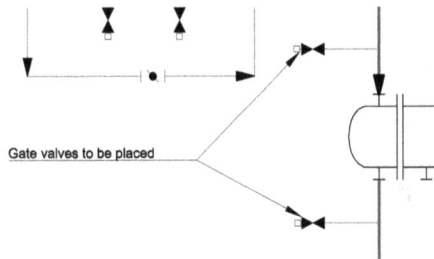

Figure 2-64 *Locations of the gate valves*

Adding Instruments to the P&ID

Next, you need to add instruments to the P&ID.

1. Choose the **Restriction Orifice** tool from the **Primary Element Symbols (Flow)** area of the **Instrument** tab in the **Tool Palettes - P&ID PIP**; you are prompted to specify the insertion point.

2. Place the restriction orifice on the steam inlet pipe of the heat exchanger, as shown in Figure 2-65.

3. Enter the values given next in the **Assign Tag** dialog box and place the annotation near the orifice.

Area: **01**
Type: **FE-FLOW ELEMENT**
Loop Number: **1002**

4. Similarly, place another restriction orifice on the pipe which is connected to the vessel, refer to Figure 2-66. Next, enter the following values in the **Assign Tag** dialog box:

Area: **01**
Type: **FE-FLOW ELEMENT**
Loop Number: **1001**

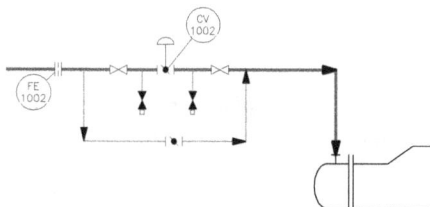

Figure 2-65 *Location of the orifice*   *Figure 2-66* *Location of the orifice*

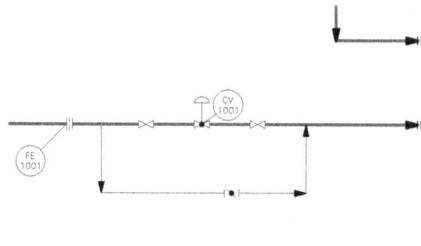

Next, you need to create instrumentation line connecting the orifice and the control valve.

5. Choose the **Electrical Signal** tool from the **Instrument Lines** area in the **Lines** tab of the **Tool Palettes - P&ID PIP**.

6. Specify the start point on the orifice located on the pipe connected to the heat exchanger and move the cursor upward upto a distance of 1.5 inches. Next, click to specify the endpoint.

    You need to make sure that the **Object Snap** and **Object Snap Tracking** options are activated at the status bar.

7. Move the cursor toward right and click when the cursor snaps to the mid point of the control valve, as shown in Figure 2-67.

8. Move the cursor downward and specify the endpoint on the control valve, refer to Figure 2-68.

9. Similarly, create an electrical signal on the pipe connected to the vessel, as shown in Figure 2-69.

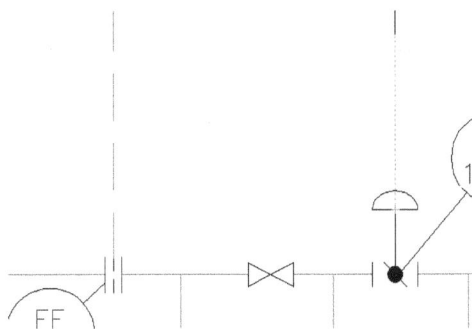

Figure 2-67 *Snapping to the midpoint of the control valve*

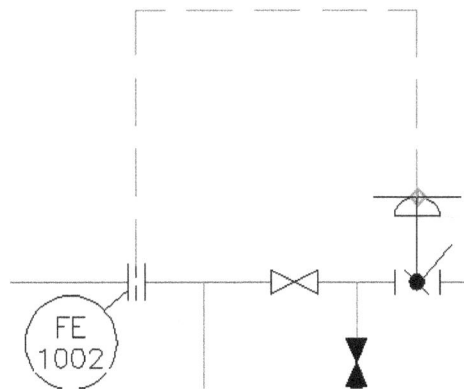

Figure 2-68 *Specifying the endpoint of the electric signal line*

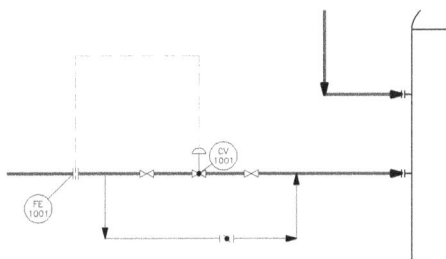

Figure 2-69 *Electric signal created on the feed line*

10. Choose **Primary Accessible DCS** from the **General Instruments** area of the **Instruments** tab in the **Tool Palettes - P&ID PIP**.

11. Place the instrument symbol on the electric signal line, as shown in Figure 2-70; the **Assign Tag** dialog box is displayed.

12. Enter the values given next in the **Assign Tag** dialog box and choose the **Assign** button.

Area	**01**
Type	**FC - FLOW CONTROLLER**
Loop Number	**1002**

13. Similarly, place another **Primary Accessible DCS** on the electric signal connected to the feed pipe, refer to Figure 2-71.

14. Assign the tag **01-FC-1001** to the instrument.

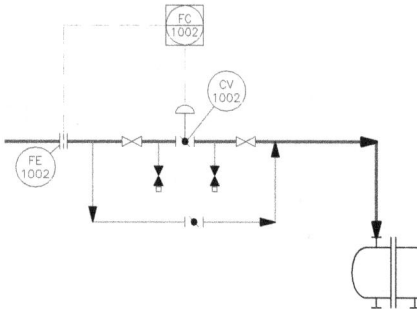

Figure 2-70 Position of the **Primary Accessible DCS**

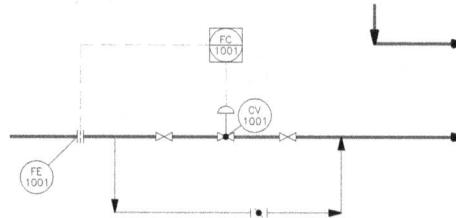

Figure 2-71 Position of the **Primary Accessible DCS** on the electric signal line of the feed line

Placing Valves and Fittings on the Lines Connecting the Pumps

Next, you need to place valves and fittings on the lines connecting the pumps.

1. Choose **Eccentric Reducer** from the **Pipe Fittings** area in the **Fittings** tab and place it on the line connecting the horizontal nozzle of the pump, as shown in Figure 2-72.

2. Place a gate valve on the same line, refer to Figure 2-72.

3. Choose the **Assign Tag** button from the **P&ID** panel and select the line connected to the reducing end of the eccentric reducer; the **Assign Tag** dialog box is invoked.

4. Modify the size of the pipe line to **8"** and choose the **Assign** button from the dialog box.

5. Similarly, place an eccentric reducer and a gate valve on the pipe connecting the other pump, as shown in Figure 2-73.

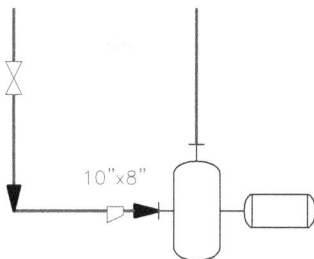

Figure 2-72 A gate valve and an eccentric reducer placed on the line connecting the first pump

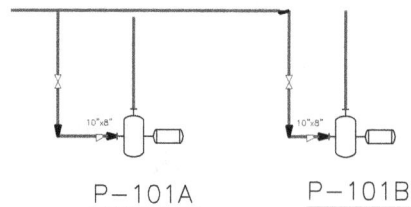

Figure 2-73 A gate valve and an eccentric reducer placed on the line connecting the second pump

6. Change the size of the pipe line connecting the reducing end of the eccentric reducer to 8".

7. Place check valves and gate valves on the outlet lines of the two pumps, refer to Figure 2-74 and 2-75.

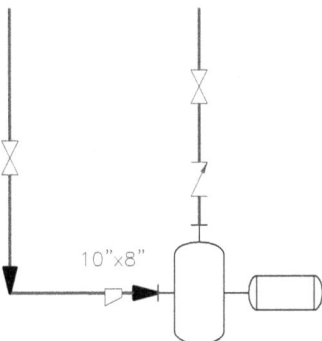

Figure 2-74 *Check valve and gate valve placed on the outlet line of the first pump*

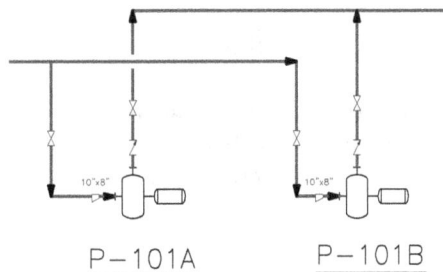

Figure 2-75 *Check valves and gate valves placed on outlet lines of both the pumps*

Next, you need to place a pressure relief valve on the pipe connecting the top nozzle of the vessel.

8. Choose **Pressure Relief Valve** from the **Relief Valves** area in the **Instruments** tab of the **Tool Palettes - P&ID PIP**.

9. Place the pressure relief valve on the secondary line of the pipe line connecting the top nozzle of the vessel, as shown in Figure 2-76. Also, assign the following tag.

Area	**01**
Type	**PSV - PRESSURE RELIEF VALVE**
Loop Number	**1002**

10. Place a gate valve on the same line, as shown in Figure 2-77.

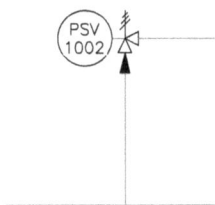

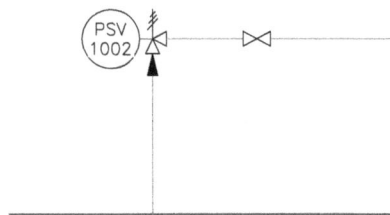

Figure 2-76 *Location of the pressure relief valve*

Figure 2-77 *Location of the gate valve*

Adding Off Page Connectors

1. Choose the **Off Page Connector** tool from the **Off Page Connectors and Tie-In**   ▭
    Symbol area of the **Non-engineering** tab in the **Tool Palettes - P&ID PIP**; you
    will be prompted to specify the insertion point.

2. Place the off page connector at the endpoint of the pipe line connecting the pumps, as
    shown in Figure 2-78.

3. Similarly, place off page connectors at the locations, as shown in Figures 2-79 and 2-80.

4. Choose the **Utility Connector** tool from the **Off Page Connectors and Tie-In**   ▤
    Symbol area of the **Non-engineering** tab in **Tool Palettes - P&ID PIP** and place
    at the locations, refer to Figures 2-80 and 2-81.

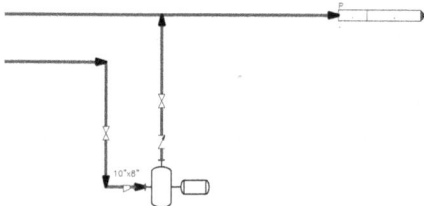

*Figure 2-78 Off page connector placed
at the endpoint of the pump outlet line*

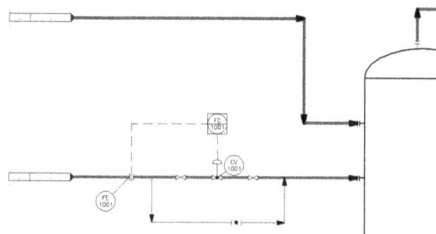

*Figure 2-79 Off page connectors placed at
the endpoints of the feed line and reflux line*

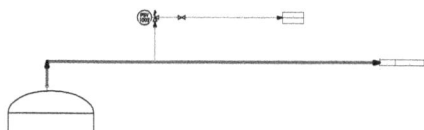

*Figure 2-80 Location of utility connector
and off page connector*

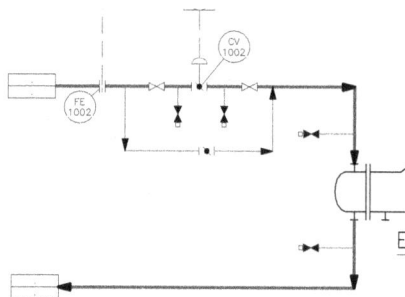

*Figure 2-81 Utility connectors placed at the
endpoints of the steam inlet line and the condensate
line*

Validating the drawing

1. Choose the **Validate Config** button from the **Validate** panel in the **Home** tab; the
    P&ID Validation Settings dialog box is displayed.

2. In this dialog box, expand the **P&ID objects** node and clear the **Unresolved off-page connectors** check box. Similarly, expand the **3D Model to P&ID checks** node and clear all the check boxes under it. Next, choose the **OK** button to close the dialog box.

Note
In this tutorial, the off page connectors are not validated because they are not connected to any other P&ID.

3. Run validation process by choosing the **Run Validation** tool from the **Validate** panel; the **Validation Summary** is displayed with a tree list of errors in the drawing.

Run
Validation

4. Right-click on any error in the **Validation Summary** and choose **Ignore** to ignore the error. Similarly, ignore all the errors.

Saving the Drawing

1. Choose **Save** from the **Application menu** or **File > Save** in the menu bar to save the drawing file *c02tut01.dwg*.

2. Choose **Close > Current Drawing** from the **Application Menu** to close the drawing file.

Self-Evaluation Test

Answer the following questions and then compare them to those given at the end of this chapter:

1. You should select the **P&ID Drawings** node in the Project tree and then choose the **New Drawing** button, to create a new P&ID file. (T/F)

2. The equipment connected to the line will also move with line. (T/F)

3. When you move a line, the inline components such as valves and balloons will also move. (T/F)

4. The reducer will be reoriented, if the diameter of the pipe line on either side of the reducer is changed. (T/F)

5. You can add a converted AutoCAD component to the **Tool Palettes - P&ID PIP** by clicking and dragging it to the Tool Palette. (T/F)

6. A nozzle will be placed automatically, when you connect a line to an equipment. (T/F)

7. You can substitute a component by invoking the _____.

8. We use _____ to indicate the continuation of the process line from one drawing sheet to another.

9. You can select the control valve body and the actuator from the _____.

10. To convert an AutoCAD component into a P&ID symbol, you need to choose the _____ option.

Review Questions

Answer the following questions:

1. The _____ dialog box is displayed, if you select the **Control Valve** tool for the first time.

2. A project file is a _____ file.

3. The **Project Manager** is used to manage the project files only. (T/F)

4. You can activate more than one project at a time. (T/F)

5. You can edit a P&ID symbol using the _____ tool.

6. The _____ option is used to attach a line to a component without physically connecting to it.

7. The _____ option is used to divide a line into two.

8. The **Make Group** tool is used to group _____.

9. The _____ button in the **Validation Summary** palette is used to run validation on selected node.

10. You can place a symbol from the **PID ISO Tool Palette**, even while working in PIP standard file. (T/F)

Exercise

Exercise 1

In this exercise, you will create the P&ID shown in Figure 2-82.

(Expected time: 1hr)

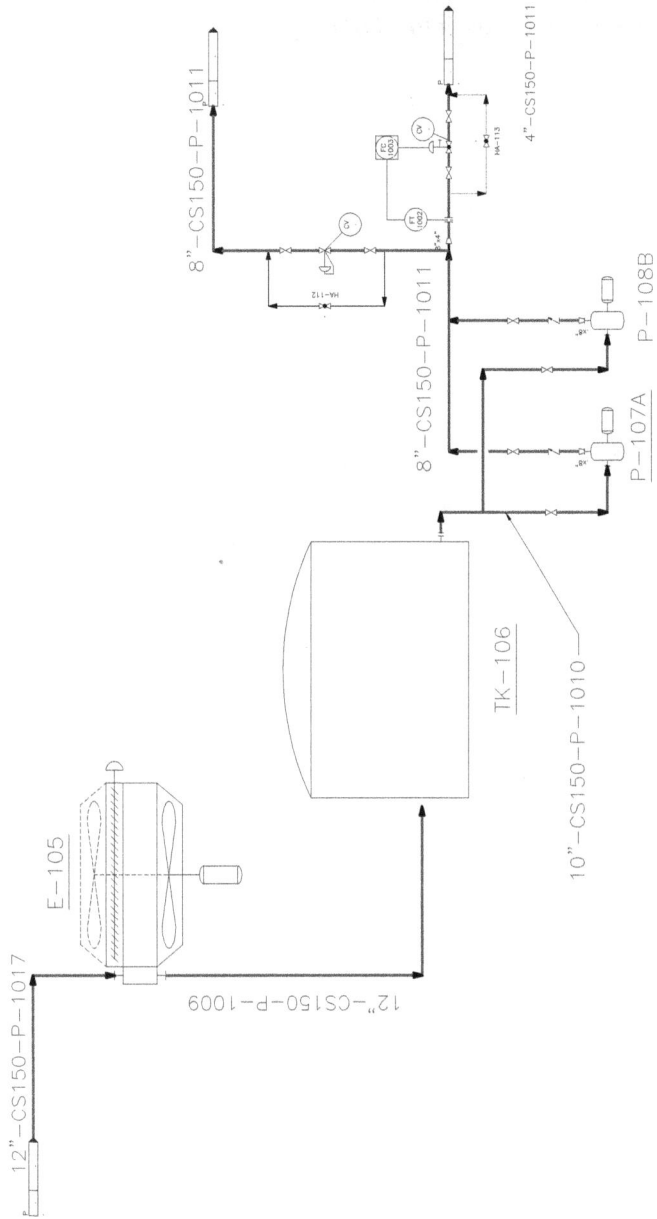

Figure 2-82 P&ID for Exercise 1

Answers to Self-Evaluation Test

1. T, **2.** F, **3.** T, **4.** T, **5.** T, **6. T**, **7.** Substitution palette, **8.** Off Page Connector, **9.** Control Valve Browser, **10.** Convert to P&ID object.

Chapter 3

Creating Structures

Learning Objectives

After completing this chapter, you will be able to:

- *Add members*
- *Configure member settings*
- *Add stairs*
- *Add ladders*
- *Add railings*
- *Add footings*
- *Modify structural members*

INTRODUCTION

In this chapter, you will learn to create a structural layout, which is an integral part of an industrial plant. To create a structural layout, first you need to create a new Plant 3D drawing using the **Project Manager**. The procedure to create a new drawing has been discussed in Chapter 2. In this chapter, the tools used to create a structural layout will be discussed. These tools are available in the **3D Piping** workspace.

CREATING A GRID

Ribbon: Structure > Parts > Grid

Grid

A grid helps you to place structural components at appropriate location. To create a grid, choose the **Grid** tool from the **Parts** panel in the **Structure** tab; the **Create Grid** dialog box will be displayed, as shown in Figure 3-1.

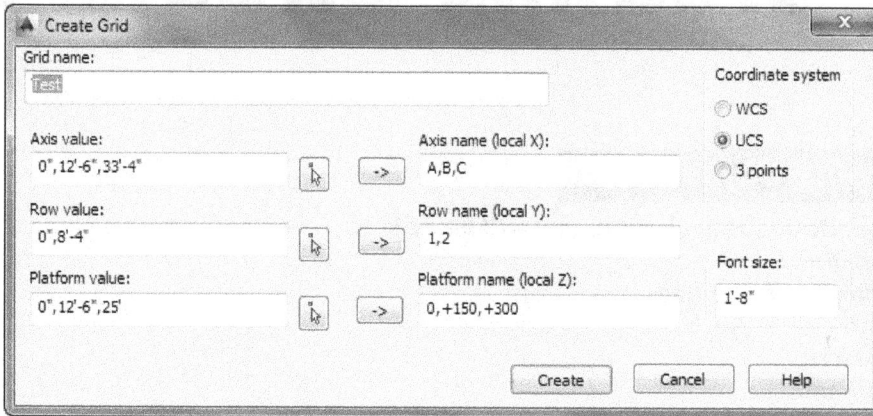

*Figure 3-1 The **Create Grid** dialog box*

In this dialog box, enter the name of the grid in the **Grid name** edit box. Next, specify the type of coordinate system to be used to set the orientation of the grid. You can specify the coordinate system by selecting anyone of the radio buttons available in the **Coordinate system** area. The **WCS** radio button is used to define the coordinate system of the grid using the World coordinate system. The **UCS** radio button is used to define a user coordinate system. If you select the **3 points** radio button, you will be prompted to specify the origin point. After specifying the origin point, you need to specify one point on the X-axis, one point on XY plane, and then one point on the Z-axis. These three points specify the directions of three axes in a grid. After specifying the coordinate system, enter values for the axis points in the **Axis value** edit box. You can use @ symbol in order to enter a value relative to the preceding value. You can also use the **Pick point** button adjacent to this edit box to directly specify the points in the workspace. Next, specify the names of the axes in the **Axis name (local X)** edit box. Similarly, enter values for the row points in the **Row value** edit box. Also, enter the names of the row points in the **Row name (local Y)** edit box. Specify the platform value points (in Z-direction) and their names in the **Platform value** and **Platform name (local Z)** edit boxes, respectively. Next, enter the font size in the **Font Size** edit box for the names displayed in the grid. Choose the **Create** button; a grid will be created in the model space, refer to Figure 3-2.

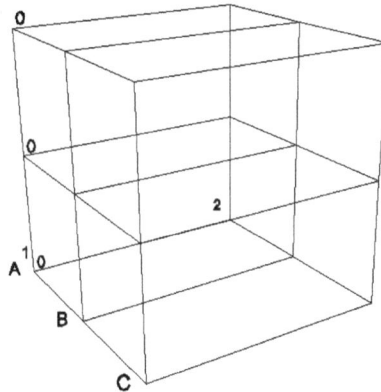

Figure 3-2 A 3D grid created

Editing Grids

To edit a grid, invoke the **Structure Edit** tool from the **Modify** panel and select the grid; the **Edit grid** dialog box will be displayed. The options in this dialog box are similar to the options in the **Create Grid** dialog box. Set the parameters in this dialog box as required and choose the **OK** button; the grid will be modified accordingly.

ADDING MEMBERS

In AutoCAD Plant 3D, a column or a beam is referred to as a structural member. A structural member can be placed in the plant 3D layout by specifying its start point and endpoint. Before adding a structural member, you need to specify the member settings such as shape, size, orientation, material standard, and so on. To add a member, choose the **Member Settings** tool from the **Settings** drop-down in the **Parts** panel in the **Structure** tab, refer to Figure 3-3; the **Member Settings** dialog box will be displayed, as shown in Figure 3-4.

Figure 3-3 The **Settings** drop-down

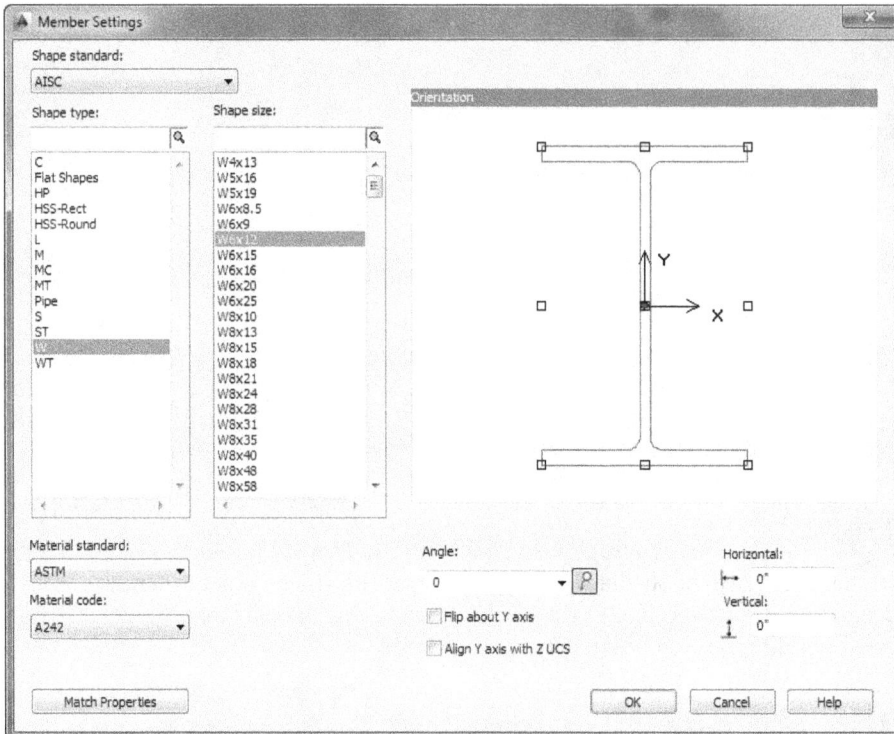

*Figure 3-4 The **Member Settings** dialog box*

In this dialog box, select the required standard for the shape of a member from the **Shape Standard** drop-down list. Next, select the shape type from the list box available in the **Shape type** area. You can search for a shape type using the search box available above the list box. After selecting the shape, you need to specify the size from the list box available in the **Shape size** area. You can search for a shape size using the search box available. Next, select the required material standard and material code from the respective drop-down lists. Specify the orientation angle using the **Angle** drop-down list. You can use the increment button available adjacent to this drop-down list to increase the angle by 90 degrees. Now, specify the justification of the member by selecting a point in the preview window. You can select the **Flip about Y axis** check box to flip the orientation about the Y axis. Also, you can select the **Align Y Axis with Z UCS** check box to align the Y axis of the member with the Z UCS. After specifying all the parameters in the **Member Settings** dialog box, choose the **OK** button to set the specified member settings.

Next, choose the **Member** tool from the **Parts** panel; you will be prompted to specify the start point of the structural member. You can do so either by clicking in the drawing area or by entering its coordinates at the Command prompt. After specifying the start point, you will be prompted to specify the endpoint of the member. Specify the endpoint of the member. At this point, you may continue to specify the points or terminate member creation by pressing ESC.

Tip: *You can align the structural member to an existing line. To do so, enter L in the Command bar; you will be prompted to select a line to align a member to. Select a line from the drawing area and then press ENTER; the member will be aligned to the selected line.*

CREATING STAIRS

To create stairs, first you need to specify the settings. To do so, choose the **Stairs Settings** tool from the **Settings** drop-down; the **Stair Settings** dialog box will be displayed, as shown in Figure 3-5. In this dialog box, you can specify the settings for creating stairs.

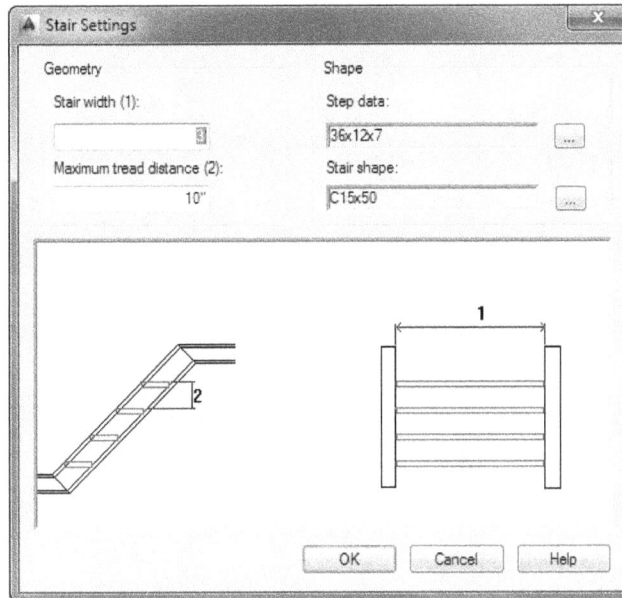

*Figure 3-5 The **Stair Settings** dialog box*

Specify the width of the stairs and the tread distance by entering desired values in the **Stair width (1)** and **Maximum tread Distance (2)** edit boxes, respectively. Next, you need to specify the step shape. To do so, click on the Browse button adjacent to the **Step data** field; the **Select Step** dialog box will be displayed. In this dialog box, select the required tread standard from the **Tread standard** list box; the tread shapes under the selected standard will be displayed in the **Tread shape** list box. Select the required shape type from the **Tread shape** list box and click **OK**; the **Select Step** dialog box will be closed. Next, you need to specify the stair shape. To do so, click on the Browse button adjacent to the **Stair shape** field; the **Select Stair Shape** dialog box will be displayed. The options in this dialog box are same as to that available in the **Member settings** dialog box. Specify the settings in this dialog box and choose the **Select** button to close this dialog box. Next, choose the **OK** button to close the **Stair Settings** dialog box.

After specifying the settings for the stairs, choose the **Stairs** tool from the **Parts** panel; you will be prompted to select the first point. Select the bottom point for the stairs;

you will be prompted to select the next point, refer to Figure 3-6. Specify the top point of the stairs and press ENTER; the stair will be created, as shown in Figure 3-7.

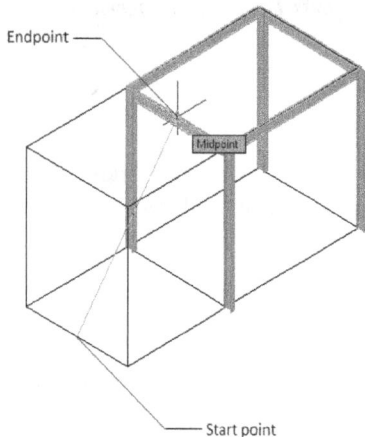

Figure 3-6 Selecting the start and endpoints of the stair

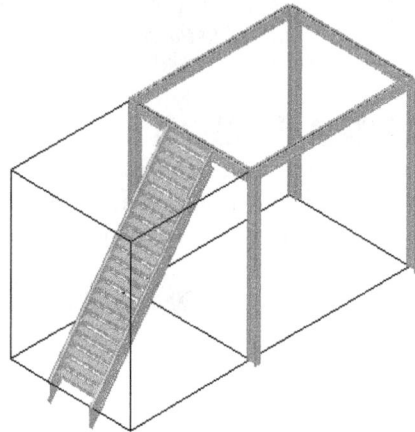

Figure 3-7 Stairs created from the selected points

Editing Stairs

To edit a stair, choose the **Structure edit** tool and then select a stair to be edited; the **Edit Stair** dialog box will be displayed. Modify the settings and choose **OK** to close the dialog box. You can also edit a stair using grips. To do so, select the stair to be edited; grips will be displayed on it, refer to Figure 3-8. You can use the center grips to change the location or length of the stairs and the side grips to increase or decrease the stair width.

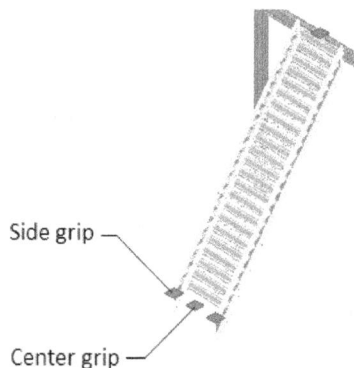

Figure 3-8 Stair with grips displayed on it

CREATING RAILING

To add railing to a stair, you need to specify parameters in the **Railing Settings** dialog box. To do so, choose the **Railing Settings** tool from the **Settings** drop-down; the **Railing Settings** dialog box will be displayed, as shown in Figure 3-9.

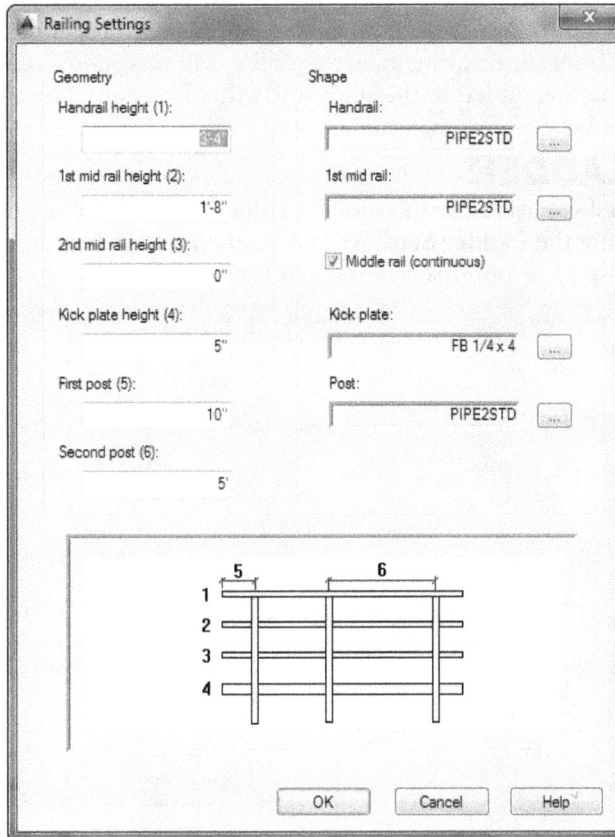

Figure 3-9 *The* **Railing Settings** *dialog box*

To specify different geometric values or parameters for the railing, enter desired values in edit boxes available in the **Geometry** area in the dialog box. Refer to the preview window in Figure 3-9 for these parameters. Next, specify the shape properties by clicking on the Browse button available adjacent to the corresponding rails. Select the **Middle rail (continuous)** check box to make the mid rail continuous. Choose the **OK** button to close the dialog box.

After defining the settings for the railing, choose the **Railing** tool from the **Parts** panel; you will be prompted to select the start point. Select the bottom point of a stair and then select the endpoint of the railing. You will notice that as soon as you specify the endpoint of the first railing, you will be prompted to select the next endpoint. Press ENTER to terminate the command.

You can also add a railing to the stairs or members by selecting them from the drawing area. To do so, enter **Object** at the prompt **Select limiting member or [cut**

Figure 3-10 *Railing added to the stairs and other structural members*

Both/Gap]: in the Command window and then press ENTER. On doing so, you will be prompted to select a stair or a structural member for applying the railing. Select the stairs or a structural member from the drawing area; the railing will be added to the selected member. Figure 3-10 shows a railing added to the stairs and other structural members.

CREATING LADDER

To create a ladder, first specify the settings for the ladder. To do so, invoke the **Ladder Settings** dialog box by choosing the **Ladder Settings** tool. In the **Ladder Settings** dialog box, choose the **Ladder** tab to display its options, as shown in Figure 3-11.

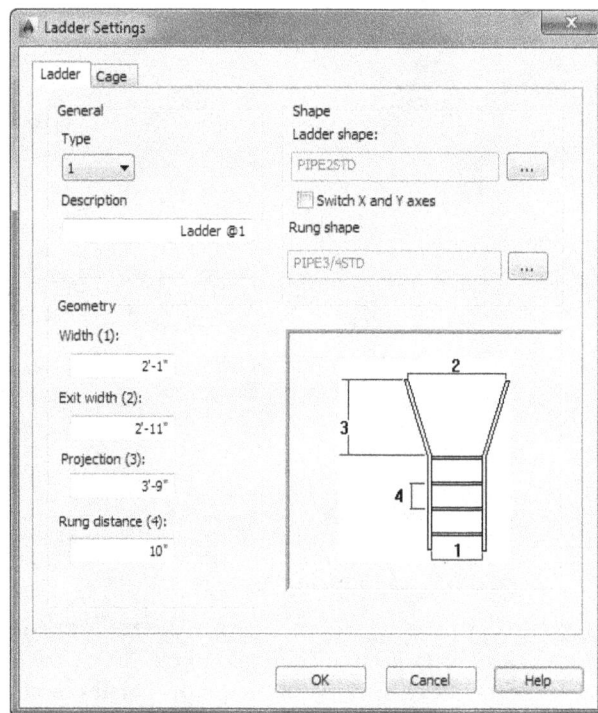

*Figure 3-11 The **Ladder Settings** dialog box with the **Ladder** tab chosen*

Ladder Tab

The **Ladder** tab has three areas, **General**, **Geometry**, and **Shape**, refer to Figure 3-11. In the **General** area, you can specify the ladder type and description using the **Type** drop-down list and the **Description** edit box, respectively.

The options in the **Shape** area allow you to specify the settings for ladder shape and rung shape. To change the ladder shape settings, choose the Browse button adjacent to the **Ladder Shape** edit box; the **Select Ladder Shape** dialog box will be displayed. Modify the settings and choose the **Select** button to close the dialog box. Also, you can select the **Switch X and Y axes** check box to rotate the ladder by 90 degrees. Similarly, you can change the rung shape settings by invoking the **Select Rung Shape** dialog box.

The **Geometry** area contains options to define the geometry of the ladder. The preview window explains the use of the options available in this area.

Cage Tab

The **Cage** tab in the **Ladder Settings** dialog box contains the **General** and **Geometry** areas. You can to preview the ladder geometry in the preview area, refer to Figure 3-12.

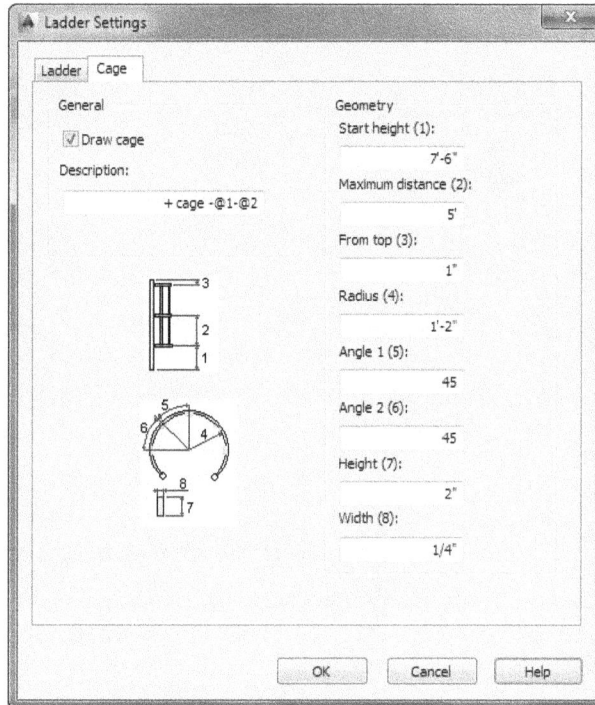

*Figure 3-12 The **Cage** tab of the **Ladder Settings** dialog box*

In the **General** area, select the **Draw cage** check box to create a ladder with a cage. You can enter the description of the cage in the **Description** edit box.

The **Geometry** area contains options to define the geometry of the cage. The preview window explains the use of the options available in this area. After specifying all the values, choose **OK**; the **Ladder Settings** dialog box will be closed.

After specifying the settings for the ladder, choose the **Ladder** tool from the **Parts** panel; you will be prompted to select the start point. Select the bottom point and then the top point as the endpoint of the ladder; you will be prompted to specify the directional distance. Move the cursor in the perpendicular direction and enter a distance at the Command prompt; the ladder will be created.

CREATING A PLATE/GRATE

Ribbon: Structure > Parts > Plate

To create a plate or a grate, choose the **Plate** tool from the **Parts** panel; the **Create Plate/Grate** dialog box will be displayed, as shown in Figure 3-13.

Plate

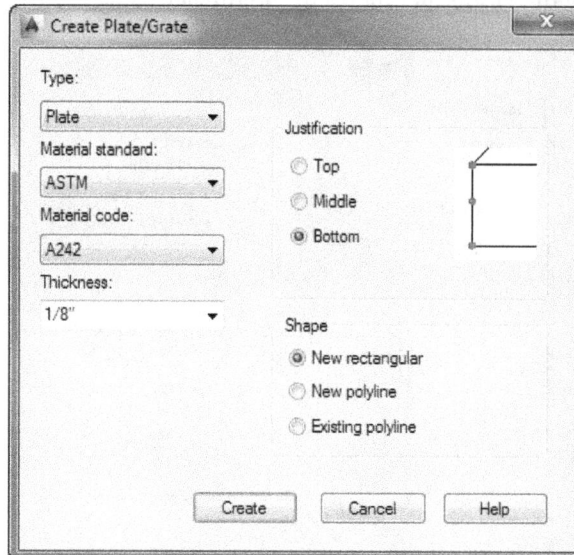

*Figure 3-13 The **Create Plate/Grate** dialog box*

Use the **Type** drop-down to specify whether you want to create a grate or a plate. Next, specify the material of the plate/grate by selecting the required material standard and material code from the respective drop-down lists. After specifying the material, specify the thickness for plate/grate from the **Thickness** drop-down list. If you have selected the **Grating** option from the **Type** drop-down list, you need to select the required hatch pattern and hatch scale of the grating from the **Hatch Pattern** and **Hatch Scale** drop-down lists, respectively.

Next, you need to specify the justification of a grating or plate using the options in the **Justification** area. Now, select an option from the **Shape** area. The **New rectangular** option is used to create a rectangular shaped plate/grating. The **New polyline** option is used to specify the shape by drawing a closed sketch using the **Polyline** tool. The **Existing polyline** option is used to select an existing polyline sketch to create the plate/grating.

After specifying the options in the **Create Plate/Grate** dialog box, choose the **Create** button; a prompt will be displayed at Command prompt depending upon the option selected in the **Shape** area. For example, if you select the **New rectangular** option, you will be prompted to specify the first corner point of the plate. Specify the first corner and then the second corner to create a plate or a grating.

CREATING FOOTING

Footing is one of the important components in a structural model. The function of footing in an industrial project is to provide support to the entire structural layout. To create a footing, first define the settings for it. To do so, invoke the **Footing Settings** dialog box by choosing the **Footing Settings** tool. The **Footing Settings** dialog box has two areas, namely **Geometry** and **Material**, and a preview window, as shown in Figure 3-14.

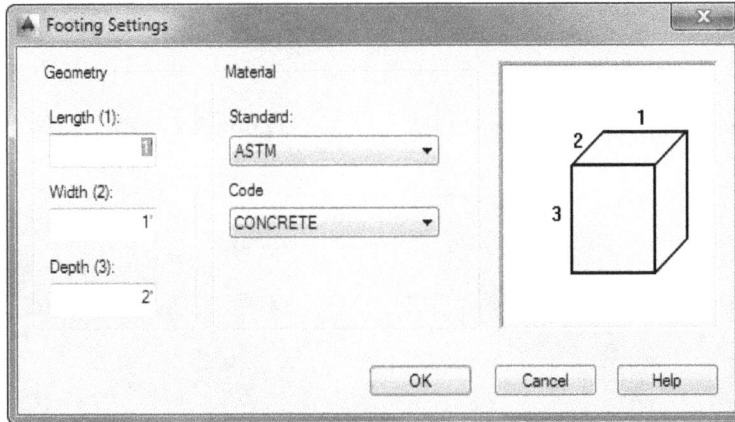

*Figure 3-14 The **Footing Settings** dialog box*

The **Geometry** area contains options to define the geometry of the footing. The preview window displays the different parameters of the **Geometry** area.

The options in the **Material** area allow you to specify the material standard and code of the footing.

After specifying the settings, choose the **Footing** tool from the **Parts** panel and specify a point to insert the footing.

SETTING THE REPRESENTATION OF THE STRUCTURAL MEMBER

You can set the representation of the structural members as required by using the tools available in the **Line Model** drop-down of the **Parts** panel, refer to Figure 3-15. The different representations that you can set for the structural members are discussed next.

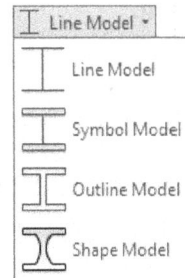

*Figure 3-15 The **Line Model** drop-down*

The Line Model is the default representation set for a member. In this mode, the members are represented as lines along with a symbol, as shown in Figure 3-16. This makes it easier to select the insertion point of a new member. In the Symbol Model mode, the members are represented as symbols, as shown in Figure 3-17. In the Outline Model mode, the members are represented as structure outlines, as shown in Figure 3-18. The Shape Model mode gives a real time look to the member with fillets added to the structural outline, as shown in Figure 3-19.

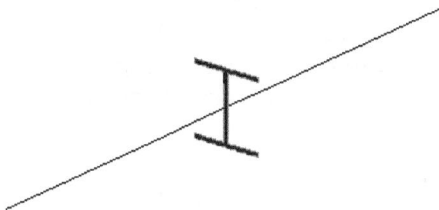

Figure 3-16 The Line Model

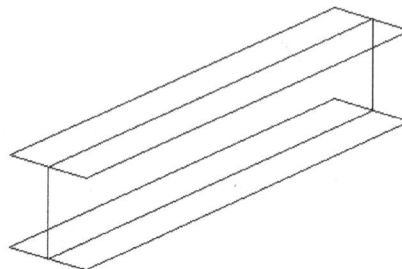

Figure 3-17 The Symbol Model

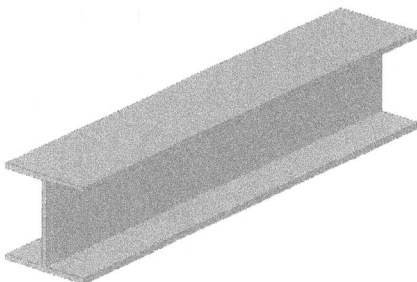

Figure 3-18 The Outline Model

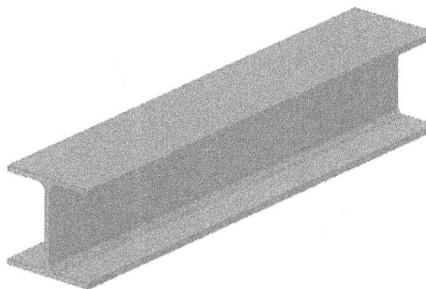

Figure 3-19 The Shape Model

EDITING THE STRUCTURAL MEMBERS

In the design process, you may need to edit the structural member. Various editing operations are discussed next.

Changing the Length of a Member

Ribbon: Structure > Cutting > Lengthen Member

You can change the length of a structural member. To do so, choose the **Lengthen Member** tool from the **Cutting** panel; you will be prompted to select a member and the following prompt sequence will be displayed at the Command prompt.

Command: **PLANTSTEELLENGTHEN**
Select a structural member or [Delta/Total]: *Select a member or enter an option*

The options in the Command prompt are discussed next.

Delta

This option allows you to enter a value of the length that is to be added to the existing length of a member. When you choose this option from the Command prompt, you will be prompted to specify a delta length. Enter a value and select a member; the length of the selected member will be changed. If you have specified a positive value, the length of the member will be increased. The length of the member will be decreased if you have entered a negative value.

Total

This option allows you to specify a value for the total length of a member.

Restoring the Member to its Original Length

Ribbon: Structure > Cutting > Restore Member

Restore Member

To restore the member to its original length, choose the **Restore Member** tool from the **Cutting** panel; you will be prompted to select the member. Select the member to restore it to its original length; the selected member will be restored to its original length.

Cutting Member at Intersections

Ribbon: Structure > Cutting > Cut Back Member

When two structural members intersect, you can remove the unwanted portion at the intersection using the **Cut Back Member** tool. To do so, invoke this tool from the **Cutting** panel; you will be prompted to select the first member. Also, the **cut Both** and **Gap** options will be displayed at the Command prompt. The **cut Both** option is used to cut both the intersecting members. The **Gap** option is used to create a gap at the intersection point. Note that you need to specify the required gap value at the Command prompt. Next, select both the intersecting members; the unwanted portion will be removed. Figure 3-20 shows a cut created using the **cut Both** option. Figure 3-21 shows a cut created by specifying a gap distance.

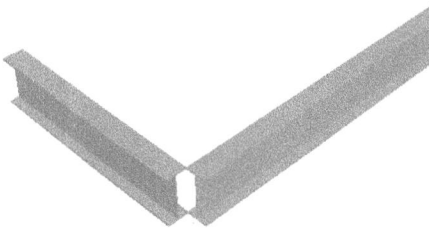

Figure 3-20 A cut created using the **cut Both** option

Figure 3-21 A cut created by specifying gap value

Creating Miter Joints

Ribbon: Structure > Cutting > Miter Cut Member

A miter joint is created at the corner of two structural members by beveling them at an angle of 45 degrees. To create a miter joint between two structural members, choose the **Miter Cut Member** tool from the **Cutting** panel; you will be prompted to select the first structural member. Also, the **Align Edges** and **Gap** options will be displayed at the Command prompt. The **Align Edges** option allows you to create a miter joint with the edges of both the members aligned, as shown in Figure 3-22. The **Gap** option is used to create a miter joint with a gap between the edges, as shown in Figure 3-23. After choosing the required option, select the two intersecting members; the miter joint will be created.

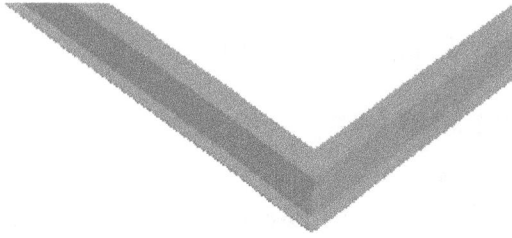

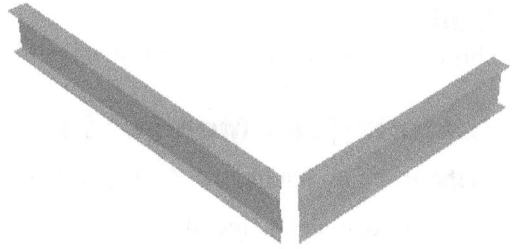

Figure 3-22 *Miter joint with aligned edges*   *Figure 3-23* *Miter joint with a gap*

Trimming/Extending a Member

Ribbon: Structure > Cutting > Trim Member/Extend Member

You can trim or extend a member upto a specified plane. To do so, choose the **Trim Member** tool from the **Cutting** panel; the **Trim to Plane** dialog box will be displayed, as shown in Figure 3-24. The options in this dialog box are discussed next.

Intersection plane Area

The options in this area are used to specify the plane at which the member is to be trimmed.

XY WCS

This radio button is chosen by default. As a result, the XY plane of the World Coordinate System is selected as the cutting plane.

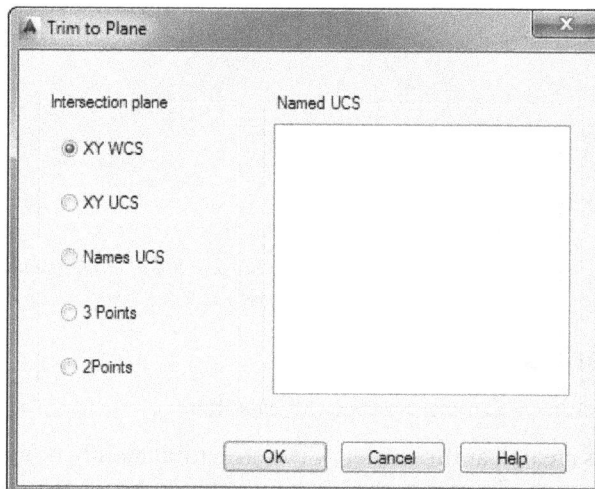

Figure 3-24 *The **Trim to Plane** dialog box*

XY UCS

Select this radio button to specify the XY plane of the User Coordinate System as the cutting plane.

Names UCS

Select this radio button to specify the XY plane of the named User Coordinate System as the cutting plane.

3 Points

Select this radio button to specify the XY plane of new coordinate system created by specifying 3 points.

2 Points

Select this radio button to create a cutting plane by drawing a line. The cutting plane is a vector of the line drawn.

Named UCS Display Box

This box displays the Named User Coordinate System.

Choose the **OK** button after selecting the required option and select a member that is to be trimmed; the selected member will be trimmed using the defined plane.

You can also extend a structural member. To do so, choose the **Extend Member** tool from the **Cutting** panel in the **Structure** tab; the **Extend to Plane** dialog box will be displayed. The options in this dialog box are the same as in the **Trim to Plane** dialog box. Select an option from the **Intersection plane** area to define the plane upto which the member should be extended, and then choose the **OK** button; the dialog box will be closed and you will be prompted to select a member to extend. Select a structural member from the drawing area; it will extend upto the specified plane.

Exploding a Structure

Ribbon: Structure > Modify > Structure Explode

You can explode structural objects into individual elements. Note that you can only explode stairs, railings, or ladders. To explode an object, choose the **Structure Explode** tool from the **Modify** panel; you will be prompted to select a structural object. Select a stair, railing, or ladder; the selected object will be exploded into individual elements that can be modified individually.

VISIBILITY OPTIONS

When you create a plant layout, whether it is large or small, you may need to toggle the visibility of various objects in the layout. You can do so by hiding the components at any stage. The method to toggle the visibility of an object is discussed next.

Hiding and Displaying Components

To hide a component placed in a plant layout, select the component from the graphics area and choose the **Hide Selected** tool from the **Visibility** panel in the **Structure** tab; the display of the component will be turned off. To hide all other components except the selected component, choose the **Hide Others** tool from the **Visibility** panel. On doing so, the selected component will be displayed while all other components will be hidden.

To show all hidden components, choose the **Show All** button; all hidden components will be displayed again in the layout.

EXCHANGING DATA WITH OTHER APPLICATIONS

Ribbon: Structure > Export > SDNF Export

SDNF allows the exchange of steel structure data between two applications. To create a SDNF report, choose the **SDNF Export** tool from the **Export** panel; the **SDNF Export** dialog box will be displayed, as shown in Figure 3-25. Specify the path for the output file by using the **Browse** button available at the right side of the **Output file** edit box in this dialog box.

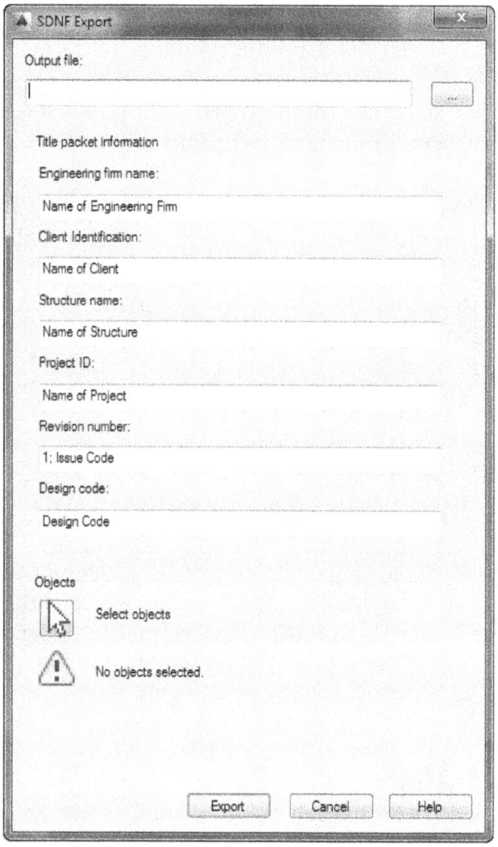

SDNF
Export

*Figure 3-25 The **SDNF Export** dialog box*

The **Title packet Information** area contains options to specify the user data. The **Select Objects** button in the **Objects** area is used to select objects from the workspace. The selection status of objects is displayed in the **Objects** area. After specifying the parameters in this dialog box, choose the **Export** button to export the structure to SDN format.

TUTORIALS

Tutorial 1

In this tutorial, you will open the **CADCIM** project created in Chapter 2 and then start a new AutoCAD Plant 3D file. In this file, you will create a pipe rack, as shown in Figure 3-26. You need to use W10 x 12 structural members to create this pipe rack. The dimensions for the model are given in Figure 3-27.   **(Expected time: 45 min)**

Note
You can also download the CADCIM project from www.cadcim.com by following the path Textbooks > CAD/CAM >AutoCAD Plant 3D > AutoCAD Plant 3D 2014 for Designers.

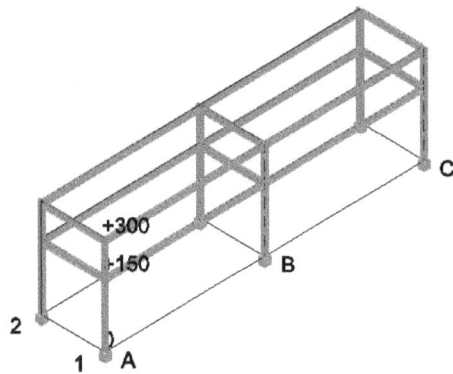

Figure 3-26 Model for Tutorial 1

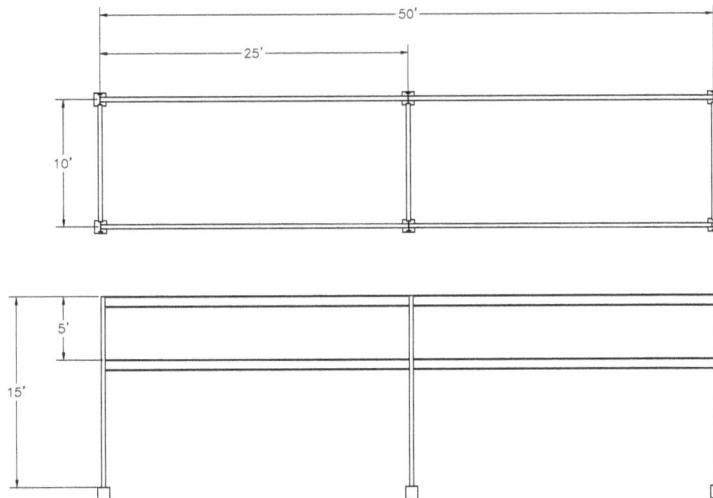

Figure 3-27 Dimensions for the Model

Footing Dimensions:

Length=12 inch
Width=12 inch
Depth= 12 inch

The following steps are required to complete this tutorial

a.   Open a new Plant 3D drawing file in the current project.
b.   Create a grid by specifying the axis, row, and platform values.
c.   Add footings at the bottom of the grid.
d.   Add columns and beams to the model.
e.   Cut members at intersections and create miters at corners.

Opening a New Plant 3D File

1.   Choose **Start > All Programs (or Programs) > Autodesk > AutoCAD Plant 3D 2014 > AutoCAD Plant 3D 2014**; AutoCAD Plant 3D starts.

     Next, you need to start a new AutoCAD Plant 3D file.

2.   Select the **CADCIM** project that you have created in **Tutorial 1** of Chapter 2 from the **Current Project** drop-down list in the **Project Manager**.

3.   Select the **Plant 3D Drawings** node in the **Project** area and choose the **New Drawing** button; the **New DWG** dialog box is displayed.

4.   Enter **Pipe_rack.dwg** in the **File name** edit box and choose **OK**; the new file is created.

5.   Select the **3D Piping** option from the **Workspace** drop-down list located in the **Quick Access Toolbar**.

Creating the Grid

1.   Choose the **Grid** tool from the **Parts** panel in the **Structure** tab; the **Create Grid** dialog box is displayed.

                                                                                              Grid

2.   In this dialog box, specify the parameters in the dialog box as given below and retain the default settings for other parameters.

     Axis value: 0, 25', 50'   **Row value: 0, 10'**
     Platform value: 0, 10', 15'

3.   Next, choose **Create**; the grid is created, as shown in Figure 3-28. Note that view orientation in the figure is set to **SW Isometric**. You can set the view orientation to **SW Isometric** by choosing the **SW Isometric** option from the **3D Navigation** drop-down list in the **View** panel of the **Structure** tab.

Creating Footings

1. Choose the **Footing Settings** tool from the **Settings** drop-down in the **Parts** panel; the **Footing Settings** dialog box is displayed.

2. Enter **12"** in the **Length(1)**, **Width(2)**, and **Depth(3)** edit boxes. Next, accept the default values in the **Material** area and choose **OK**; the settings of the footings are changed and the **Footings Settings** dialog box is closed.

3. Choose the **Footing** tool from the **Parts** panel; you are prompted to select the insertion point. Place the footings at the bottom grid points, as shown in Figure 3-29.

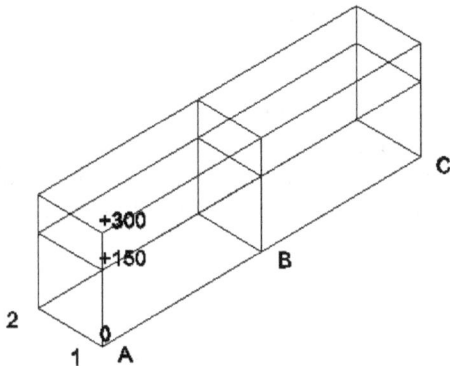

Figure 3-28 Model after creating the grid

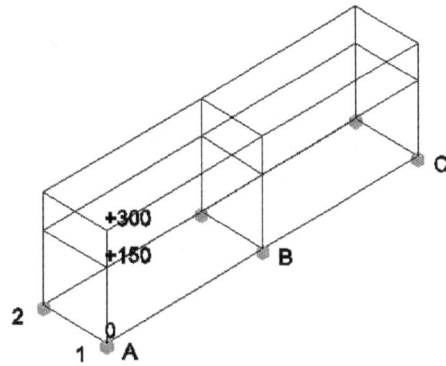

Figure 3-29 Model after adding footings

Creating Columns and Beams

To create structural members, first you need to set the properties of the member. Also, you need to turn on the object snap and 3D snap in order to make the grid points easier.

1. Choose the **Member Settings** tool from the **Settings** drop-down in the **Parts** panel; the **Member Settings** dialog box is displayed.

2. Select **W** and **W 10x12** from the **Shape Type** and **Shape Size** list boxes, respectively. Next, accept the default values of the other options and choose **OK**; the settings of the structural member are changed and the **Member Settings** dialog box is closed.

3. Choose the **Member** tool from the **Parts** panel; you are prompted to select the start point of the member.

    Make sure that the **Ortho Mode** is turned on at the Status Bar.

4. Select the start point, as shown in Figure 3-30; you are prompted to select the endpoint.

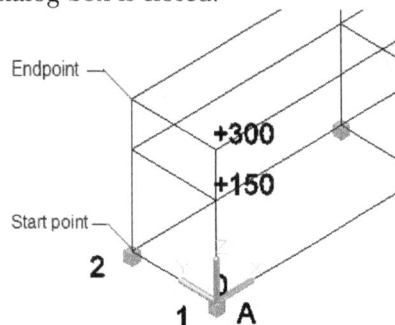

Figure 3-30 Selecting the start point and the endpoint of the column

5. Move the cursor vertically upward and snap to the grid point, refer to Figure 3-30. Next, click to select it; the column is created between the two selected points. Similarly, place columns on other footings, as shown in Figure 3-31. You can also use the **Copy** command to create rest of the columns.

    Next, you need to add beams. But before that, you need to change the justification of the member to top.

6. Choose the **Member Settings** tool from the **Settings** drop-down in the **Parts** panel; the **Member Settings** dialog box is displayed.

7. In the **Member Settings** dialog box, choose the top center justification point in the **Orientation** window. Next, choose **OK** to close the dialog box.

8. Invoke the **Member** tool; you are prompted to select the start point. Select the start point and the end point, as shown in Figure 3-32; a beam is created between the two selected points and a rubber-band line is displayed between the cursor and the specified point. Also, you are prompted to select the end point of a new member. Similarly, create rest of the beams, as shown in Figure 3-33.

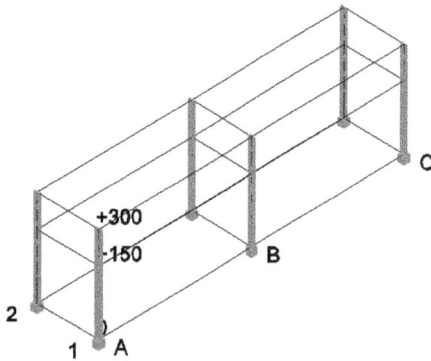

Figure 3-31 *Model after creating columns*

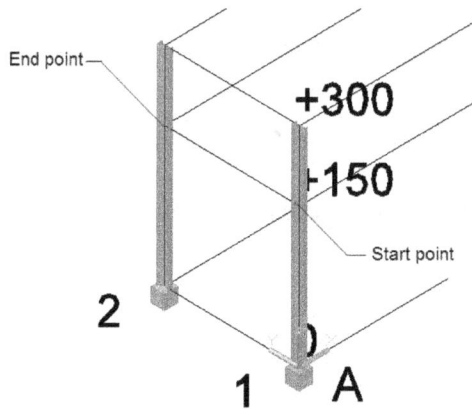

Figure 3-32 *The start point and end point of the beam*

Note

You can hide columns such that only the grid points are visible while specifying the start and end points of the beam.

9. Similarly, create the second level. The model after placing all the beams is shown in Figure 3-34.

Tip

*You can select all the beams at the first level and then use the **COPY** command to create the second level.*

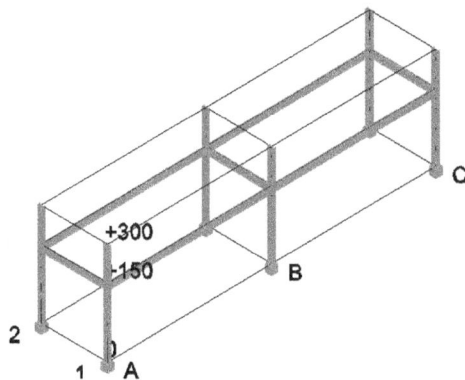

Figure 3-33 *Model after creating the lower level*

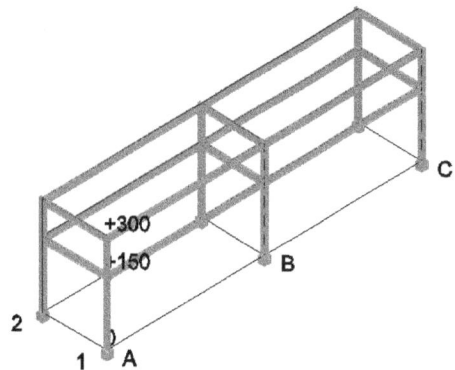

Figure 3-34 *Model after adding two levels of beams*

Next, you need to cut intersections between members.

10. Choose the **Cut Back Member** tool from the **Cutting** panel; you are prompted to select the limiting member.

11. Select the first column on the left side of the grid; you are prompted to select the structural member to be cut.

12. Select the anyone of the beam that is intersecting with the column; the beam is cut at the intersection. Also, you are prompted to select the limiting member. Similarly, cut the beams at the intersections by using columns as the limiting members.

Saving the Model

1. Choose the **Save** tool from the **Quick Access Toolbar**; the file is saved.

2. Choose **Application Button > Close** to close the file.

Tutorial 2

In this tutorial, you will open the CADCIM project created in Chapter 2 and then start a new AutoCAD Plant 3D file. In this file, you will create a structural model, as shown in Figure 3-35. The layout uses W10 x12 structural members. The dimensions of the model are shown in Figure 3-36.   **(Expected time: 45 min)**

Note

You can also download the CADCIM project from www.cadcim.com by following the path Textbooks > CAD/CAM >AutoCAD Plant 3D > AutoCAD Plant 3D 2014 for Designers.

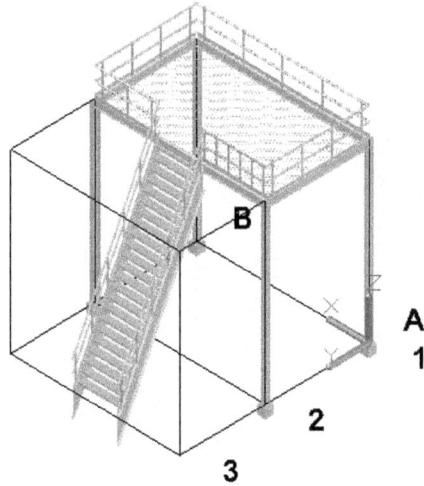

Figure 3-35 *Structural model for Tutorial 2*

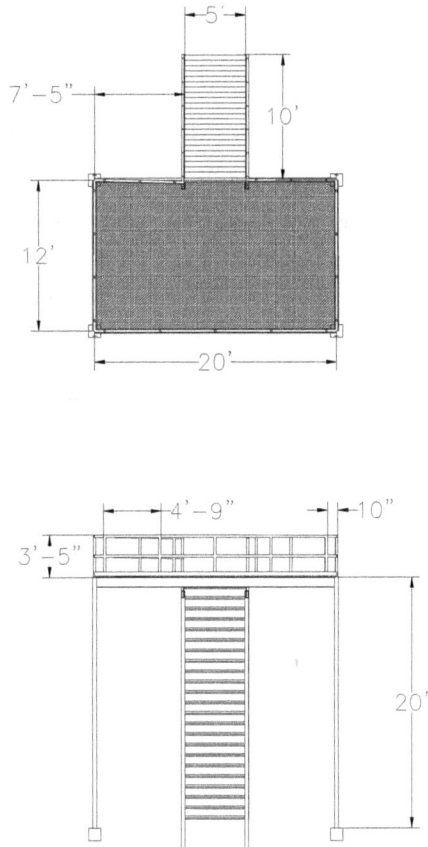

Figure 3-36 *Dimensions of the model*

The other specifications of the model to be created are as follows:

Grating specifications

Material standard:	ASTM
Material code:	A242
Thickness:	1"
Hatch pattern:	ZIGZAG
Hatch scale:	10"
Justification:	Bottom
Shape:	New rectangular

Footing dimensions:

Length=12 inch
Width=12 inch
Depth= 12 inch

Railing specifications:

Handrail Height:	40 inch
1st mid rail height:	20 inch
2nd mid rail height:	0
Kick plate height:	5 inch
First post:	10 inch
Second post:	5 inch
Handrail Shape:	PIPE2STD
Kick plate Shape:	FB 1/4x4
Post Shape:	PIPE2STD

Stair specifications:

Stair width:	60 inch
Maximum tread distance:	10 inch
Tread shape:	36x12x7
Stair Shape:	C15x50

The following steps are required to complete this tutorial

a. Open a new Plant 3D drawing file in the current project.
b. Create a grid by specifying the axis, row and platform values.
c. Add footings at the bottom of the grid.
d. Add columns and beams to the model.
e. Cut members at intersections.
f. Add stairs to the model.
g. Create a platform by using **Plate** tool.
h. Add railings to stairs and platform.

Opening a New Plant 3D file

1. Choose **Start > All Programs (or Programs) > Autodesk > AutoCAD Plant 3D 2014 > AutoCAD Plant 3D 2014**; AutoCAD Plant 3D starts.

   Next, you need to start a new AutoCAD Plant 3D file.

2. Select the CADCIM project that you created in chapter 2, from the **Current Project** drop-down list in the **Project Manager**.

3. Select the **Plant 3D Drawings** node in the **Project** area and choose the **New Drawing** button available; the **New DWG** dialog box will be displayed.

4. Enter **c03tut02.dwg** in the **File name** edit box and choose **OK**; the new file will be created.

Creating the Grid

1. Choose the **Grid** tool from the **Parts** panel in the **Structure** tab; the **Create Grid** dialog box is displayed.

2. In this dialog box, specify the values, as shown in Figure 3-37:

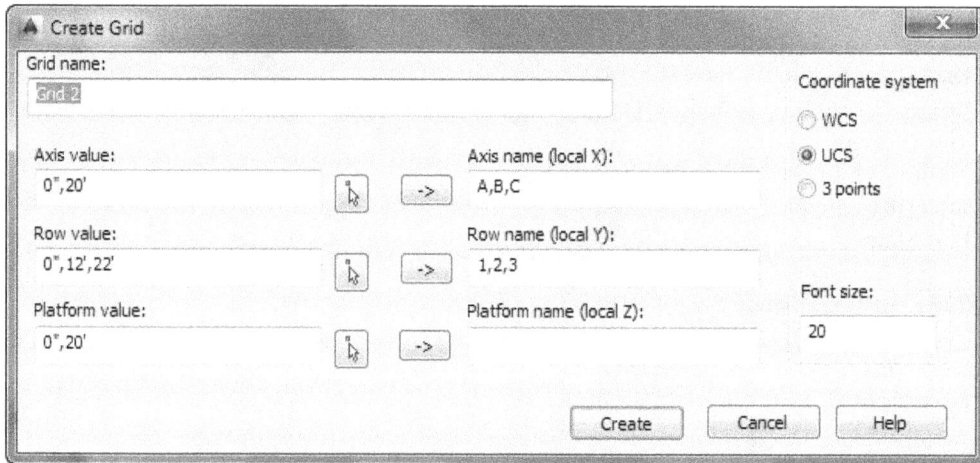

*Figure 3-37 The values in the **Create Grid** dialog box*

3. Next, accept default values for other options and choose the **Create** button; the grid is created, as shown in Figure 3-38. Next, change the view orientation to **SW Isometric**.

Creating Footings

1. Choose the **Footing Settings** tool from the **Settings** drop-down in the **Parts** panel; the **Footing Settings** dialog box is displayed.

2. In the **Footing Settings** dialog box, specify the following values:

   Length(1): 1'   **Width(2): 1'**   **Depth(3): 1'**

Next, accept the default values in the **Material** area and choose **OK** to close the dialog box.

3. Invoke the **Footing** tool and place the footings at the bottom grid points, as shown in Figure 3-39.

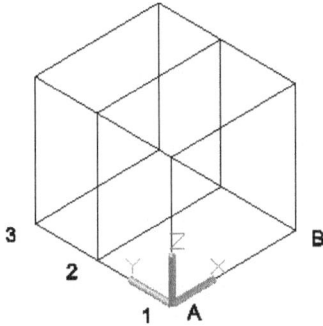

Figure 3-38 Model after creating the grid

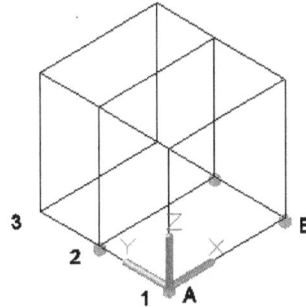

Figure 3-39 Model after adding footings

Creating Columns and Beams

To create structural members, first you need to set the properties of the member and then place a member. Also, you need to turn on the object snap and 3D snap so that you can easily select the grid points.

1. Choose the **Member Settings** tool from the **Settings** drop-down in the **Parts** panel; the **Member Settings** dialog box is displayed.

2. In the **Member Settings** dialog box, select **W** and **W 10x12** from the **Shape Type** and **Shape Size** list boxes, respectively. Next, choose the middle center justification point in the **Orientation** window and accept default values of the other options. Choose **OK**; the settings of the structural member are changed and the **Member Settings** dialog box is closed.

3. Choose the **Member** tool from the **Parts** panel and then place columns on all footings, as shown in Figure 3-40. Use grid points for the precise placement of columns.

   Next, you need to add beams. But before that, you need to change the justification of the member to top.

4. Invoke the **Member Settings** dialog box and choose the top center justification point in the **Orientation** window. Next, choose **OK** to close the dialog box.

5. Choose the **Member** tool and place beams on the grid. The model after placing all the beams is shown in Figure 3-41.

6. Trim the intersections between the members.

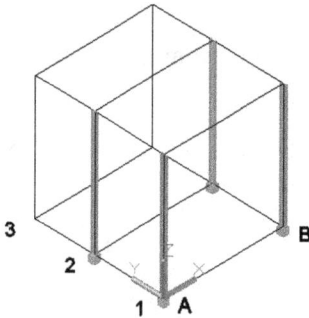

Figure 3-40 *Model after adding columns*

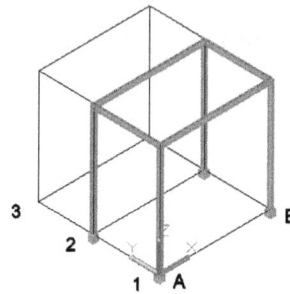

Figure 3-41 *Model after adding columns and beams*

Creating Stairs

Next, you need to add stairs to the model. To do so, first you need to set the stair settings and then place stairs.

1. Choose the **Stair Settings** tool from the **Settings** drop-down in the **Parts** panel; the **Stair Settings** dialog box is displayed.

2. Enter **5'** in the **Stair width(1)** edit box and **10"** in the **Maximum tread distance(2)** edit box. Next, choose the **OK** button in the **Stair Settings** dialog box to close it.

    Next, you need to create stairs. Make sure that the **Object Snap** and the **3D Object snap** are turned ON.

3. Choose the **Stairs** tool from the **Parts** panel; you are prompted to select the first point of the stair.

4. Rotate the model and select the midpoint on the top edge of the member, as shown in the Figure 3-42; you are prompted to select the second point.

5. Select the midpoint of the bottom grid line, as shown in Figure 3-43; a line is placed between the two specified points. Press ENTER to create the stairs on the specified line, as shown in Figure 3-44.

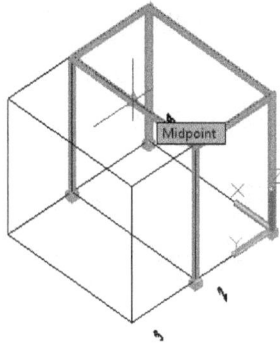

Figure 3-42 *Selecting the first point*

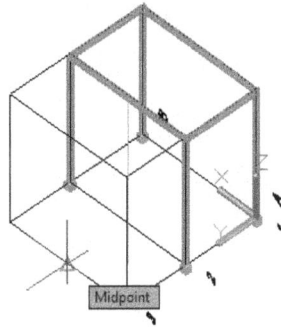

Figure 3-43 *Selecting the second point*

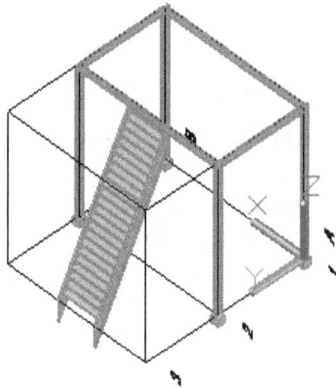

Figure 3-44 *Model after creating stairs*

Adding Grating
You need to create platform by adding grating.

1. Choose the **Plate** tool from the **Parts** panel; the **Create Plate/Grate** dialog box is displayed.

2. Specify the parameters for the grating as given below and then choose the **Create** button; you are prompted to specify the first corner of the grate.

Type:	**Grating**
Material standard:	**ASTM**
Material code:	**A242**
Thickness:	**1"**
Hatch pattern:	**ZIGZAG**
Hatch scale:	**10"**
Justification:	**Bottom**
Shape:	**New rectangular**

3.   Select the corner points, as shown in Figure 3-45; a rectangular grating is created on the platform 3-46.

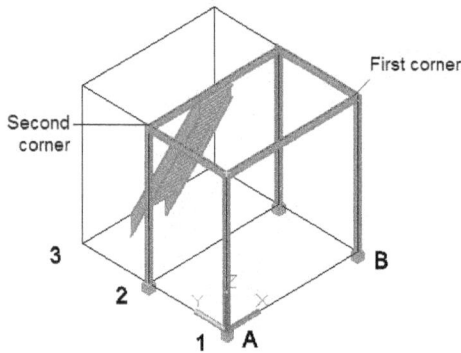

Figure 3-45 *Selecting corner points to create a grating*

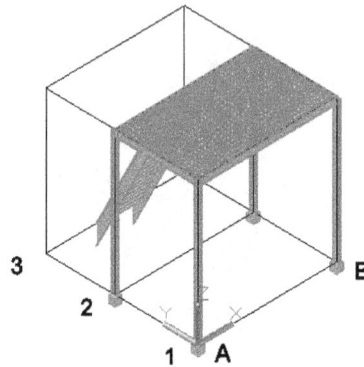

Figure 3-46 *Selecting corner points to create other grating*

Adding Railing

Next, you need to add railing to the stairs and the platforms.

1.   Choose the **Railing** tool from the **Parts** panel; you are prompted to select the start point of the railing. Enter **Object** at the Command prompt; you are prompted to select an object to align railing.

2.   Select the stairs; a railing is added to the stairs, as shown in Figure 3-47. Next, you need to add railing to the beams.

3.   Invoke the **Railing** tool again and select the points, as shown in Figure 3-48; a railing is created between the selected points. Next, press ENTER.

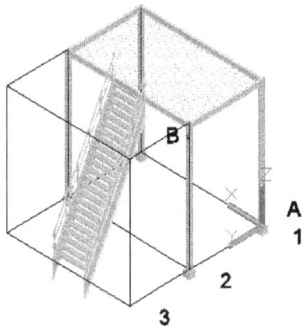

Figure 3-47 *Railing added to the stairs*

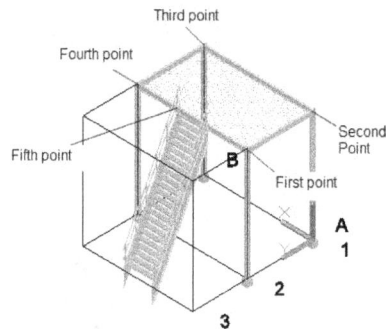

Figure 3-48 *Points to be selected to create the railing*

4. Invoke the **Railing** tool again and select the start and end point, as shown in Figure 3-49. The model after adding the railing is shown in Figure 3-50.

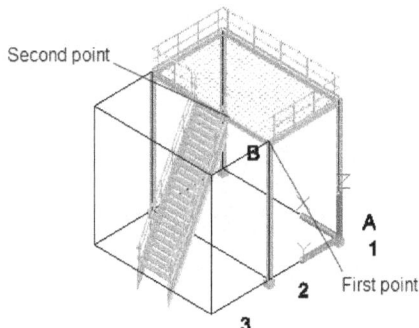

Figure 3-49 *Points to be selected to create the railing*

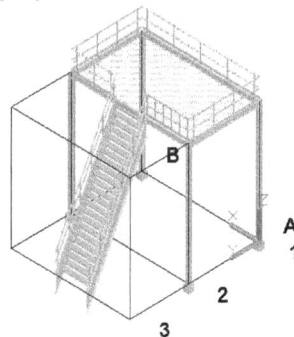

Figure 3-50 *Model after adding railing*

Saving the Model

1. Choose the **Save** tool from the **Quick Access Toolbar**; the file is saved.

2. Choose **Application Button > Close** to close the file.

Self-Evaluation Test

Answer the following questions and then compare them to those given at the end of this chapter:

1. In the _____ mode, the members are represented as lines with a shape symbol.

2. The _____ mode gives a 3D look to the members with fillets added to the structural outline.

3. You can add a railing to a structural member or a stair by invoking the _____option and selecting them directly.

4. The _____ allows you to exchange steel structure data between two applications.

5. You need to specify _____ to place a ladder at some distance from the specified location.

6. The _____ option creates a miter joint with a gap between the edges.

7. You need to select the _____check box in the **Ladder Settings** dialog box to create a ladder with a cage.

8. You can use the _____ button to match the properties of an already existing member with a new member.

9. You can modify individual elements of a stair, railing, or a ladder after exploding them. (T/F)

10. You can also convert a line into a structural member by invoking the **Member** tool and selecting it. (T/F)

Review Questions

Answer the following questions:

1. You can make the middle rail continuous by selecting the _____ check box in the **Railing Settings** dialog box.

2. The _____ are used to increase or decrease the width of the stairs.

3. You can hide all other components except the selected one by using the _____ button.

4. You can select an existing polyline to convert it into a grating or a plate, if the _____ radio button is chosen in the **Create Plate/Grate** dialog box.

5. The _____ check box in the **Member Settings** dialog box is used to orient the Y axis in the opposite direction.

6. You can place a stair with railing aligned to it. (T/F)

7. You can cut members at their intersections using the **Trim/Extend** tool. (T/F)

Exercise

Exercise 1

In this exercise, you will create the model shown in Figure 3-51. Its orthographic views are given in the drawing shown in Figure 3-52. The layout uses W10 x12 structural members.

(Expected time: 30 min)

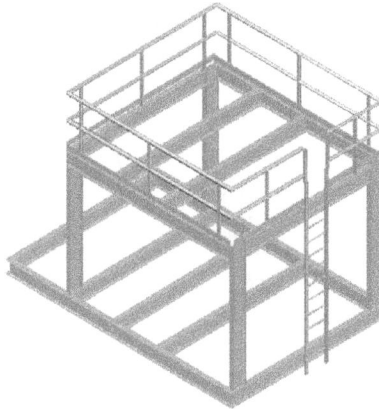

Figure 3-51 *The model for Exercise 1*

Railing specifications:

Handrail Height:	40 inch
1st mid rail height:	20 inch
2nd mid rail height:	0
Kick plate height:	5 inch
First post:	10 inch
Second post:	5 inch
Handrail Shape:	PIPE2STD
Kick plate Shape:	FB 1/4x4
Post Shape:	PIPE2STD

Ladder specifications:

Width:	25 inch
Exit width:	35 inch
Projection:	45 inch
Rung Distance:	10 inch
Ladder shape:	PIPE2STD
Rung Shape:	PIPE3/4STD

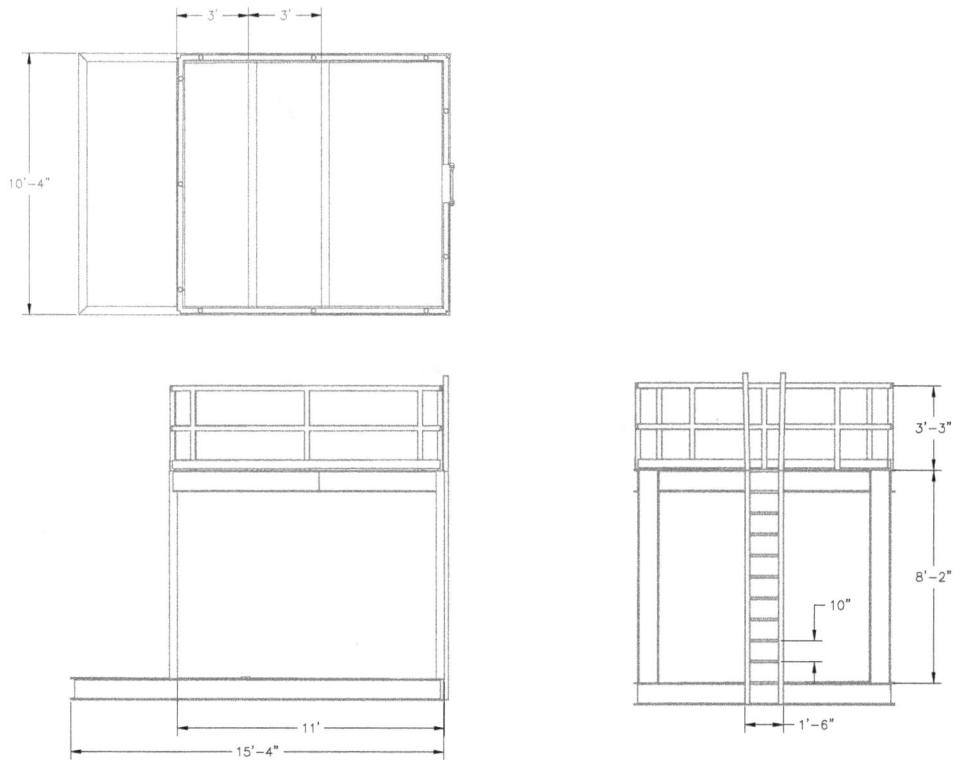

Figure 3-52 *Orthographic views for Exercise 1*

Answers to Self-Evaluation Test
1. Line Model, **2. Shape Model**, **3. Object**, **4. SDNF**, **5.** Directional distance, **6. Gap**, **7. Draw cage**, **8. Match Properties**, **9.** T, **10.** T

Chapter 4

Creating Equipment

Learning Objectives

After completing this chapter, you will be able to:
- *Create equipment*
- *Create equipment supports*
- *Create custom equipment*
- *Modify equipment*
- *Convert solid models into equipment*

INTRODUCTION

In this chapter, you will learn to create and place equipment in the **3D Piping** workspace. In addition to that, you will learn to convert solid models into equipment.

CREATING EQUIPMENT

You can create equipment in AutoCAD Plant 3D by choosing the **Create** tool from the **Equipment** panel in the **Home** tab. On choosing the **Create** tool the **Create Equipment** dialog box will be displayed, as shown in Figure 4-1. The options in this dialog box are discussed next.

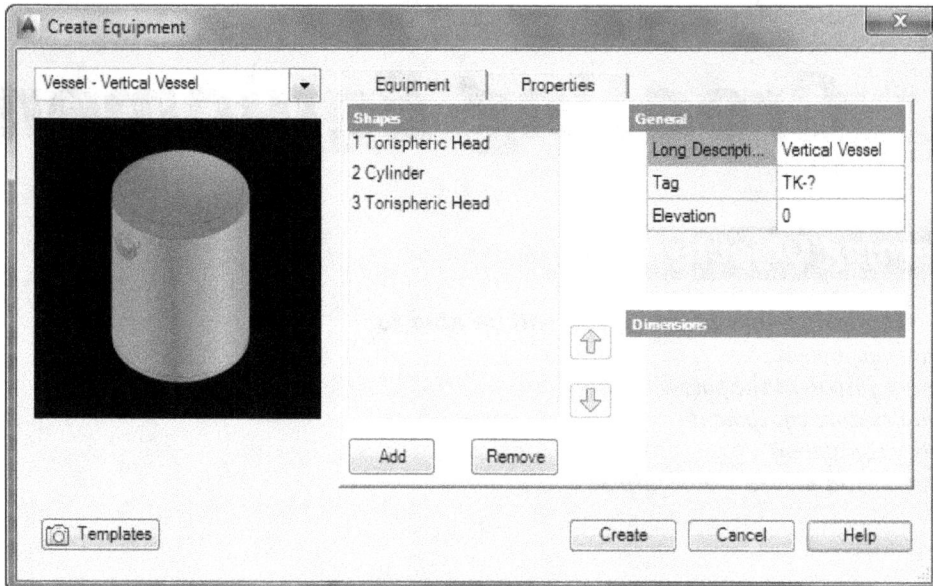

*Figure 4-1 The **Create Equipment** dialog box*

Equipment Drop-down List

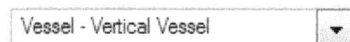

This drop-down list displays a list of equipment class. On selecting an equipment class, a flyout will be displayed with a list of predefined equipment types and options to create a new custom equipment. Note that the predefined equipment models are available only for heater, heat exchanger, pump, vessel, tank and strainer.

Equipment Tab

The **Equipment** tab displays options to determine the shape and size of the equipment. In addition to that, you can also specify the general data. The options in this tab are discussed next.

Shapes

The **Shapes** area displays a list of shapes piled up to create a piece equipment. You can add shapes to an existing equipment and also remove shapes. Also, you can change the stacking order by using the **Up Arrow** and **Down Arrow** buttons. This can only be done while adding a new shape to the existing equipment.

General

There are three fields available in this area: **Long Description**, **Tag** and **Elevation**. You can enter description and elevation of the equipment in its respective fields. To assign a tag, click in the **Tag** field; the **Assign Tag** dialog box will be displayed. Enter the desired text in the edit boxes available in this dialog box, and then choose the **Assign** button; the tag will be assigned to the equipment.

Dimensions

This area displays dimensions of the equipment. You can modify them as required. The preview window displays the annotated view of the equipment.

Properties Tab

The **Properties** tab displays properties of nozzles and fields to specify the data for the new equipment models. There are two areas in this tab, which are discussed next.

Nozzles

This area displays the properties of the nozzle such as size, pressure class, description and so on. Note that you cannot modify the data under this area.

Data

This area displays the data related to the equipment such as manufacturer, material, material code, and so on.

Template

This button displays the options to load the existing templates and save the new settings as a template.

PLACING EQUIPMENT

To place an equipment in the drawing area, select the required equipment from the **Equipment** drop-down list and then choose the **Create** button from the **Create Equipment** dialog box; the selected equipment will be attached to the cursor and you will be prompted to specify an insertion point in the drawing area. On specifying the insertion point, the equipment will be placed in the drawing area and a compass tool will be displayed at the bottom of the equipment. Use the compass tool to specify the orientation angle of the equipment. The procedures to add various equipment are discussed next.

Adding a Vessel

To add a vessel, choose **Vessel > Horizontal Vessel/Vertical Vessel** option from the **Equipment** drop-down list; the options related to the vessel are displayed in the **Equipment** tab. You can increase or decrease the length or the diameter of the vessel by specifying values in the **Dimensions** area. You can also add more cylinders to increase the length of the vessel. Next, click in the **Tag** edit box to assign tag to the vessel. After assigning tag, you can specify the elevation in the **Elevation** edit box. Next, choose the **Create** button and place the vessel in the geometry area; the compass will be displayed, refer to Figure 4-2 and you will be prompted to specify the rotation angle. Specify the rotation angle to orient the vessel according to the requirement.

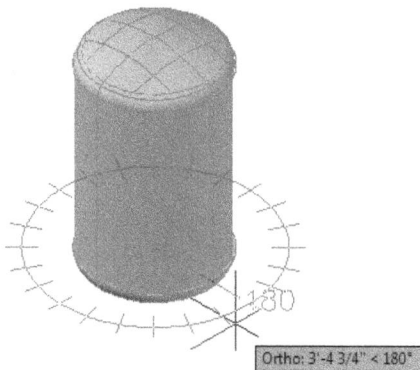

Figure 4-2 A vessel placed in the drawing area

Adding a Heat Exchanger

Heat exchangers are an important part of a process plant. They maintain the heat balance in the plant through addition or removal of heat. They exchange heat either with outside sources such as cooling towers or with streams of fluids operating at different temperatures. The heat exchanger can be classified into a cooler, exchanger, reboiler, condenser, heater, or a chiller. There are five predefined heat exchangers available in the **Create Equipment** dialog box. To create a heat exchanger, select the **Heat Exchanger** option from the **Equipment** drop-down list and then choose the desired predefined heat exchanger type. Next, click on the shapes in the **Shapes** area; the dimensions of the shapes will be displayed in the **Dimensions** area. You can edit their dimensions, if required. Choose the **Create** button and place the heat exchanger in the drawing area, refer to Figure 4-3. You can also create a new heat exchanger. The procedure to create an user-defined equipment is discussed later in this chapter.

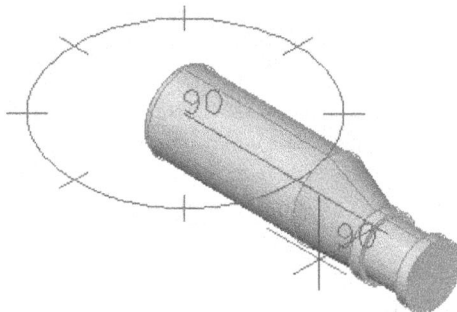

Figure 4-3 A heat exchanger placed in the drawing area

Adding a Pump

Pump is a mechanical device that is used to supply fluid to the desired point by using mechanical force. There are eight types of predefined pumps available in the **Create Equipment** dialog box. Choose any of the pumps; the preview will be displayed in the preview window, refer to Figure 4-4. You can modify the dimensions in the **Dimensions** area as per your requirement and choose the **Create** button; the pump will be attached to the cursor and you will be prompted to specify the insertion point in the drawing area. Place the pump and orient it using the compass tool, refer to Figure 4-5.

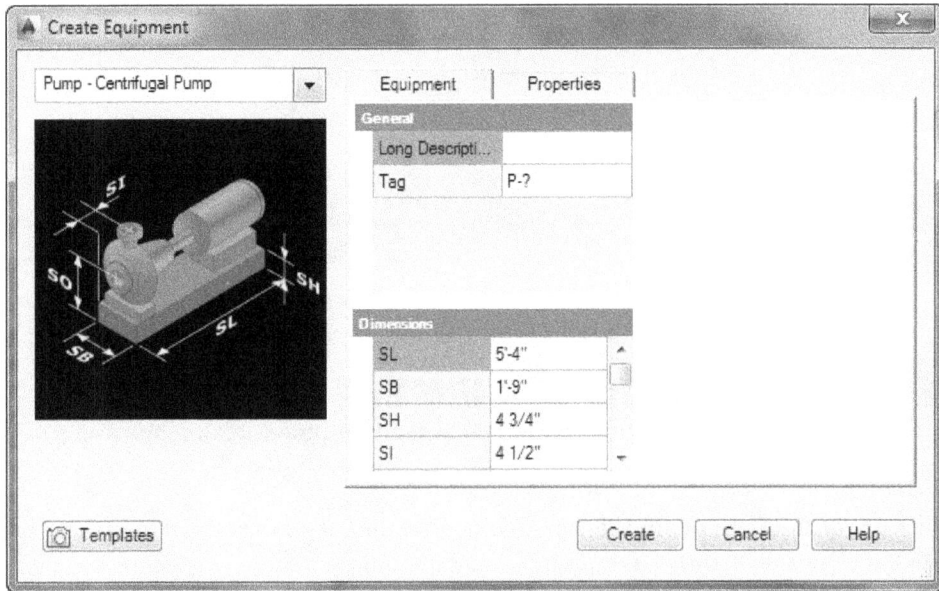

*Figure 4-4 The **Create Equipment** dialog box displaying a pump*

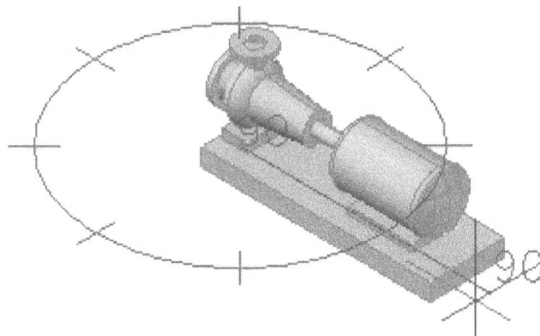

Figure 4-5 A centrifugal pump placed

Adding a Heater

Heaters are used to raise the temperature of a fluid to meet specific process requirement. There are two types of predefined heaters available, the box type heater and the cylindric heater.

To add a heater to a plant model, invoke the **Create Equipment** dialog box. Next, select the **Heater** option from the **Equipment** drop-down list and then choose any of the predefined heater types. Edit the dimensions, if required and then choose the **Create** button; you will be prompted to specify the insertion point. Specify the insertion point and then orient it using the compass, refer to Figure 4-6.

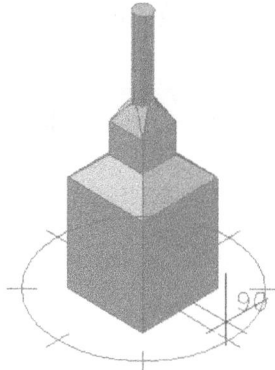

Figure 4-6 A heater placed in the drawing area

CREATING A CUSTOM EQUIPMENT

To create a customized equipment, invoke the **Create Equipment** dialog box and follow the steps given below.

1. Click on the **Equipment** drop-down list and move the cursor over the required equipment type; a flyout will be displayed.

2. Choose the **New Horizontal** or **New Vertical** equipment type from the flyout.

3. Choose the **Add** button at the bottom of the **Shapes** area; a flyout will be displayed with a list of shapes, as shown in Figure 4-7.

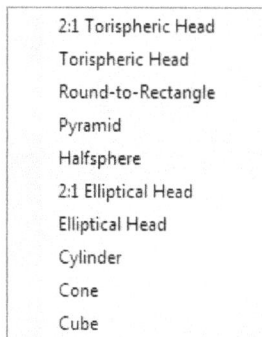

Figure 4-7 Flyout showing the list of shapes

4. Select the desired shape from the available options in this flyout; the selected shape will be added under the **Shapes** area.

5. Similarly, add other shapes and arrange them in a systematic manner using the **Up Arrow** and **Down Arrow** buttons. The shapes should be compatible with each other and form a definite shape.

6. Modify the dimensions of the shapes from the **Dimension** area. Also, assign a tag, add description and elevation in the respective fields.

7. Choose the **Create** button and place the model in the drawing area.

8. Set the orientation of the model using the compass tool. Figure 4-8 shows a custom equipment.

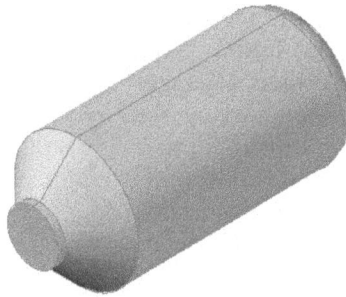

Figure 4-8 *A custom equipment*

MODIFYING EQUIPMENT

To modify an equipment, choose the **Modify Equipment** tool from the **Equipment** panel; you will be prompted to select an equipment. Select the equipment to be modified; the **Modify Equipment** dialog box will be displayed. The options in this dialog box are same as in the **Create Equipment** dialog box. Modify the parameters of the equipment and choose the **OK** button; the dialog box will be closed and the equipment will be modified accordingly. Figures 4-9 and 4-10 show a cylindrical heater before and after modification.

Figure 4-9 *A cylindrical heater before modifications*

Figure 4-10 *The cylindrical heater after modifications*

CONVERTING SOLID MODELS INTO EQUIPMENT

To convert a solid model into an equipment, you need to create a solid model by using the tools that are available in the **Modeling** tab of the ribbon. You can also convert solid models into equipment from other AutoCAD drawing files. To do so, right-click on the **Plant 3D Drawings** folder in the **Project Manager** tree and choose the **Copy Drawing to Project** option from the shortcut menu displayed. Next, select the drawing file which contains the solid model; the file will be added to the **Plant 3D Drawing folder**. Now, double-click on the file in the **Project Manager** and choose the **Convert Equipment** tool from the **Equipment** Panel; you will be prompted to select a solid model. Select the solid model and press ENTER; the **Convert to Equipment** dialog box will be displayed, as shown in Figure 4-11. Select an equipment type from the **Equipment** tree and choose the **Select** button; you will be prompted to specify the insertion base point. Select a point on the model by using snaps; the **Modify Equipment** dialog box will be displayed, as shown in

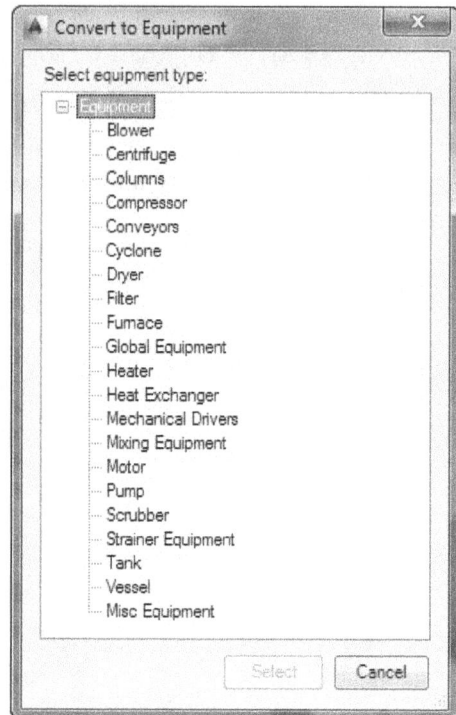

Figure 4-11 *The Convert to Equipment dialog box*

Figure 4-12. In the **General** area, enter the long description and assign tag by clicking in the **Tag** field. In the **Properties** tab, enter data in the fields related to manufacturer, material, size and so on. You can also save the equipment as a template by choosing the **Template** button. Choose the **OK** button to close the **Modify Equipment** dialog box. Figure 4-13 shows an AutoCAD object converted into an equipment.

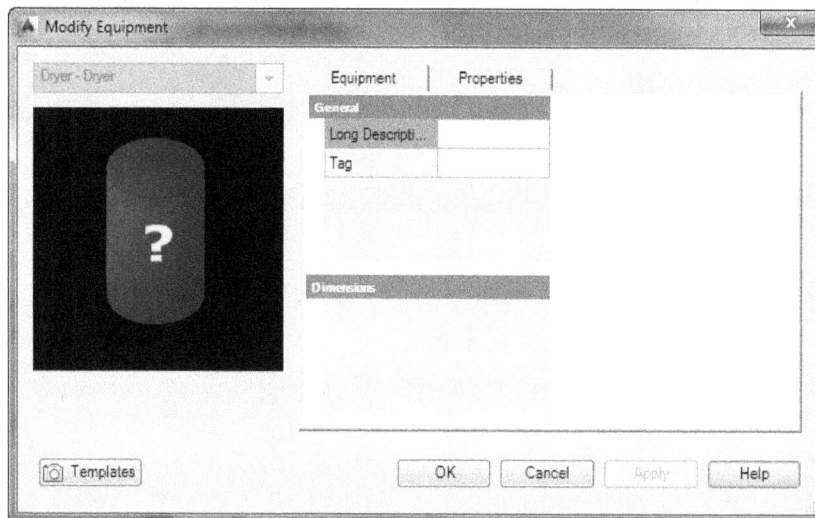

Figure 4-12 *The Modify Equipment dialog box*

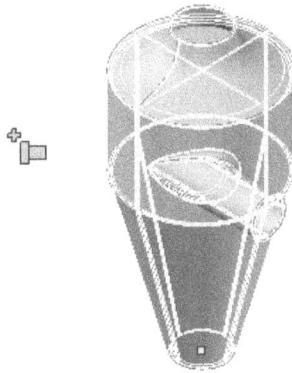

Figure 4-13 *AutoCAD object after*
converting into an equipment

ATTACHING OBJECTS TO AN EQUIPMENT

You can attach solid objects to an equipment. To do so, choose the **Attach Equipment** tool from the **Equipment** panel; you will be prompted to select an equipment. Select the required equipment, refer to Figure 4-14; you will be prompted to select the objects to be attached. Select the objects from the graphics window, refer to Figure 4-14 and press ENTER; the selected objects will be attached to the equipment.

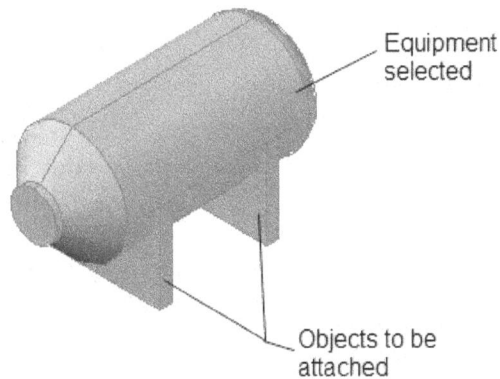

Equipment
selected

Objects to be
attached

Figure 4-14 *Equipment and objects to be selected*

DETACHING OBJECTS FROM AN EQUIPMENT

To detach the previously added objects from an equipment, choose the **Detach Equipment** tool; you will be prompted to select the equipment to which the objects have been attached. On doing so, you will be asked to detach all the attached objects or not. Choose **Yes** from the Command prompt to detach all the components.

ADDING NOZZLES TO A CUSTOM EQUIPMENT

You can add a nozzle to an equipment after creating or modifying it. To do so, first select the equipment, and click on the nozzle grip that is displayed; a nozzle preview will be displayed on the equipment, refer to Figure 4-15. Also, the **Add or Modify Nozzles** dialog box will be displayed with **Change Location** tab chosen, refer to Figure 4-16. The options in this dialog box are discussed next.

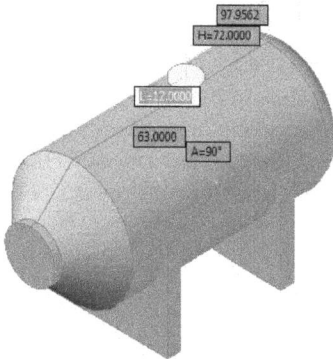

Figure 4-15 *The preview of the nozzle displayed on the equipment*

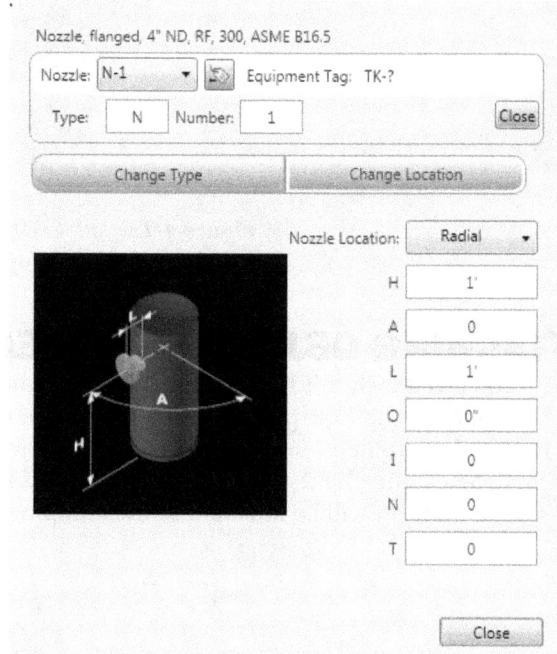

Figure 4-16 *The **Add or Modify Nozzles** dialog box with the **Change Location** tab chosen*

Change Location Tab

This tab displays options to define the location of the nozzle. The options in this tab are discussed next.

Nozzle

This drop-down list contains the total number of nozzles on an equipment, although it displays the currently selected nozzle. You can select any nozzle from this drop-down list and modify it.

Tag Icon

On choosing this icon, the **Add or Modify Nozzles** dialog box will be expanded. You can edit the nozzle type and number in this dialog box.

Nozzle Location

This drop-down list displays types of nozzle locations that can be specified on an equipment.

You can specify the location using the **Top, Bottom, Radial,** and **Line** options. On choosing the **Line** option; you are prompted to select a line from the equipment. The nozzle will be created in line with the selected equipment.

Preview Window
The preview window explains the usage of options available in the **Change location** tab.

Radius(R)
This edit box is available for the **Top** and **Bottom** location options. You can specify the radial distance of the nozzle from the center of the vessel in this edit box.

Angle(A)
This edit box is used to specify the radial angle between the initial location and the new location of the nozzle.

Length(L)
This edit box is used to specify the length of the nozzle

Offset(O)
This edit box is used to specify an offset value of the nozzle. The nozzle will be placed at the offset distance.

Perpendicular(P)
On selecting this check box, the nozzle will be created perpendicular to the equipment. Note that this check box is available only if you select the **Top** or **Bottom** option.

Inclination(I)
This edit box is used to specify the angle of inclination to the nozzle axis.

Rotation(N)
This edit box is used to specify the rotation angle of the nozzle on the horizontal plane.

Twist(T)
This edit box is used to specify the twist angle along the vertical plane.

Change Type Tab
This tab contains a nozzle list and the options to filter it, refer to Figure 4-17. You can choose the required nozzle by filtering various parameters such as the nozzle type, size, units, and pressure class. The options in this tab are discussed next.

Nozzle Type Area
There are four types of nozzles available in this area: **Straight Nozzle, Bent Nozzle, Vent Nozzle** and **Manway**. Choose a nozzle type; different sizes available for the selected nozzle type are displayed.

Size

This drop-down list is used to specify sizes for the chosen nozzle type. On selecting a size, the **Select Nozzle** list is filtered to display nozzles of the selected size.

Units

This drop-down list is used to select a unit type for the nozzle. When you select a unit type from this drop-down list, the **Select Nozzle** list displays nozzles available under the selected unit.

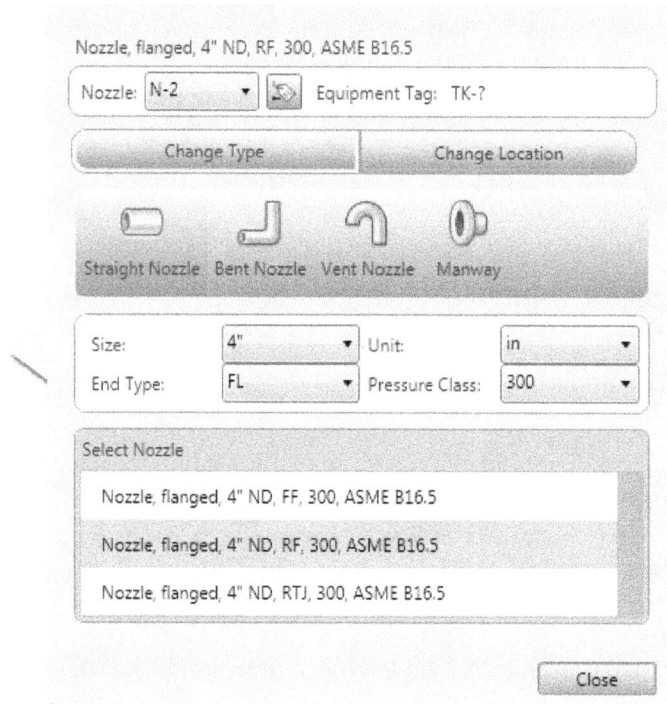

*Figure 4-17 The **Change Type** tab of the **Add or Modify Nozzles** dialog box*

End Type

This drop-down list displays various end types. When you select the **Flange (FL)** end type, the **Select Nozzle** list will be filtered to display nozzles with flange end type.

Pressure Class

On selecting a pressure class from the **Pressure Class** drop-down list, the **Select Nozzle** list will be filtered to display the nozzles with the selected pressure class.

Pressure Class

The nozzles available for the selected nozzle type will be displayed in the **Select Nozzle** list. Specify the parameters such as size, end type, unit and, pressure class; the nozzle list gets filtered. Select the required nozzle from the list and choose the **Close** button; the nozzle will be added at the specified location.

ADDING NOZZLES TO A CONVERTED EQUIPMENT

After converting a solid model into an equipment, you can add nozzles to it. To do so, select the equipment; a nozzle grip will be displayed on it. Click on the nozzle grip; you are prompted to specify the center of the nozzle. To specify the center of the nozzle, snap to the location where you want to add the nozzle and then select a point, refer to Figure 4-18. On doing so, a rubber band line will be attached to the cursor and you are prompted to specify the direction by selecting a second point, refer to Figure 4-19. Specify the second point; the **Add or Modify Nozzles** dialog box will be displayed. Specify the nozzle type using the options in this dialog box and then close the dialog box.

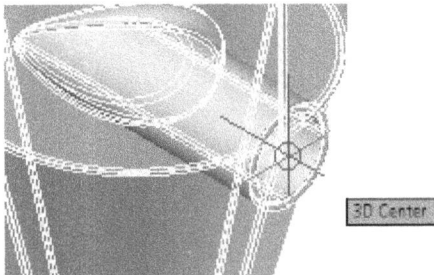

Figure 4-18 *Selecting a point to add a nozzle*

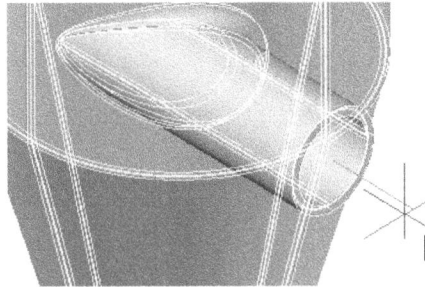

Figure 4-19 *Specifying the direction of the nozzle*

MODIFYING NOZZLES

To modify the type and location of the nozzles, you need to click on the nozzle while holding the CTRL key. On doing so, a pencil grip will be displayed on it, as shown in Figure 4-20. Release the CTRL key and click on the pencil symbol; the **Add or Modify Nozzles** dialog box will be displayed. Now, you can change the location as well as the type of nozzle, refer to Figure 4-21

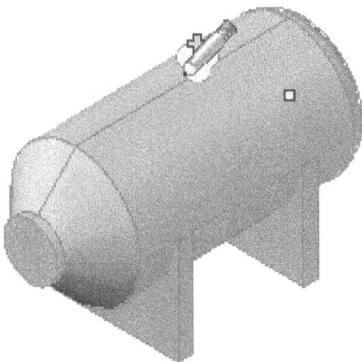

Figure 4-20 *Pencil grip displayed on the nozzle*

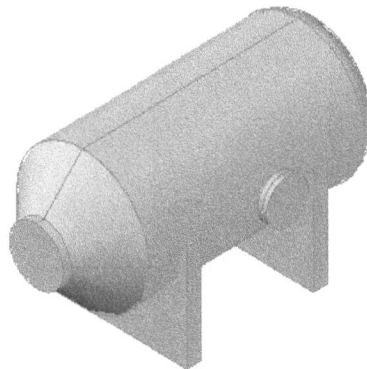

Figure 4-21 *Nozzle after modifying the location and type*

TUTORIALS

Tutorial 1

In this tutorial, you will open the **CADCIM** project created in Chapter 2 and then start a new AutoCAD Plant 3D file. In this file, you will create equipment, as shown in Figure 4-22. The dimensions and location of the equipment are shown in Figures 4-23 through 4-26.

(Expected time: 1 hr)

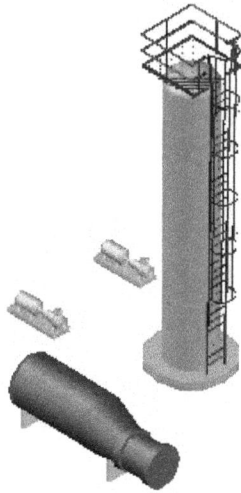

Figure 4-22 The model for Tutorial 1

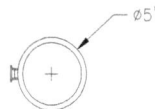

Figure 4-23 Dimensions of the Vertical vessel

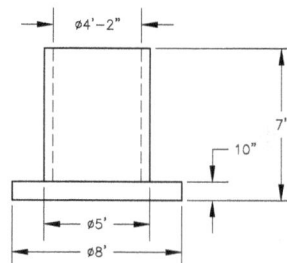

Figure 4-24 Dimensions of the support

Grating Specifications

Material standard:	ASTM
Material code:	A242
Thickness:	1"
Hatch pattern:	ZIGZAG
Hatch scale:	10"
Justification:	Top
Shape:	New rectangular
Size:	70"x70"

Railing Specifications:

Handrail Height:	40 inch
1st mid rail height:	20 inch
2nd mid rail height:	0
Kick plate height:	5 inch
First post:	10 inch
Second post:	5 inch
Handrail Shape:	PIPE2STD
Kick plate Shape:	FB 1/4x4
Post Shape:	PIPE2STD

Ladder Specifications:

Width:	25 inch
Exit width:	35 inch
Projection:	45 inch
Rung Distance:	10 inch
Ladder shape:	PIPE2STD
Rung Shape:	PIPE3/4STD

Cage Specifications:

Start Height:	7'-6"
Maximum distance:	5'
Distance From top:	1 inch
Radius:	14 inch

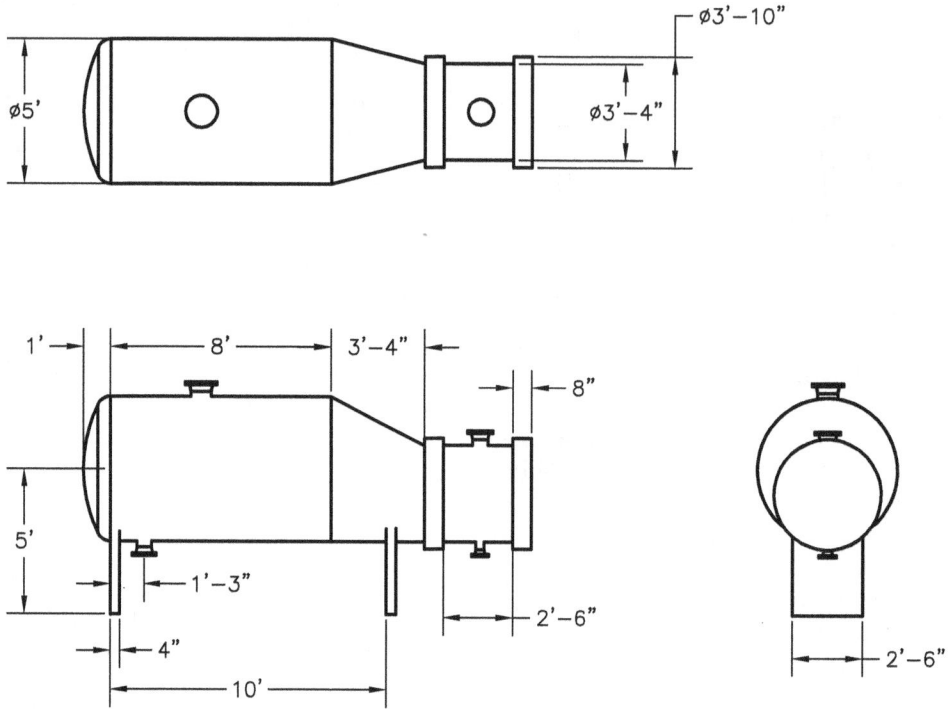

Figure 4-25 *Dimensions of the Reboiler*

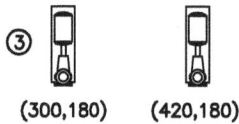

S.NO.	EQUIPMENT NAME	QTY
1	Vertical Equipment	1
2	Reboiler	1
3	Centrifugal Pump	2

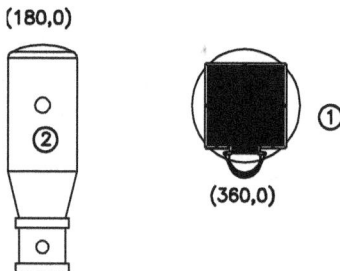

Figure 4-26 *Coordinate dimensions of the model*

The following steps are required to complete this tutorial:

a. Open a new Plant 3D drawing file in the current project.
b. Create a vertical vessel and place it in the model space.
c. Create a grating at the top of the vertical vessel and add a ladder and a railing.

    d. Place the reboiler.
    e. Create supports and attach to the reboiler.
    f. Place centrifugal pumps.
    g. Save the model.

Creating a New AutoCAD Plant 3D File

1. Choose **Start > All Programs (or Programs) > Autodesk > AutoCAD Plant 3D 2014 > AutoCAD Plant 3D 2014**; AutoCAD Plant 3D starts.

    Next, you need to start a new AutoCAD Plant 3D file.

2. Select **CADCIM** from the **Current Project** drop-down list in the **Project Manager**.

3. Select the **Plant 3D Drawings** node in the **Project** area and choose the **New Drawing** button available; the **New DWG** dialog box will be displayed.

4. Enter **Piping Model.dwg** in the **File name** edit box and choose **OK**; the new file will be created.

Creating a Support for the Vessel

You need to create a support before adding the vessels.

1. Create a support for the vessel using the modeling tools, as shown in Figure 4-27. The dimensions of the support are given in Figure 4-24. You can also download this model from www.cadcim.com by following the link Home > Textbooks > CAD/CAM > AutoCAD Plant 3D > AutoCAD Plant 3D 2014 for Designers.

Figure 4-27 *Support for the vertical vessel*

2. Place it at the coordinate point **360, 0**.

3. Change the view orientation to **SW Isometric**.

Creating a Vertical Vessel

1. Choose the **Create** tool from the **Equipment** panel in the **Home** tab; the **Create Equipment** dialog box is displayed.

2. Choose **Vessel > Vertical Vessel** option from the drop-down list located at the top left of the dialog box; a list of shapes stacked together to form the vessel are displayed in the **Shapes** area.

3. Order the shapes in the **Shapes** area, as given below.

    Torispheric Head
    Cylinder
    Torispheric Head

4. Select the **Torispheric Head** in the **Shapes** area; its diameter will be displayed in the **Dimensions** area. Change the diameter to **60 inch**.

5. Similarly, change the diameter of the other torispheric head to **60 inch**.

6. Select the **Cylinder** from the **Shapes** area and change its dimensions as given next.

    D   60"
    H   240"

    Next, you need to assign a tag to the vessel.

7. Click in the field next to the **Tag** box in the **General** area; the **Assign Tag** dialog box will be displayed.

8. Enter **TK** in the **Type** edit box and **101** in the **Number** edit box and choose the **Assign** button; a tag will be assigned to the vessel.

9. Enter **84** as the value of elevation in the field next to the **Elevation** box and choose the **Create** button; the vessel will be attached to the cursor and you will be prompted to select an insertion point.

10. Specify the insertion point of the vessel as 360,0 in the Command prompt; the compass tool is displayed at the bottom of the vessel. Set the orientation of the vessel to 0 degree; the vessel will be placed at the specified point.

Adding a Platform and a Ladder to the Vessel

Next, you need to add a platform and a ladder to the vessel.

1. Choose the **Plate** tool from the **Parts** panel in the **Structures** tab of the **Ribbon**; the **Create Plate/Grate** dialog box is displayed.

2. In this dialog box, select the **Grating** option from the **Type** drop-down list. Next, select the **New rectangular** radio button from the **Shapes** area.

3. Select the **Top** radio button from the **Justification** area.

4. Specify the settings given at the start of the tutorial and choose the **Create** button; you are prompted to specify the first corner of the rectangular grating.

    Make sure that the Dynamic Input is turned off.

5. Enter **325,-35,340** at the Command prompt to specify the first corner; you are prompted to specify the second corner.

6. Enter **395,35,340** at the Command prompt; the rectangular grating is created, as shown in Figure 4-28.

7. Invoke the **Ladder** tool from the **Parts** panel in the **Structures** tab and then select the first point, refer to Figure 4-28.

8. Select the second point, as shown in Figure 4-29. Next, move the cursor in the -Y direction and enter 6" and press ENTER; the ladder is placed, as shown in Figure 4-30.

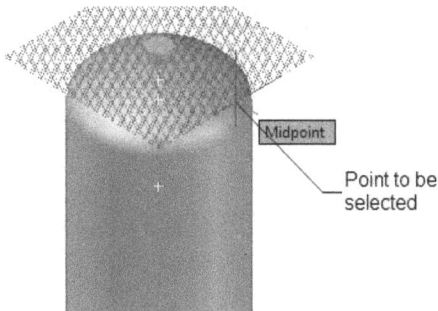

Figure 4-28 Selecting the first point of the ladder

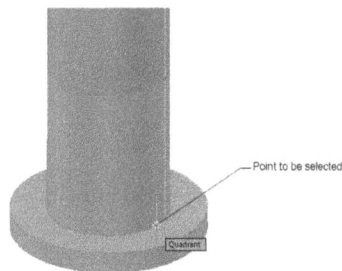

Figure 4-29 Selecting the second point of the ladder

9. Invoke the **Railing** tool from the **Parts** panel and create the railing on the edges of the grating, as shown in Figure 4-31. For dimensions, refer to the specifications given in the tutorial description.

Placing the Reboiler

1. Invoke the **Create Equipment** dialog box and choose **Heat Exchanger > Reboiler** option from the drop-down list located at the top left of the dialog box; a list of shapes is displayed in the **Shapes** area, as given next.

    Cylinder
    Cylinder
    Cylinder

       Cone
       Cylinder
       Torispheric Head

2.   Select the **Cylinder** located at the top in the **Shapes** area; the **Dimensions** area is displayed in the **Create Equipment** dialog box.

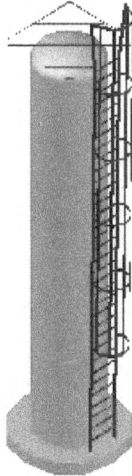

Figure 4-30 Vertical vessel after placing the ladder

Figure 4-31 Vertical vessel after adding railing to the grating

3.   Enter the following values in the **Dimensions** area:

       D:46"   **H:8"**

4.   Similarly, specify the values for the other shapes, as given next.

       Cylinder2
       D: 40"   **H: 30"**

       Cylinder3
       D: 46"   **H: 8"**

       Cone
       Orientation: Upwards   **D1: 60"**   **D2:40"**
       H: 40"   **E: 10"**   **A: 0**

       Cylinder4
       D: 60"   **H: 96"**

Torispheric Head
D: 60"

Next, you need to assign a tag to the reboiler.

5. Click in the **Tag** field in the **General** area of the **Create Equipment** dialog box; the **Assign Tag** dialog box is invoked. Enter **101** in the **Number** edit box and choose the **Assign** button; the **Assign Tag** dialog box is closed. Next, enter **48** in the **Elevation** field in the **Create Equipment** dialog box.

6. Choose the **Create** button and insert the reboiler at the point 180, 40. Next, set the orientation to -90 degrees using the Compass tool. Figure 4-32 shows the model after placing the reboiler.

Figure 4-32 *The model after placing the heat exchanger*

Next, you need to create supports and attach them to the reboiler.

7. Invoke the **Box** tool from the **Modeling** panel in the **Modeling** tab of the **Ribbon**; you are prompted to specify the first corner. You need to make sure that the **Dynamic Input** is turned off.

8. Enter **165,36** at the Command Prompt; you are prompted to specify the second corner of the box. Enter **195,40** at the Command prompt; you are prompted to specify the length of the box.

9. Enter **50** at the Command prompt; the rectangular box is created, as shown in Figure 4-33.

10. Copy the box and select a base point, refer to Figure 4-33. Move the cursor along Y-direction and enter **-10'** at the Command prompt; the box will be copied at the specified location. Note that **Ortho Mode** should be turned ON.

11. Invoke the **Attach Equipment** tool from the **Equipment** panel in the **Home** tab; you are prompted to select a single equipment item.

12. Select the reboiler from the model space; you are prompted to select objects to be attached to the selected equipment.

13. Select the boxes below the reboiler. Next, press ENTER; the boxes are attached to the reboiler. Figure 4-34 shows the reboiler after attaching supports.

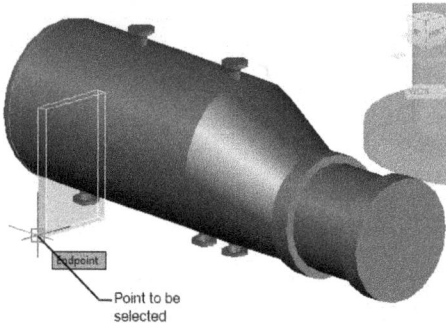

Figure 4-33 *Support created for the reboiler*

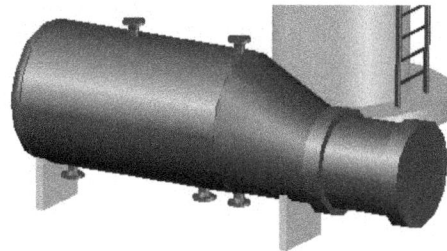

Figure 4-34 *Reboiler after attaching supports*

Placing Centrifugal Pumps

Next, you need to add centrifugal pumps to the model.

1. Invoke the **Create Equipment** dialog box and choose **Pump > Centrifugal Pump** option from the drop-down list located at the top left of the dialog box; the **Equipment** tab along with the preview image of the centrifugal pump is displayed in the **Create Equipment** dialog box.

2. Click in the **Tag** field and enter **101A** in the **Number** edit box in the **Assign Tag** dialog box. Choose the **Assign** button to close the dialog box. Next, enter the long description as **Centrifugal Pump**.

3. Choose the **Create** button; the **Create Equipment** dialog box will be closed and you are prompted to select an insertion point. Enter the insertion point as 25',15' and orient the pump to 90 degrees, as shown in Figure 4-35.

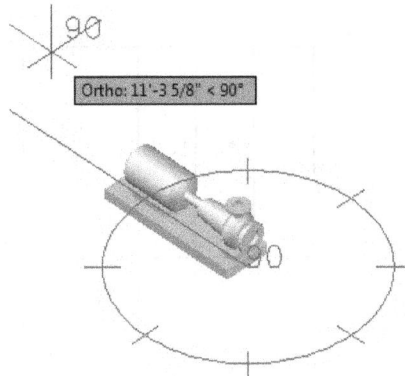

Figure 4-35 *Setting the orientation of the centrifugal pump*

4. Similarly, place another pump with the same orientation at the insertion point 35',15'.

Note
The pump should be assigned the tag P-101B.

The model after placing pumps is shown in Figure 4-36.

Figure 4-36 *The model after placing pumps*

Saving the Model

1. Choose the **Save** button from the **Quick Access Toolbar**; the file will be saved.

2. Choose **Close > Current Drawing** on the **Application Menu**; the file is closed.

Tutorial 2

In this tutorial, you will add nozzles to the equipment created in Tutorial 1. The nozzle list is given next.

(Expected time: 45 min)

Nozzle List of the Vertical Vessel

S.No	Nozzle Location	Long Description	Height (H)	Angle (A)	Length (L)
1	Radial	Nozzle, flanged 6" ND, RF, 150, ASME B16.5	10'	180	6"
2	Radial	Nozzle, flanged 8" ND, RF, 150, ASME B16.5	5'	180	6"
3	Bottom	Nozzle, flanged 10" ND, RF, 150, ASME B16.5		180	6"

Nozzle List of the Reboiler

S.No	Nozzle Location	Long Description	Height (H)	Angle (A)	Length (L)
1	Radial	Nozzle, flanged 8" ND, RF, 150, ASME B16.5	3'-4"	90	6"
2	Radial	Nozzle, flanged 6" ND, RF, 150, ASME B16.5	1'-3"	270	6"
3	Radial	Nozzle, flanged 6" ND, RF, 150, ASME B16.5	13'-4"	90	6"
3	Radial	Nozzle, flanged 3" ND, RF, 150, ASME B16.5	13'-4"	270	6"

Nozzle List of the Centrifugal Pump

S.No	Nozzle Type	Long Description
1	Inlet Nozzle	Nozzle, flanged 10" ND, RF, 150, ASME B16.5
2	Outlet Nozzle	Nozzle, flanged 8" ND, RF, 150, ASME B16.5

a. Open the **Piping_Model.dwg** file created in the previous tutorial.
b. Add nozzles to all vertical vessel.
c. Modify nozzles of the reboiler.

d. Modify nozzles of the pumps
e. Save the model.

Opening the File
1. Choose **Start > All Programs (or Programs) > Autodesk > AutoCAD Plant 3D 2014 > AutoCAD Plant 3D 2014**; AutoCAD Plant 3D is started and the welcome screen is displayed.

2. Choose *CADCIM > Plant 3D Drawings > Piping_Model* from the **Project Manager;** the Plant 3D model will be opened.

Adding Nozzles to the Vertical Vessel
1. Click on the vertical vessel; the **Add Nozzle** grip will be displayed on it.

2. Click on the **Add Nozzle** grip; a preview of the nozzle and the **Add or Modify** Nozzles dialog box will be displayed.

3. Select the **Change Location** tab from the dialog box displayed, if not already selected; the options to change the location of the nozzle are displayed.

4. Select the **Radial** option from the **Nozzle Location** drop-down list. Next, enter the following values in the edit boxes displayed in this tab:

   H: 60"   **A: 180**   **L: 6"**

5. Choose the **Change Type** tab and filter the **Select Nozzle** list by setting the following options:

   Size: 8"   **Unit: in**   **End Type: FL**
   Pressure Class:150

6. Select **Nozzle, flanged 8" ND, RF, 150, ASME B16.5** from the **Select Nozzle** list. Next, choose the **Close** button to close the **Add or Modify Nozzles** dialog box; the nozzle is added.

7. Invoke the **Add or Modify Nozzles** dialog box again and select the **Radial** option from the **Nozzle Location** drop-down list in the **Change Location** tab. Next, enter the following values in the respective edit boxes:

   H: 120"   **A: 180**   **L: 6"**

8. Choose the **Change Type** tab and filter the **Select Nozzle** list by setting the following options:

   Size: 6"   **Unit: in**   **End Type: FL**
   Pressure Class: 150

9. Select **Nozzle, flanged 6" ND, RF, 150, ASME B16.5** from the **Select Nozzle** list. Next, choose the **Close** button to close the **Add or Modify Nozzles** dialog box; the nozzle is added.

    Next, you need to modify the nozzle at the bottom of the vessel. To do so, first you need to hide the support at the bottom.

10. Choose the **Hide Selected** button from the **Visibility** panel in the **Home** tab of the **Ribbon**; you are prompted to select the object to be hidden.

11. Select the support of the vertical vessel and press ENTER; it is hidden.

12. Press CTRL key and click on the nozzle located at the bottom; a pencil symbol is displayed on it.

13. Click on the pencil symbol; the **Add or Modify Nozzles** dialog box is displayed. Choose the **Change Location** tab from this dialog box.

14. Select the **Bottom** option from the **Nozzle Location** drop-down list and enter **180** in the **T** edit box and **6** in the **L** edit box.

15. Choose the **Change Type** tab and then filter the **Select Nozzle** list by setting the following options:

Size	**10"**
Unit	**in**
End Type	**FL**
Pressure Class	**150**

16. Select **Nozzle, flanged, 10" ND, RF, 150, ASME B16.5** from the **Select Nozzle** list. Next, choose the **Close** button from the **Add or Modify Nozzles** dialog box; the bottom nozzle is modified.

Deleting a Nozzle on the Reboiler

1. Press CTRL and click on the nozzle located on the bottom of the reboiler, refer to Figure 4-37.

2. Press DELETE to delete the selected nozzle.

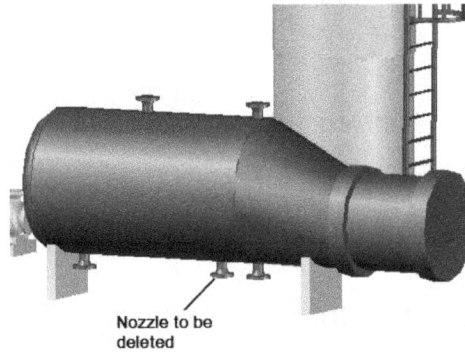

Figure 4-37 Nozzles to be deleted

Modifying Nozzles of the Reboiler

1. Press CTRL and click on **Nozzle 1** located on the top of the reboiler, refer to Figure 4-38; a pencil grip is displayed on it, as shown in Figure 4-39. Note that the nozzles are numbered in the figure for reference only.

2. Click on the pencil symbol displayed on the nozzle; the **Add or Modify Nozzles** dialog box is displayed.

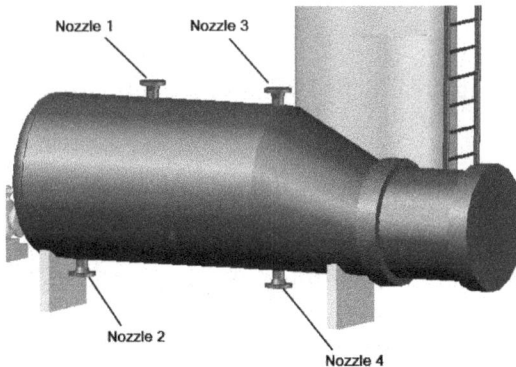

Figure 4-38 Nozzles displayed on the reboiler

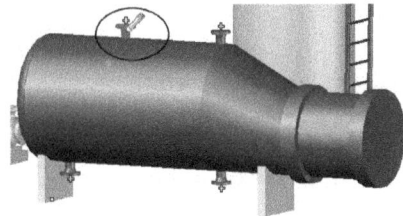

Figure 4-39 Pencil symbol displayed on the nozzle

3. Choose the **Change Type** tab from the dialog box and filter the **Select Nozzle** list by specifying the following parameters:

Size	**8"**
Unit	**in**
End Type	**FL**
Pressure Class	**150**

4. Select **Nozzle, flanged 8" ND, RF, 150, ASME B16.5** from the **Select Nozzle** list and close the dialog box.

5. Similarly, modify the other nozzles of the reboiler as given next. For nozzle sequence, refer to Figure 4-38.

Nozzle 2
Type: Nozzle, flanged 6" ND, RF, 150, ASME B16.5

Location
H: 15"   A: 270   L: 6"

Nozzle 3
Type: Nozzle, flanged 6" ND, RF, 150, ASME B16.5

Location
H: 160"   A: 90   L: 6"

Nozzle 4
Type: Nozzle, flanged 3" ND, RF, 150, ASME B16.5

Location
H: 160"   A: 270   L: 6"

Modifying the Pump Nozzles

Next, you need to modify the pump nozzles.

1. Select a centrifugal pump from the drawing area; pencil symbols are displayed on the inlet and outlet nozzles of the pump.

2. Click on the pencil symbol displayed on the inlet nozzle; the **Add or Modify Nozzles** dialog box is displayed.

3. Choose the **Change Type** tab from the dialog box and filter the **Select Nozzle** list by specifying the following parameters:

Size: 10"   Unit: in   End Type: FL
Pressure Class: 150

4. Select **Nozzle, flanged 10" ND, RF, 150, ASME B16.5** from the **Select Nozzle** list and close the dialog box.

   Next, you need to modify the outlet nozzle.

5. Select the pump and then click on the pencil symbol displayed on the outlet; the **Add or Modify Nozzles** dialog box is displayed.

6. Choose the **Change Type** tab from the dialog box and filter the **Select Nozzle** list by specifying the following parameters.

Size: 8"   Unit: in   End Type: FL
Pressure Class: 150

7. Select **Nozzle, flanged 6" ND, RF, 150, ASME B16.5** from the **Select Nozzle** list and close the dialog box.

Similarly, modify the inlet and outlet nozzles of the other centrifugal pump.

Saving the Model

1. Choose the **Save** button from the **Quick Access Toolbar**; the file is saved.

2. Choose **Close > Close All** on the **Application Menu**; the file is closed.

Self-Evaluation Test

Answer the following questions and then compare them to those given at the end of this chapter:

1. On choosing the _____ button in the **Create Equipment** dialog box, a flyout will be displayed with a list of shapes.

2. The _____ area displays a list of shapes piled up to create a piece equipment.

3. The _____ edit box in the **Add or Modify Nozzles** dialog box is used to specify the value to create a nozzle at an inclination to the nozzle axis.

4. The _____ edit box in the **Add or Modify Nozzles** dialog box is used to specify a twist angle to the nozzle.

5. On selecting the _____ check box, the nozzle will be created perpendicular to the equipment.

Review Questions

Answer the following questions:

1. You can create a new equipment and then save it as a template. (T/F)

2. You can add shapes to a pump. (T/F)

3. The **Convert Equipment** tool is used to convert a solid model into an equipment.

4.   You can attach objects to an equipment using the _____ tool.

5.   The _____ grip is displayed on selecting an equipment from the model space.

Answers to Self-Evaluation Test
1. **Add**, 2. **Shapes**, 3. **I** (Inclination), 4. **T** (Twist), 5. **P** (Perpendicular).

Chapter 5

Editing Specifications and Catalogs

Learning Objectives

After completing this chapter, you will be able to:

- *Create a New Spec File*
- *Add parts to a Spec*
- *Edit parts added to a Spec*
- *Add Notes to a part*
- *Modify a Spec*
- *Update a Spec*
- *Assign a Branch table*
- *Export and Import Spec Data*
- *Create a new Catalogue*
- *Add a new part to a Catalogue*

INTRODUCTION

In this chapter, you will learn to create and edit the specifications and catalog files. A catalog is a list of parts that can be used while creating a piping model. The specifications file contains specifications of the parts present in the catalog. When you are creating a 3D piping model, the specifications file provides the required specifications for pipes, fittings, fasteners, and so on.

GETTING STARTED WITH AutoCAD Plant 3D Spec Editor

You can start **AutoCAD Plant 3D Spec Editor** by double-clicking on its shortcut icon on the desktop of your computer. Alternatively, choose **Start > All Programs (or Programs) > Autodesk > AutoCAD Plant 3D 2014 > AutoCAD Plant 3D Spec Editor 2014**; the **Autodesk AutoCAD Plant 3D Spec Editor** will be displayed, as shown in Figure 5-1. Close the Welcome window by using the **Close** button in the top right corner of the window.

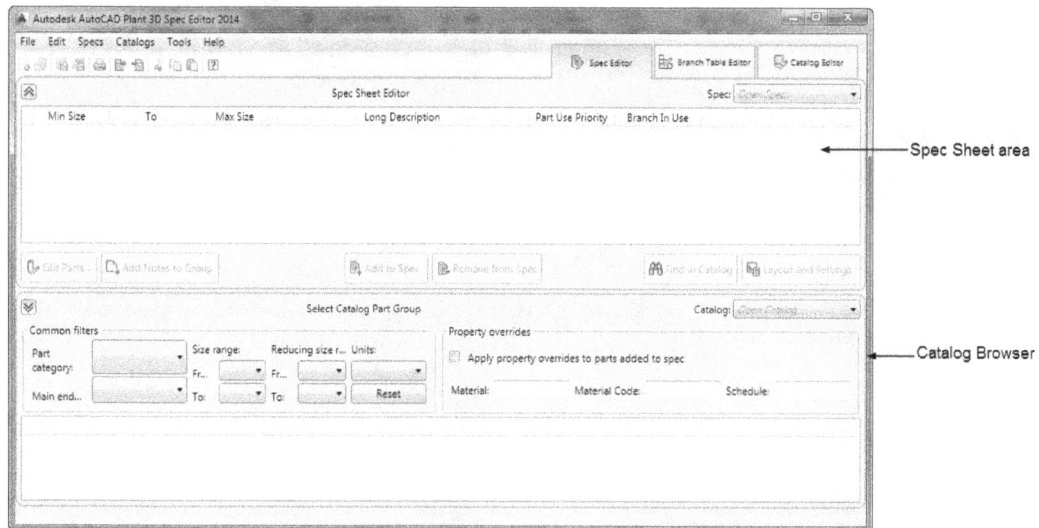

*Figure 5-1 The **Autodesk AutoCAD Plant 3D Spec Editor 2014** window*

The screen contains three tabs: **Spec Editor**, **Branch Table Editor**, and **Catalog Editor**. You can use the **Spec Editor** tab to edit the spec sheet of the parts. The **Branch Table Editor** tab is used to assign a fitting for a branch size. The assigned fitting will be used while creating a branch. The **Catalog Editor** tab is used to add or remove parts from a catalog. Also, you can edit the general properties and size parameters of a part.

CREATING A NEW SPEC FILE

You can create a new spec file by choosing **New > Create Spec** from the **File** menu; the **Create Spec** dialog box will be displayed, as shown in Figure 5-2. Next, enter the spec name in the **New Spec name** edit box. You can set the path of the file by using the **Browse** button next to the edit box. Enter a description of the spec in the **Spec description** text box. Next, specify the part catalog which you want to load for creating the spec from the **Load catalog** drop-down list. Then, choose the **Create** button; a new spec file will be created. The newly created spec file will be displayed in the **Spec sheet** area. The parts of the selected catalog are loaded in the Catalog Browser. You can also load additional catalogs in the Catalog Browser by using the **Open Catalog** option from the **Catalog** drop-down list.

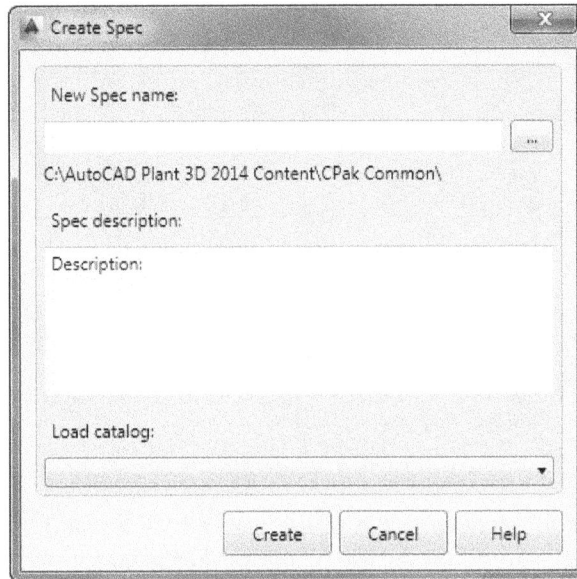

Figure 5-2 The ***Create Spec*** *dialog box*

CREATING A NEW SPEC FILE FROM AN EXISTING SPEC

You can create a new spec file from the already existing specs. On doing so, the parts of the already existing spec file will be loaded to the new spec file and you can add more parts to it. To do so, choose **New > Create Spec from Existing** from the **File** menu; the **Create Spec From Existing Spec** dialog box will be displayed. Choose the Browse button next to the **Source Spec name** edit box; the **Open** dialog box will be displayed. Browse to the location of the existing spec file and choose the **Open** button; the file path will be displayed in the **Source Spec name** and the **New Spec name** boxes, refer to Figure 5-3.

You can modify the name and path of the new spec in the **New Spec name** edit box. It is optional to enter the description in the **Spec description** text box. Next, choose the **Create** button; the **Create Spec From Existing Spec** dialog box will be closed and the new spec file will be created and displayed in the **Spec Editor** tab. The path of the new spec file is displayed in the **Spec Sheet** area, as shown in Figure 5-4. The options in the **Spec Editor** tab are discussed next.

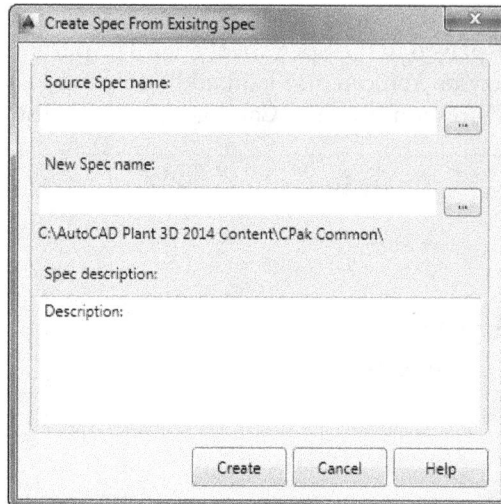

*Figure 5-3 The **Create Spec From Existing Spec** dialog box*

Spec Sheet

The **Spec Sheet** area displays properties of the parts which are added to a spec file, refer to Figure 5-4. It displays the properties such as **Minimum size**, **Maximum size**, **Long Description**, **Part Use Priority**, and **Branch In Use**.

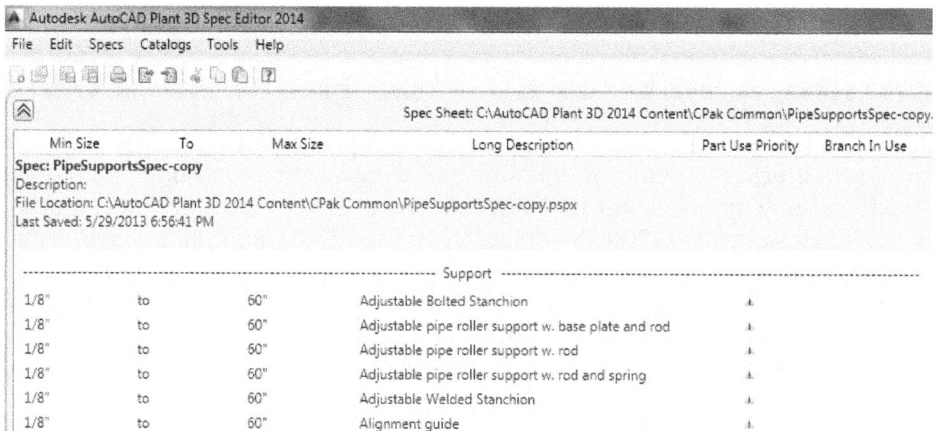

*Figure 5-4 The **Spec Sheet** area*

Catalog Browser

The **Catalog Browser** area consists of two areas namely, **Common Filters** and **Property overrides**, and a table which displays the data related to the catalog, refer to Figure 5-5. The **Common Filters** area contains options to filter the display of the catalog data, whereas in the **Property overrides** area, you can override the property values of the part copied from the catalog to the spec sheet. In AutoCAD Plant 3D, you can only override Material, Material Code, and Schedule.

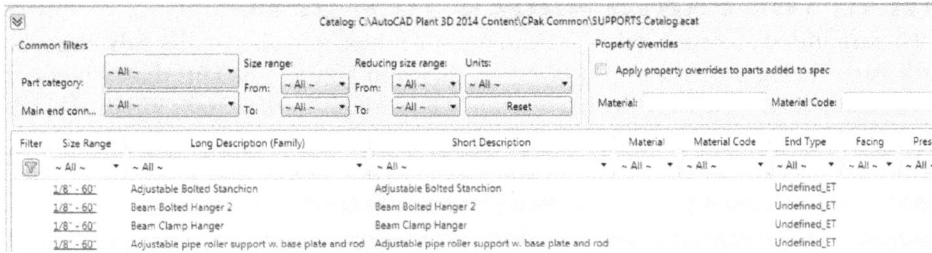

*Figure 5-5 The **Catalog Browser** area*

ADDING PARTS TO THE SPEC SHEET

You can add parts from a catalog to a newly created spec or an existing spec. For example, to add a pipe of size ranging from 3" to 10", select **Pipe** from the **Part category** drop-down list in the **Common Filters** area of the Catalog Browser, refer to Figure 5-6. Set the size range using the **From** and **To** drop-down lists in the **Size range** group; pipes under the specified range will be displayed in the table. Select a pipe from the table and enter the property values (Material, Material Code and Schedule) in the **Property overrides** area, refer to Figure 5-7. Next, select the **Apply property overrides to parts added to spec** check box and choose the **Add to Spec** button; the pipe group will be added to the spec sheet. Similarly, you can add fittings that you want to use while routing a pipe. Also, you can add multiple parts by loading multiple catalogs using the **Catalog** drop-down list.

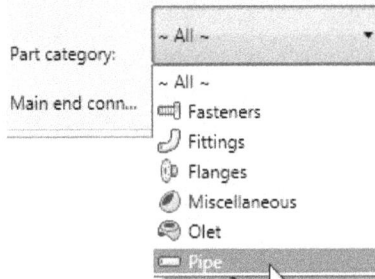

*Figure 5-6 Selecting an option from the
Part category drop-down list*

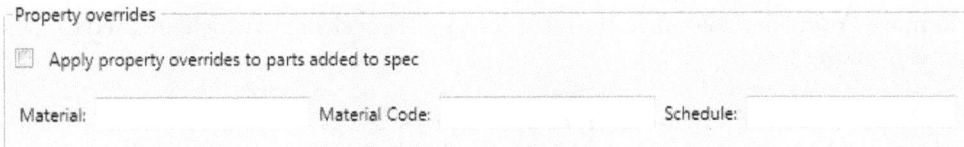

*Figure 5-7 The **Property overrides** area*

To remove a part from a spec, select the part from that **Spec Sheet** area and choose the **Remove from Spec** button; the part will be removed from the spec sheet.

EDITING THE PARTS ADDED TO A SPEC

To edit a part added to a spec, select it from the spec sheet and choose the **Edit Parts** button; the **Edit Parts** dialog box will be displayed, as shown in Figure 5-8. This dialog box contains the **Parts List** tab and the **Edit Properties** tab. The options in these tabs are discussed next.

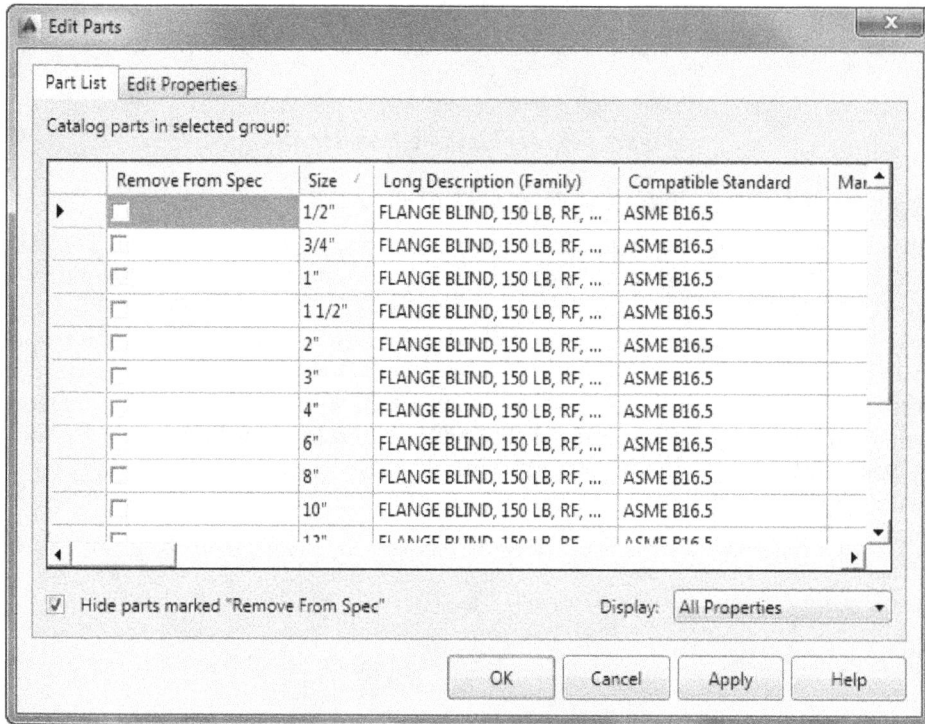

Figure 5-8 The **Edit Parts** dialog box

Part List Tab

This tab consists of a table that displays the properties of the parts. You can modify the display of part properties by using the options in the **Display** drop-down list. You can display the catalog properties, added properties, or all properties.

The **Remove From Spec** column in the table consists of check boxes which are used to remove the parts from the spec.

The **Hide parts marked "Remove From Spec"** check box is selected by default. When you clear this check box, the parts which are not included in the selected group will be displayed. The properties of these parts are read-only. You can clear the check boxes to include them into the spec file.

Edit Properties Tab

You can add properties to the selected part using the options in this tab. Figure 5-9 shows the options in the **Edit Properties** tab. The **Property definition** area contains the options which are used to define a new property. The **Add Property to** drop-down list in this area is used to specify whether to add the property to the current part or to all parts.

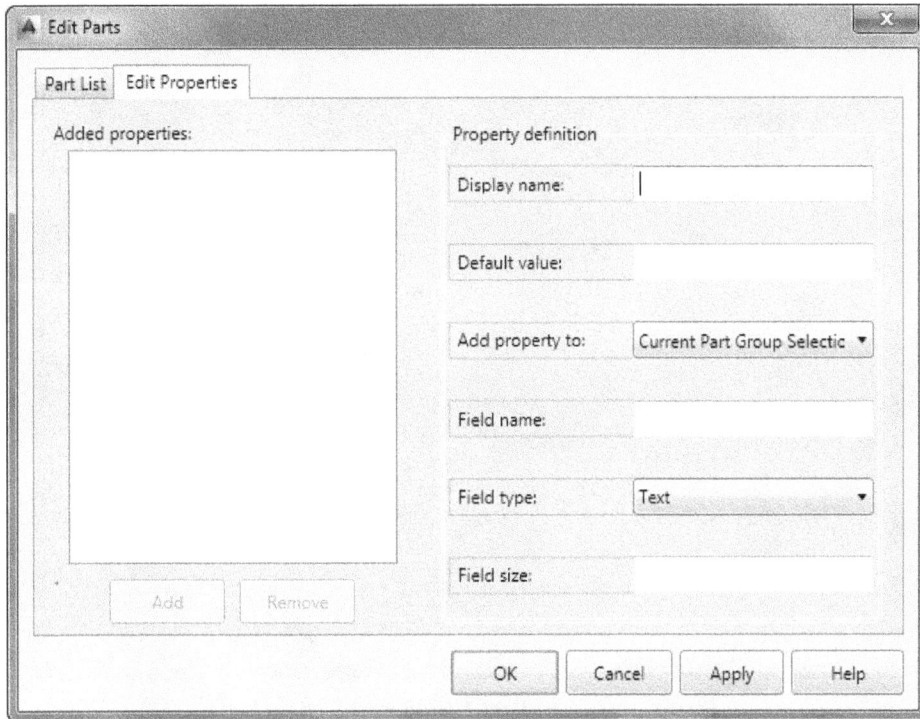

Figure 5-9 *The* **Edit Properties** *tab in the* **Edit Parts** *dialog box*

The **Added properties** list box displays the newly added properties. The **Add** button below this list box is used to add the properties defined in the **Property definition** area. Note that the **Add** button will be active only after entering values in all the edit boxes of the **Property definition** area. You can remove the newly added properties by selecting them from the **Added properties** list box and then choosing the **Remove** button. Choose the **OK** button to accept the changes and close the **Edit Parts** dialog box.

SETTING THE PART USE PRIORITY

After assigning sizes to the parts in the spec sheet, there may be a case in which there are more than one part of the same size in a spec file. In such a case, the system is unable to decide the part to be used first and it displays an error symbol in the spec sheet. Therefore, you need to manually set the priority of the parts. To do so, click on the error symbol in the spec sheet; the **Part Use Priority** dialog box will be displayed, as shown in Figure 5-10. Follow the steps given next to set the priority in this dialog box.

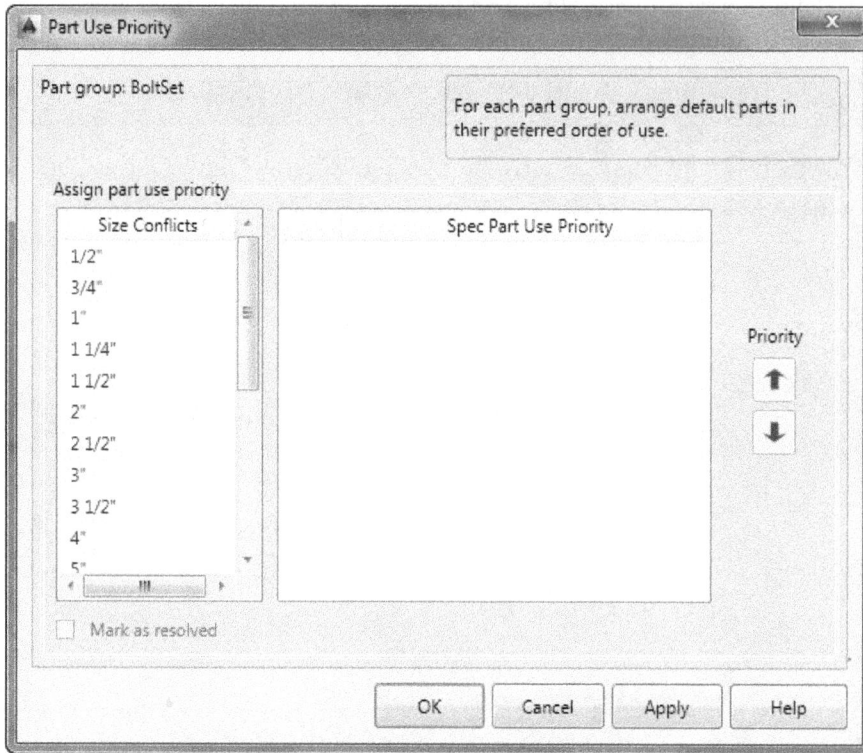

Figure 5-10 *The Part Use Priority dialog box*

1. Select a size from the **Size Conflicts** list in the **Assign part use priority** area; the parts under the selected size will be displayed in the **Spec Part Use Priority** list box.

2. Select a part from the **Spec Part Use Priority** list and move it up and down in the list, using the up and down arrows, respectively.

3. Select the **Mark as resolved** check box and choose the **OK** button to close the dialog box.

ADDING NOTES TO A GROUP

You can add notes to a part group that you have added to a spec sheet. To do so, select a part from the spec sheet and choose the **Add Notes to Group** button; the **Add Notes To Group** dialog box will be displayed, as shown in Figure 5-11. This dialog box consists of a text box in which you can enter notes. The entered note will be displayed when the spec sheet is printed.

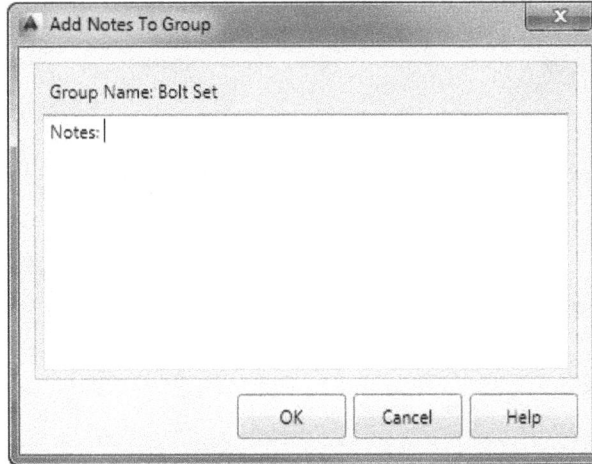

*Figure 5-11 The **Add Notes To Group** dialog box*

EDITING THE LONG DESCRIPTION STYLES

In **AutoCAD Plant 3D Spec Editor**, you can customize the long description styles of the parts as per the requirement of the project. For example, a default long description style of a Y-Type Strainer is shown in Figure 5-12. You can modify the long description, as shown in Figure 5-13.

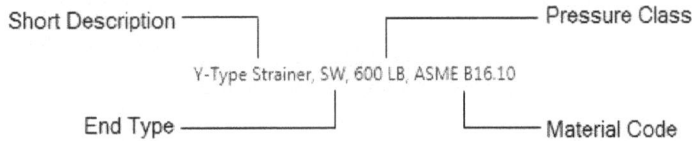

Figure 5-12 A default long description style of a Y Type Strainer

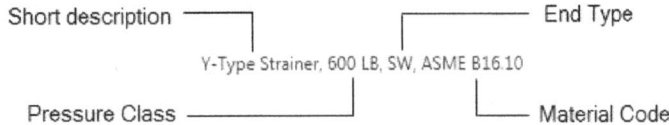

Figure 5-13 Modified long description Style

To modify the long description style of a piping component, choose the **Layout and Settings** button in the **Spec Editor**; the **Spec Editor Layout and Settings** dialog box will be displayed, as shown in Figure 5-14. In this dialog box, choose the **Edit long description styles** button; the **Edit Long Description Style** dialog box will be displayed, as shown in Figure 5-15.

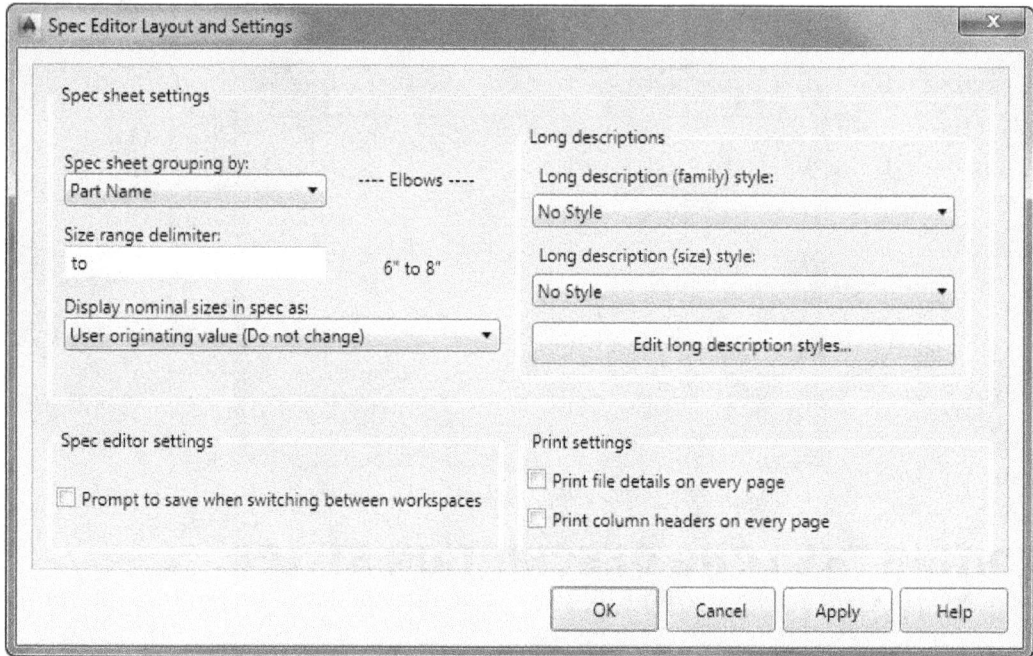

Figure 5-14 The **Spec Editor Layout and Settings** *dialog box*

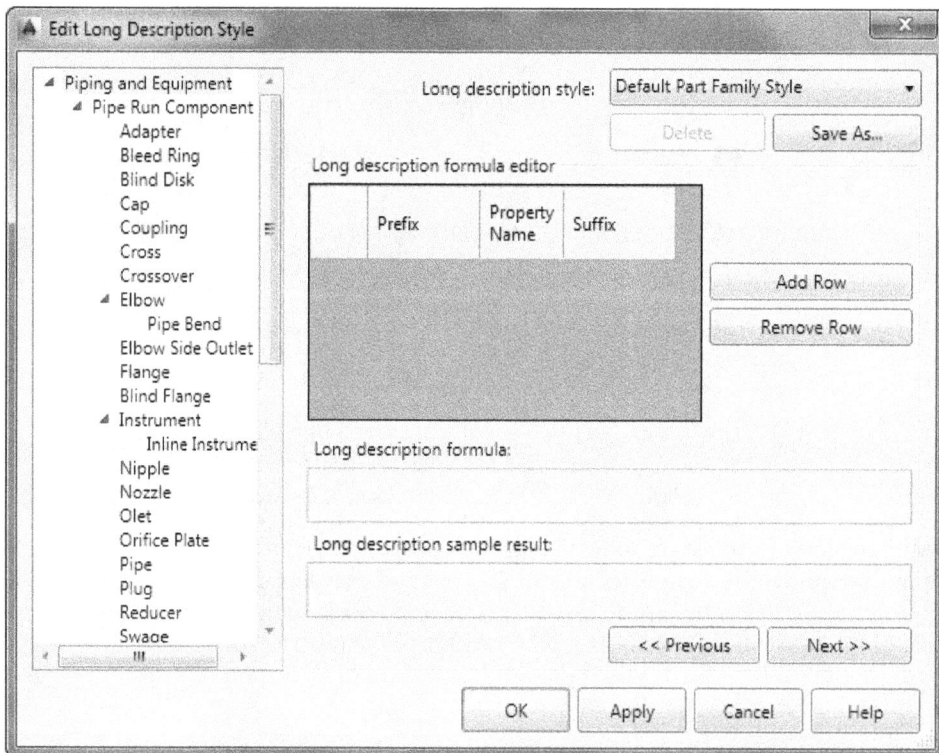

Figure 5-15 The **Edit Long Description Style** *dialog box*

Follow the steps given next to edit the long description.

1. Select a component from the **Part** list available at the left side; the default long description settings will be displayed.

2. Select the default long description style from the **Long description style** drop-down list; the editor table in the **Long description formula editor** area displays the property names arranged.

3. Modify the formula editor table by modifying the Property name values and adding new prefixes and suffixes. Also, you can add a new row to the formula editor table by choosing the **Add Row** button. The resultant long description style will be displayed in the **Long description style sample result** box.

4. Choose the **Save as** button located below the **Long description styles** drop-down list; the **Save As New Style** dialog box will be displayed.

5. Enter the name in the **Style name** edit box and choose the **Create** button; a new long description style will be created.

6. Choose the **OK** button; the **Edit Long Description Style** dialog box will be closed and the **Spec Editor Layout and Settings** dialog box will be activated.

7. Select the newly created long description style from the **The Long Description (Family) Style** or the **Long Description (Size) Style** drop-down list and choose the **OK** button; the selected long description style will be assigned to the part.

ASSIGNING A LONG DESCRIPTION STYLE TO MULTIPLE SPECS

You can assign a long description style to multiple specs at a time. To do so, choose **Spec > Batch Assign Long Description Styles** from the menu bar; the **Batch Assign Long Description Styles** dialog box will be displayed, as shown in Figure 5-16. Select a long description style from the **Long Description (family) style** or **Long Description (size) style** drop-down lists, or both. Next, choose the **Add** button from the **Select Spec file** area; the **Open** dialog box will be displayed. Browse to the spec file location and double-click on it; the spec file will be added to the **Apply long description styles to these pipe specifications** table, refer to Figure 5-16. Similarly, you can add more spec files to the table. You can also remove a spec file from the table by selecting it and choosing the **Remove** button. Choose the **OK** button after adding all the spec files to which you want to assign a long description style; the long description style will be assigned and a message box will be displayed with the message **Batch assignment of Long Description Styles has successfully completed**.

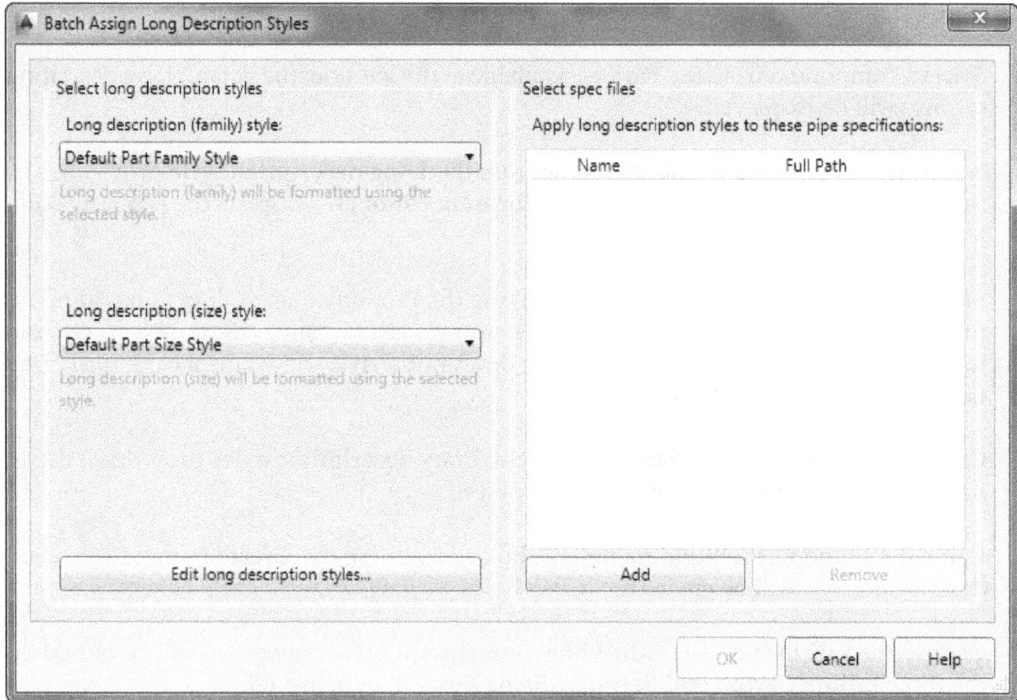

*Figure 5-16 The **Batch Assign Long Description Styles** dialog box*

ASSIGNING OPERATORS (ACTUATORS) TO VALVES

Actuators are assigned to valves by default but they are not displayed in the spec sheet. To assign a different actuator to a valve, right-click on the valve in the spec sheet and choose the **Edit Valve Operator** option from the shortcut menu displayed, refer to Figure 5-17; the **Edit Valve Operators** dialog box will be displayed, as shown in Figure 5-18.

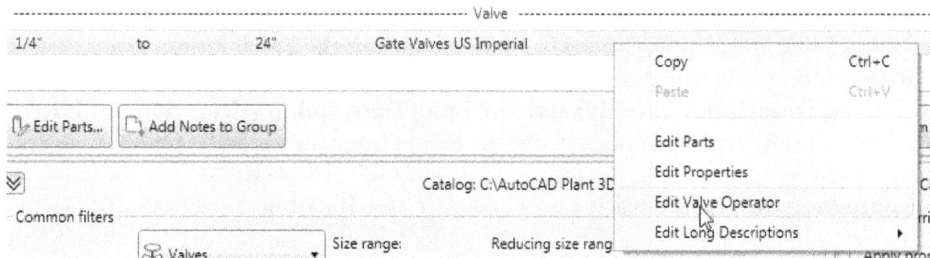

*Figure 5-17 Choosing the **Edit Valve Operator** option*

In this dialog box, select the required valve from the **Valves in current spec** tree; the default operators of the selected valve will be displayed in the **Default Valve Operators** table. Select the required size from the **Default Valve Operators** table and choose the **Valve Operator Shape Browser** button; the **Valve Operator Shape Browser** dialog box will be displayed, as shown in Figure 5-19. Select the required valve operator from the **Select Operator Shape** area of this dialog box and choose the **OK** button; the

Valve Operator Shape Browser dialog box will be closed. Next, choose the **OK** button from the **Edit Valve Operators** dialog box; the new operator will be assigned to the selected valve.

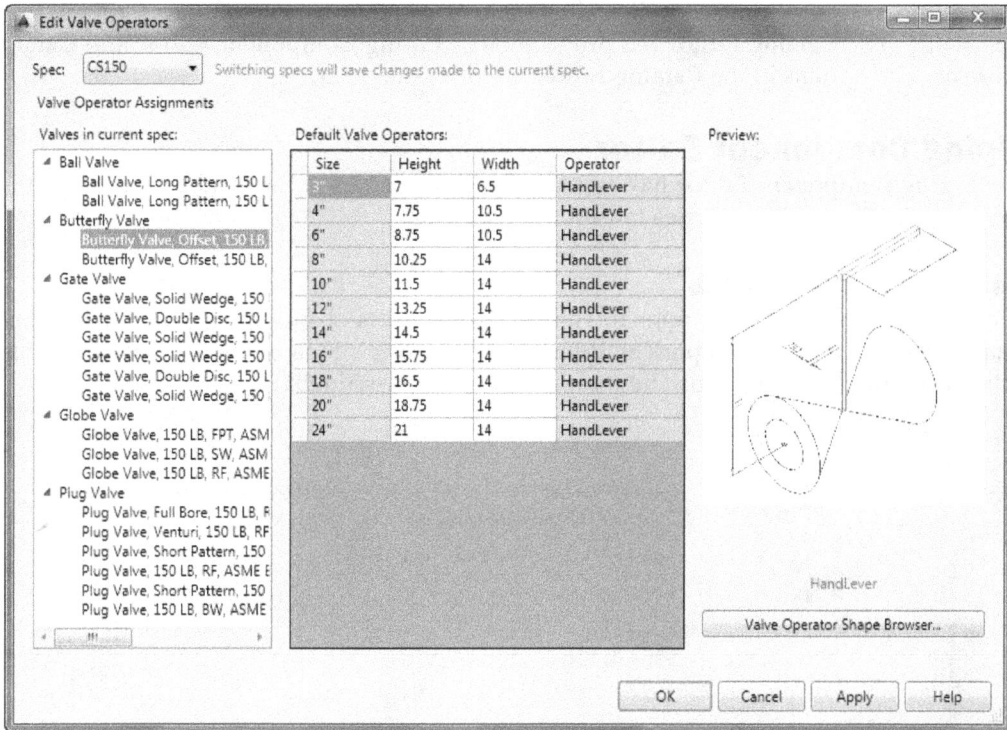

Figure 5-18   The **Edit Valve Operators** *dialog box*

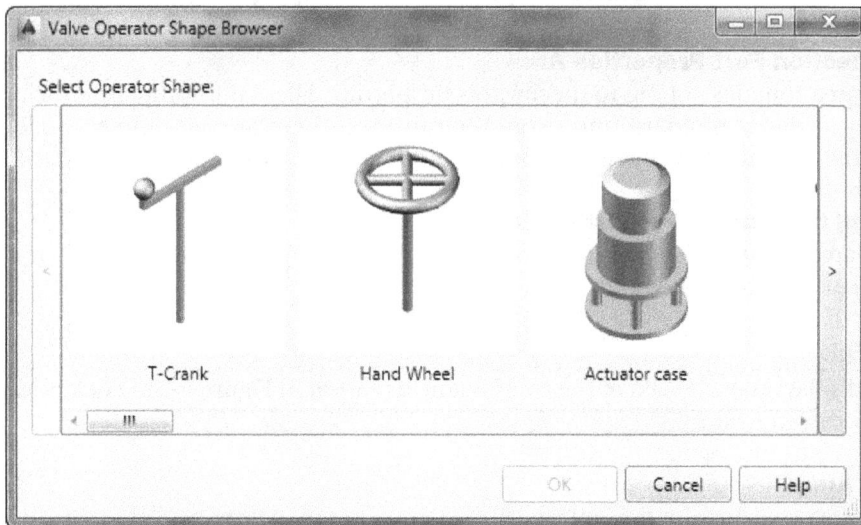

Figure 5-19 The **Valve Operator Shape Browser** *dialog box*

WORKING WITH THE CATALOG EDITOR

The **Catalog Editor** is used to modify a catalog and customize it according to the requirement. You can add or remove parts as well as change the size and properties of the existing parts in the catalog. The **Catalog Editor** has two windows - **Piping Component Editor** and **Catalog Browser**. The options in the **Catalog Editor** are discussed next.

Piping Component Editor

The **Piping Component Editor** has options to edit properties of the component. It contains two tabs - the **General Properties** tab and the **Sizes** tab.

General Properties Tab

The **General Properties** tab, refer to Figure 5-20, consists of two areas - **Connection Port Properties** and **Piping Component Properties**. A preview area of the component is also available on the left of this tab. These two areas are discussed next.

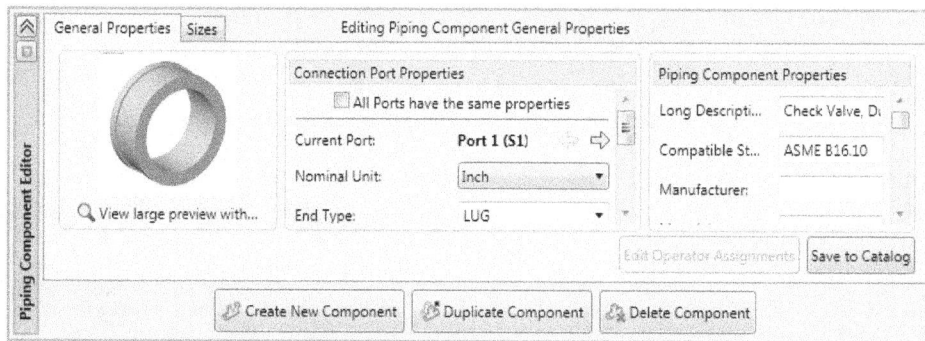

*Figure 5-20 The **General Properties** tab in the **Piping Component Editor***

Connection Port Properties Area

This area contains options to specify port properties. These port properties include units, end type, flange standard, and so on. You can apply these properties to multiple ports by selecting the **All Ports have the same properties** check box.

Piping Component Properties Area

This area is used to set the property values of the components such as material, design standards, valve details, and so on.

Sizes Tab

This tab displays the size data of the component, as shown in Figure 5-21. The options in this tab are discussed next.

Size Area

This area consists of a list box which displays a range of sizes available for the selected component. You can select the required size for the selected component from the list box. Also, you can add or remove a size from the list box using the **Add Size** and **Remove Size** buttons. Also, you can duplicate a size using the **Duplicate size** button.

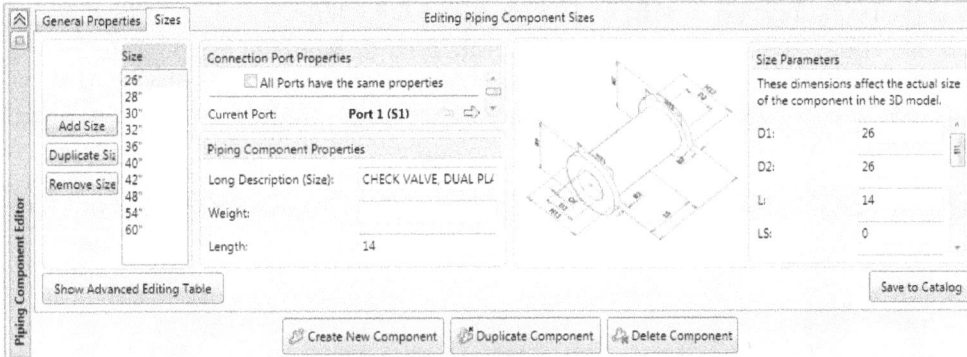

*Figure 5-21 The **Sizes** tab in the **Piping Component Editor***

Size Parameters Area

This area displays dimensional parameters of the selected component. The preview area explains the parameters in the **Size Parameters** area.

The **Show Advanced Editing Table** button is used to display the component size properties in the form of a table.

Catalog Browser

This is similar to **Catalog Browser** in the **Spec Editor** tab. But, there are some additional options to filter the display of parts in this browser. You can display an individual part size or all part sizes using the **Show All Part Sizes** and the **Show All Part Families** buttons.

CREATING A NEW CATALOG FROM AN EXISTING CATALOG

You can create a new catalog from an existing one and modify the catalog content as per your requirement. To do so, choose **New > Create Catalog from Existing** from the **File** menu; the **Create Catalog From Existing Catalog** dialog box will be displayed, as shown in Figure 5-22. Select an existing catalog file by using the **Browse** button next to the **Source catalog name** edit box and then choose the **Create** button; the copy of the selected catalog will be created.

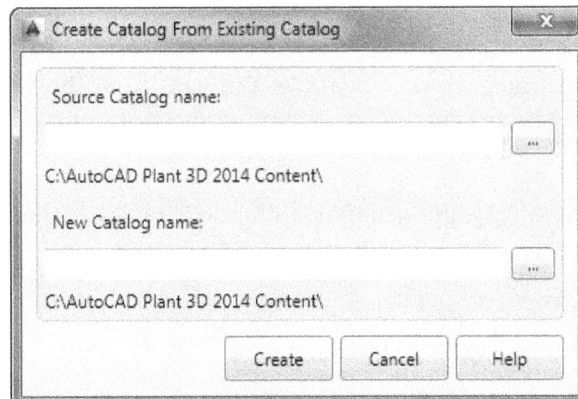

*Figure 5-22 The **Create Catalog From Existing Catalog** dialog box*

ADDING A NEW PART TO A CATALOG

To add a new part to a catalog, choose the **Create New Component** button from the **Piping Component Editor**; the **Create New Component** dialog box will be displayed, as shown in Figure 5-23. Using the options in this dialog box, you can create a new component using two different methods, which are discussed next.

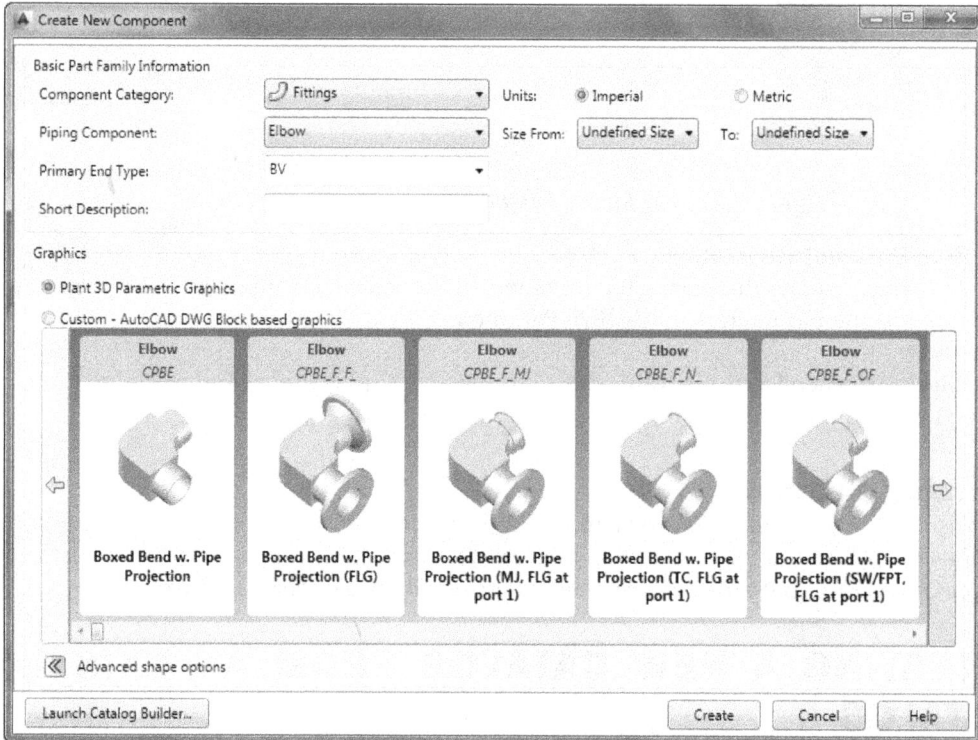

Figure 5-23 *The* ***Create New Component*** *dialog box*

Creating a New Component Using Parametric Graphics

To create a component using parametric graphics, follow the steps given next.

1. Select the **Plant 3D Parametric Graphics** radio button from the **Graphics** area.

2. Select a component category (For example: **Fittings**) from the **Component Category** drop-down list; the **Piping Component** drop-down list displays a list of component types available for the selected component type.

3. Select the desired piping component type from the **Piping Component** drop-down list.

4. Select the primary end type (For example: **FL** for flanged end) from the **Primary End Type** drop-down list.

5. Enter a short description about the component in the **Short Description** edit box.

6. Select the required component from the **Graphics** area.

7. Specify units (For example: **Imperial**) by selecting the respective radio buttons.

8. Select the size range by using the **Size From** and **To** drop-down list available below the **Units** radio buttons.

9. Choose the **Create** button; the component will be added to the **Catalog Editor**.

10. Select the **Sizes** tab from the **Piping Component Editor**.

11. Enter the outer diameter value in the **Matching Pipe OD** edit box in the **Connection Port Properties** area. Next, select the **All Ports have the same properties** check box from the **Connection Port Properties** area, if the connection port properties of all ports are same. If not, select the right arrow and enter values of the matching pipe OD for all the other ports.

12. Enter the long description of the component in the **Long Description** edit box.

13. Choose the **Save to Catalog** button; the component will be added to the catalog.

Creating a New Component Using Block Based Graphics

To create a new component using blocks, first you need to create a block and convert it into an AutoCAD Plant 3D component. Create a valve, elbow, flange, or any other piping component by using AutoCAD modeling tools and convert it into a block. Next, invoke the **PLANTPARTCONVERT** command and select the block; the prompt **Select a port operation [Add/Delete/Move/eXit]** will be displayed at the Command prompt. Select the **Add** option at the Command prompt; you are prompted to specify a port location. Specify a point and create a port, as shown in Figure 5-24. Similarly, add

Figure 5-24 *Creating a port on an AutoCAD block*

ports on the other side of the block. Select the **eXit** option after adding all ports. Save the file in which the block is created.

Switch to the **Catalog Editor** and choose the **Create New Component** button; the **Create New Component** dialog box will be displayed. In this dialog box, select the **Use Custom Geometry** radio button and follow the steps given next.

1. Select a component category (For example: **Valves**) from the **Component Category** drop-down list.

2. Select the desired piping component type from the **Component** drop-down list.

3. Enter a short description about the component in the **Short Description** edit box.

4. Select the primary end type from the **Primary End Type** drop-down list.

5. Specify the number of ports available on the block in the **Number of Connection Ports** drop-down list, refer to Figure 5-25.

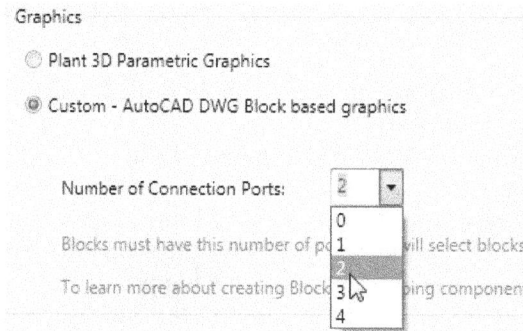

Figure 5-25 Specifying the number of ports

6. Specify units (For example: **Imperial**) by selecting the respective radio buttons.

7. Select the size range by using the **Size From** and **To** drop-down lists available below the **Units** radio button.

8. Choose the **Create** button; the dialog box will be closed.

9. Choose the **Sizes** tab in the **Piping Component Editor**.

10. In the **Size** area, select sizes that you do not want and choose the **Remove size** button. You can also add sizes to the list by using the **Add size** button.

11. Choose the **Select Model** button from the **Size** area; the **Open** dialog box will be displayed. Open the file containing the block; the **Select Block Definition** dialog box will be displayed, as shown in Figure 5-26.

12. Select the block that has ports specified using the **PLANTPARTCONVERT** command and choose the **OK** button; the selected block will be displayed in the **Piping Component Editor**, as shown in Figure 5-27.

13. Enter the outer diameter value in the **Matching Pipe OD** edit box in the **Connection Port Properties** area. Next, select the **All Ports have the same properties** check box from the **Connection Port Properties** area, if the connection port properties of all ports are same. If not, select the right arrow and enter values of the matching pipe OD for all the other ports.

14. Enter the long description of the component in the **Long Description** edit box.

15. Choose the **Save to Catalog** button; the component will be added to the catalog.

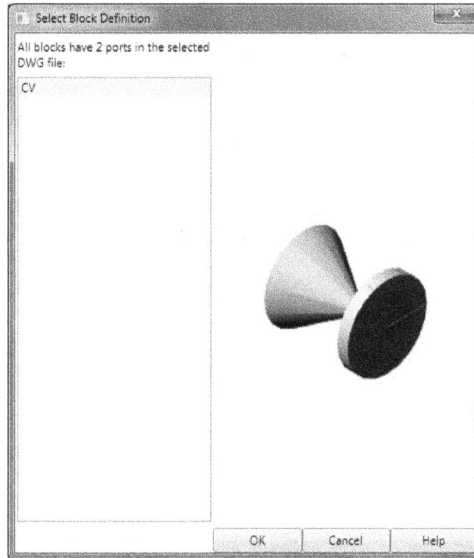

*Figure 5-26 The **Select Block Definition** dialog box*

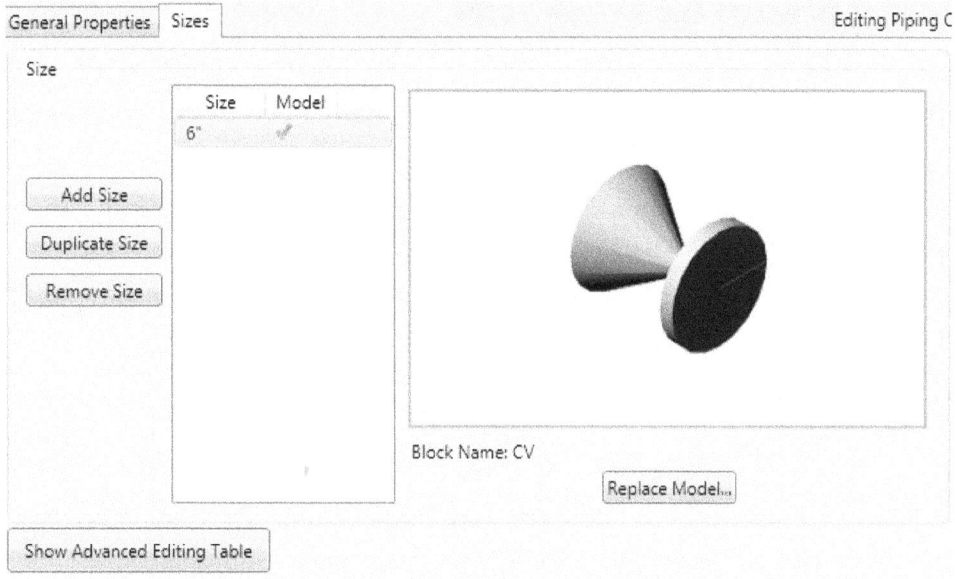

*Figure 5-27 The selected block displayed in the **Piping Component Editor***

MODIFYING THE BRANCH TABLE

The branch table determines the type of fitting to be used while joining a header pipe with a branch pipe. The branch table is prepared by arranging the header sizes in the columns and the branch sizes in rows, as shown in Figure 5-28. While routing a pipe in AutoCAD Plant 3D, you can use the branch table information to cite the header size with the branch size and insert an appropriate fitting connecting them. In the branch table, the cell where the header size column and the branch size row meet displays a legend symbol representing the type of fitting, refer to the highlighted cell shown in Figure 5-28, where the header size 24" and the branch size 2" meet. It displays two legend symbols representing the fittings (S001,O001) to be used while connecting the header pipe of 24" with the branch pipe of 2". You can also assign more symbols to the branch cell.

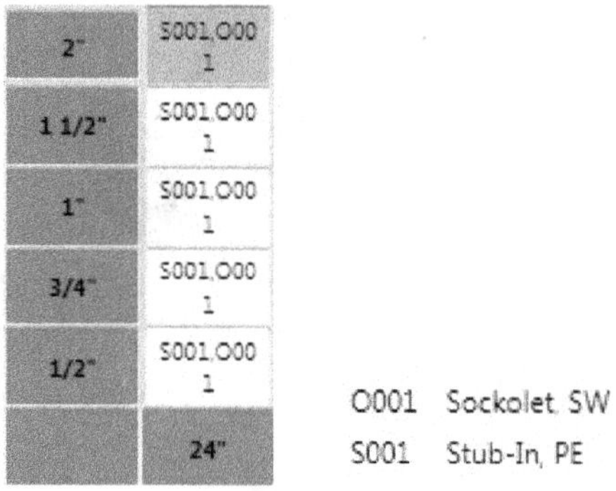

Figure 5-28 *Cells displaying fittings to be used for different branches*

Creating Branch Table Legends

To create a legend (symbol) for a branch table, first you need to identify all branch fittings available in a spec. Next you need to decide the type of fittings to be used for each type of branch and then create their legends. Follow the steps given next to create and add a new legend to a branch table.

1. Select the **Edit Legend** button from the **Legends** pane in the **Branch Table Editor**; the **Branch Table Setup** dialog box will be displayed, refer to Figure 5-29.

2. Choose the **Add Branch** button from the **Branch Connection Part Setup** area; a new branch row will be added to the part setup table.

3. Select a desired fitting from the **Part Type** drop-down list in the newly created row.

4. Select a part from the **Spec Part** drop-down list. If you select the **Use preferred part from the Spec** option, the part used will depend on the priority set in the spec sheet.

5. Double-click in the **Legend Symbol** field and enter a symbol (for example: T001). Next, choose **OK**.

You can also add a reducer along with the fitting to connect the branch pipe. To do so, select the check box in the corresponding field in the **Add Reducer** column of the part setup table; a new row will be added. You first need to select a reducer type and then the reducer part.

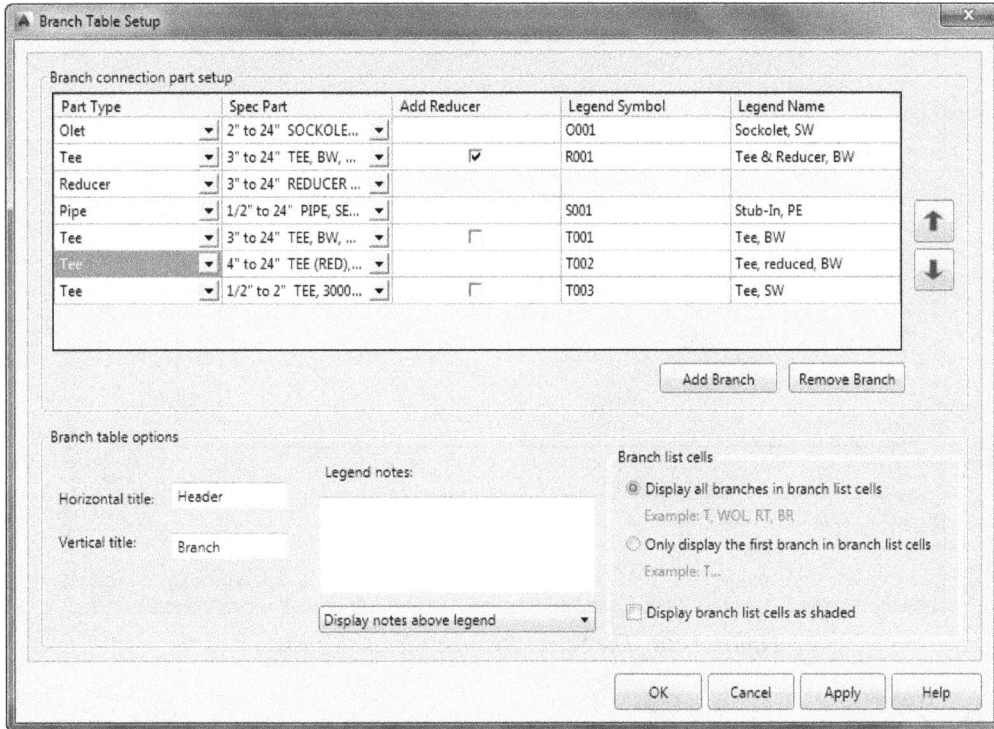

Figure 5-29 The **Branch Table Setup** *dialog box*

Assigning Legends to a Branch Table

You can assign more than one branch fitting to a branch. Also, you can change the default branch fitting and assign a new one. To do so, open the **Branch Table Editor** and follow the steps given next.

1. Select a cell that is intersecting with the header size and branch size.

2. Right-click on the selected cell and choose the **Multi Branch Selection** option from the shortcut menu displayed; the **Select Branch List** dialog box will be displayed, refer to Figure 5-30.

3. The table in this dialog box displays the available branch symbols. In the table, select the check box adjacent to the branch fitting; the fitting will be added to the selected cell in the **Branch Table Editor** and will be used while routing.

4. Use the up and down arrows to set the priority of the fitting. Next, choose the **OK** button to close the dialog box.

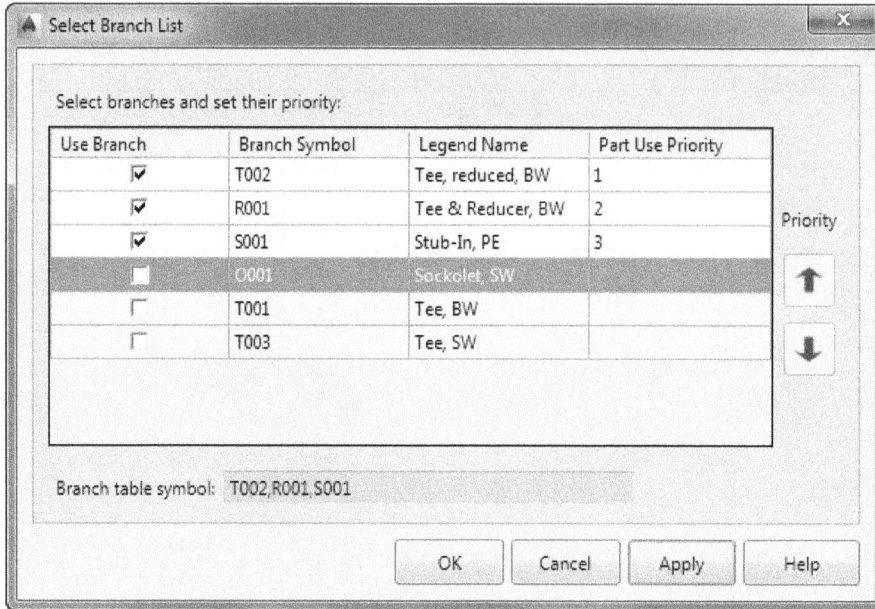

Figure 5-30 The *Select Branch List* dialog box

Note

You can also assign fittings to multiple branches at the same time. To do so, you need to select multiple cells. To select multiple cells which are in a sequential order, click on a cell. Next, press and hold the SHIFT key and keep clicking on the cells to be selected.

To select multiple cells that are not in sequence, press and hold the CTRL key and then click on multiple cells.

TUTORIALS

Tutorial 1

In this tutorial, you will create a spec file and add parts to it, as shown in Figure 5-31. You will also modify the part properties.   **(Expected time: 45 min)**

The following steps are required to complete this tutorial:

a.   Open the AutoCAD Plant 3D Spec Editor.
b.   Create a new spec file.
c.   Add parts to the spec sheet.
d.   Edit the part properties.
e.   Set part use priorities.
f.   Edit long description styles.
g.   Save the spec file.

```
------------------------------------------------------ Cross ----------------------------------------
2"      to      8"          CROSS, 125 LB, FF, ASME B16.1
3/4"    to      1 1/2"      CROSS, BW, ASME B16.9

------------------------------------------------------ Olet -----------------------------------------
2"      to      8"          SOCKOLET, 3000 LB, BWXSW, 9/16" LG, ASME B16.11

------------------------------------------------------ Pipe -----------------------------------------
3/4"    to      1 1/2"      PIPE, SEAMLESS, 10S, PE, ASTM A312
2"      to      8"          PIPE, SEAMLESS, 5S, PE, ASTM A312
2 1/2"  to      8"          PIPE, SEAMLESS, 80, PE, ASTM A106

----------------------------------------------------- Reducer ---------------------------------------
2"      to      8"          REDUCER (ECC), 125 LB, FF, ASME B16.1

------------------------------------------------------ Tee ------------------------------------------
2"      to      8"          TEE, 125 LB, FF, ASME B16.1
3/4"    to      1 1/2"      TEE, BW, ASME B16.9
```

*Figure 5-31 The **Spec Sheet** after adding parts to it*

Creating a New Spec File

1. Click on the shortcut icon of **AutoCAD Plant 3D Spec Editor 2014** on the Desktop; **AutoCAD Plant 3D Spec Editor 2014** is opened and the welcome window is displayed.

2. Choose the **Create** option under the **Spec** category from the welcome window; the **Create Spec** dialog box is displayed.

3. Enter **CADCIM** in the **New Spec name** edit box and **Sample Spec** in the **Description** editor.

4. Choose **ASME Pipes and Fittings Catalog** from the **Load catalog** drop-down list and then choose the **Create** button; the new spec file is created.

Adding Parts to the Spec Sheet and Editing Their Part Properties

In this section, you will add parts to the spec sheet. To do so, you need to filter the parts in **Catalog Browser** by using filter options available in it.

1. Select the **Pipe** option from the **Part Category** drop-down list from the **Common filters** area and apply the following filters:

   Size range: **0.75 to 1.5**
   Unit: **in**

2. Select the **PIPE, SEAMLESS** option from the **Short Description** drop-down list in the **Catalog Browser** to show seamless pipes in it.

3. Select **PIPE, SEAMLESS, 10S, PE, ASTM A312** parts from the **Catalog Browser** and choose the **Add to Spec** button from the **Spec Sheet** area; the part is added to the spec sheet.

    Next, you need to add a larger pipe to the spec sheet.

4. Select the **Pipe** option from the **Part Category** drop-down list, if it is not already selected, and apply the following filters:

    Size range: **2 to 8**
    Unit: **in**
    Short Description: PIPE, SEAMLESS

5. Select the **Apply property overrides to parts added to spec** check box from the **Property overrides** area and enter the following values:

    Material: **CS**
    Material Code: **A106**
    Schedule: **100**

6. Press the CTRL key on the keyboard and select **PIPE, SEAMLESS, 80, PE, ASTM A106** and **PIPE, SEAMLESS, 5S, PE, ASTM A312** from the **Catalog Browser**. Next, choose the **Add to Spec** button; the parts are added to the spec sheet.

7. Select the **Fittings** option from the **Part Category** drop-down list and apply the following filters:
    Size range: **2 to 8**
    Main end connection: **FL**
    Short Description: **REDUCER (ECC)**

8. Select **REDUCER (ECC), 125 LB, FF, ASME B16.1** from the **Catalog Browser** and choose the **Add to Spec** button from the **Spec Sheet** pane; the reducer will be added to the spec sheet.

    Similarly, add other parts to the spec sheet by filtering the **Catalog Browser** using the following filter options:

    Part Category: **Fittings**
    Main end connection: **FL**
    Size range: **2 to 8**

    Select the following parts from the **Catalog Browser**:

    TEE, 125 LB, FF, ASME B16.1

CROSS, 125 LB, FF, ASME B16.1

Next, you need to add an olet to the spec sheet.

9. Select the **Olet** option from the **Part category** drop-down list and apply the following filters:

Main end connection:	**ALL**
Size range:	**2 to 8**
Short Description:	**SOCKOLET**

10. Select **SOCKOLET, 3000 LB, BWXSW, 3/8" LG, ASME B16.11** from the **Catalog Browser** and choose the **Add to Spec** button; the part will be added to the spec sheet.

11. Add another **Tee** and **Cross** to the spec sheet. Use the following filter options to filter the **Catalog Browser**:

Part Category:	**Fittings**
Main end connection:	**BV**
Size range:	**0.75 to 1.5**

    Select the following parts from **Catalog Browser**:

    TEE, BW, ASME B16.9
    CROSS, BW, ASME B16.9

    Figure 5-32 shows the spec sheet after adding parts to it

-- Cross ----------------------------------			
2"	to	8"	CROSS, 125 LB, FF, ASME B16.1
3/4"	to	1 1/2"	CROSS, BW, ASME B16.9
-- Olet -----------------------------------			
2"	to	8"	SOCKOLET, 3000 LB, BWXSW, 9/16" LG, ASME B16.11
-- Pipe -----------------------------------			
3/4"	to	1 1/2"	PIPE, SEAMLESS, 10S, PE, ASTM A312
2"	to	8"	PIPE, SEAMLESS, 5S, PE, ASTM A312
2 1/2"	to	8"	PIPE, SEAMLESS, 80, PE, ASTM A106
--- Reducer ---------------------------------			
2"	to	8"	REDUCER (ECC), 125 LB, FF, ASME B16.1
-- Tee ------------------------------------			
2"	to	8"	TEE, 125 LB, FF, ASME B16.1
3/4"	to	1 1/2"	TEE, BW, ASME B16.9

Figure 5-32 *The spec sheet after adding parts to it*

Setting the Part Use Priority

In this section, you will set the part priority for parts with conflicts. You will notice that an error symbol is displayed next to the parts with conflicts, refer to Figure 5-32. This is because of the system not being able to assign part usage priority to parts with the same size.

1. Click on the error symbol displayed in the **Part Use Priority** column in the **Spec Sheet**; the **Part Use Priority** dialog box is displayed.

2. Select the **2 1/2"** size from the **Size Conflicts** list in the dialog box; the parts for the selected size are displayed in the **Spec Part Use Priority** list.

3. Move **PIPE, SEAMLESS, 80, PE, ASTM A106** to top in the **Spec Part Use Priority** list and select the **Mark as resolved** check box. Similarly, move **PIPE, SEAMLESS, 80, PE, ASTM A106** to top for all sizes and choose the **OK** button; the dialog box is closed and a green dot is displayed in the **Part Use Priority** column. The green dot indicates that the conflict is solved.

Editing Long Description Styles

In this section, you will edit the long description style of the pipe.

1. Select **PIPE, SEAMLESS, 80, PE, ASTM A106** from the spec sheet and right-click on it; a shortcut menu is displayed. Choose **Edit Long Descriptions > Assign Long Description Styles to Spec** option from the shortcut menu; the **Spec Editor Layout and Settings** dialog box is displayed.

2. Choose the **Edit long description style** button from the dialog box; the **Edit Long Description Styles** dialog box is displayed.

3. Select **Pipe** from the Component list available on the left side in the **Edit Long Description Styles** dialog box; the default long description settings will be displayed.

4. Select the **Default Part Size Style** option from the **Long description style** drop-down list; the editor table in the **Long description formula editor** area displays the property names arranged by default, refer to Figure 5-33.

5. Modify the formula editor table by deleting some rows and modifying the **Property Name** values and adding new prefixes and suffixes. Figure 5-34 shows the formula editor table after modification.

6. Choose the **Save as** button located below the **Long description style** drop-down list; the **Save As New Style** dialog box is displayed.

7. Enter **Long description1** in the **Style name** edit box and choose the **Create** button; a new long description style will be created.

8. Choose the **OK** button; the **Edit Long Description Style** dialog box will be closed and the **Spec Editor Layout and Settings** dialog box will be activated.

	Prefix	Property Name		Suffix
▶		Short Description	▾	
	,	Size	▾	ND
	,	End Type	▾	
	,	Facing	▾	
	,	Pressure Class	▾	LB
	,	Compatible Standard	▾	
	,	Material Code	▾	
	, SCH	Schedule	▾	

Figure 5-33 *The formula editor table*

	Prefix	Property Name		Suffix
▶		Short Descri...	▾	
	,	Size	▾	ND
	, SCH	Schedule	▾	
	,	End,Type	▾	
	,	Compatible ...	▾	

Figure 5-34 *The formula editor table after modification*

9. Select the newly created long description style from the **Long Description (Size) Style** drop-down list and choose the **OK** button; the selected long description style will be assigned to the pipe component.

Saving the Spec File

1. Choose the **Save** tool from the main toolbar to the save the file.

2. Choose **File > Exit** from the main menu to close the spec file.

Tutorial 2

In this tutorial, you will create a new catalog from the existing one, add a part with two different size parameter. Figure 5-35 shows the part to be added to the catalog. Figure 5-36 and 5-37 show two different size parameters of the part.   **(Expected time: 45 min)**

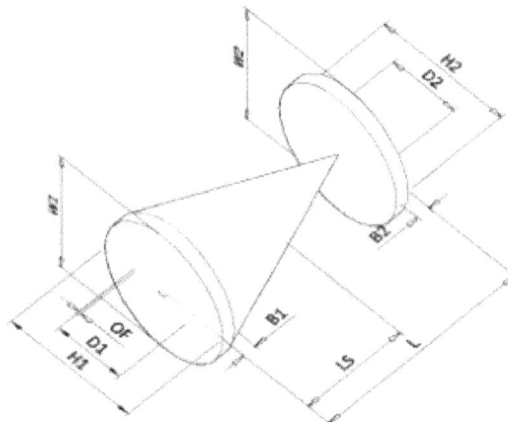

Figure 5-35 *Part to be added to the catalog*

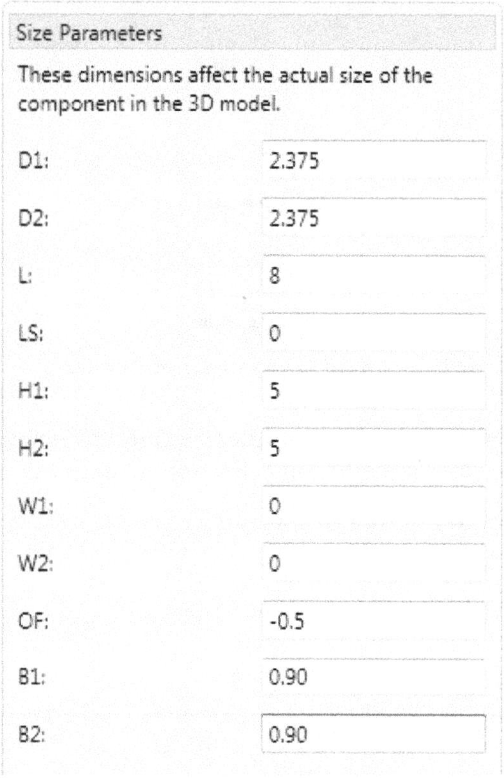

Size Parameters	
These dimensions affect the actual size of the component in the 3D model.	
D1:	2.375
D2:	2.375
L:	8
LS:	0
H1:	5
H2:	5
W1:	0
W2:	0
OF:	-0.5
B1:	0.90
B2:	0.90

Size Parameters	
These dimensions affect the actual size of the component in the 3D model.	
D1:	8.625
D2:	8.625
L:	18
LS:	0
H1:	12.5
H2:	12.5
W1:	0
W2:	0
OF:	-0.5
B1:	1.5
B2:	1.5

Figure 5-36 Size parameters for 2" nominal diameter

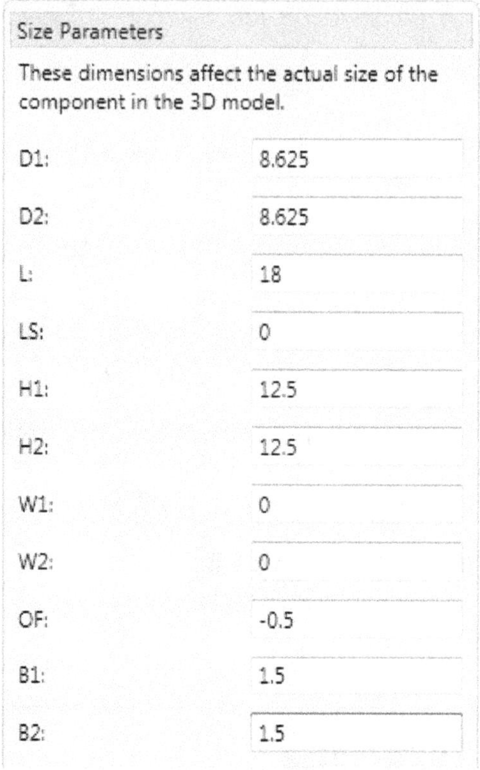

Figure 5-37 Size parameters for 8" nominal diameter

The following steps are required to complete this tutorial:

a. Create a new catalog file.
b. Add parts to the catalog.
c. Specify the general properties.
d. Specify the size parameters.
e. Save part to the catalog.

Creating a New Catalog from the Existing Catalog

1. Click on the shortcut icon of **AutoCAD Plant 3D Spec Editor 2014**; the **AutoCAD Plant 3D Spec Editor 2014** starts and the welcome window is displayed.

2. Close the welcome screen and choose **File > New > Create Catalog From Existing** in the Menu bar; the **Create Catalog From Existing Catalog** dialog box is displayed.

3. Choose the **Browse** button next to the **Source Catalog Name** edit box; the **Open file** dialog box is displayed. Browse to the location *C:\ AutoCAD Plant 3D 2014 Content\CPak ASME* and double-click on *ASME Valves Catalog.pcat*; the selected catalog file is displayed in the **Source Catalog Name** edit box.

4. Choose the **Browse** button next to the **New Catalog name** edit box, browse to the location *C:\Documents\CADCIM*. Now enter **c05tut02** in the **File name** edit box.

5. Choose the **Create** button from the **Create Catalog From Existing Catalog** dialog box; the dialog box is closed and a new catalog file is created.

Adding a New Part to the Catalog

In this section, you will create a new part and add it to the catalog. The part will be created using the parametric graphics.

1. Choose the **Catalog Editor** tab from the top right corner of the window.

2. Choose the **Create New Component** button from the **Piping Component Editor** window; the **Create New Component** dialog box is displayed.

3. Select the **Plant 3D Parametric Graphics** radio button from the **Graphics** area, if not selected already.

4. Select the **Valves** option from the **Component Category** drop-down list; the **Piping Component** drop-down list displays the **Valve** and **Valve Body** options.

5. Select **Valve** from the **Piping Component** drop-down list.

6. Select **FL** from the **Primary End Type** drop-down list.

7. Browse the valve images in the **Graphics** area and select **Inline Valve, Check Valve Style (FLG/BW/PE)** from it.

8. Enter **Check Valve** in the **Short Description** edit box.

9. Select the **Imperial** radio button and specify the size range as **2"** to **8"**.

10. Choose the **Create** button; the component is displayed in the **Piping Component Editor** pane.

11. Select the **General Properties** tab.

12. Specify the connection port properties, as shown in Figure 5-38.

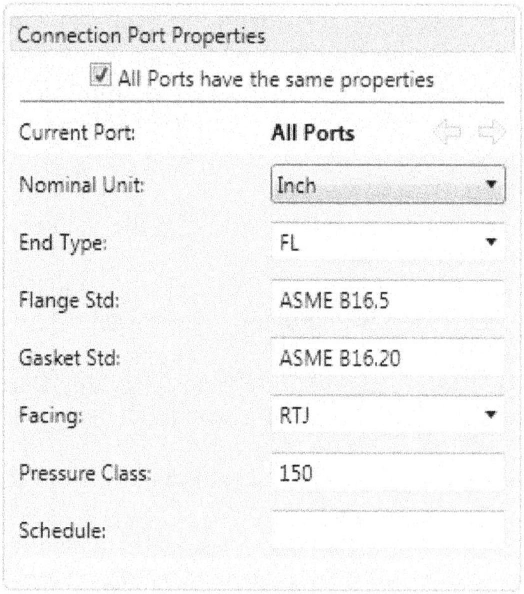

Figure 5-38 *The **Connection Port Properties** area (Family)*

Next, you need to specify the piping component properties.

13. Enter **Check Valve, Lift, 150 LB, RTJ, ASME B16.10** and **ASME B16.10** in the **Long description (Family)** and **Compatible Standard** edit boxes in the **Piping Component Properties** area. Also, specify other property values in this area, refer to Figure 5-39.

14. Choose the **Sizes** tab and remove sizes between 2" and 8" from the **Size** list.

15. Select **2"** from the **Size** list and enter **2.375** in the **Matching Pipe OD** edit box in the **Connection Port Properties** area.

16. Enter **CHECK VALVE, LIFT, 2" ND, 150 LB, RTJ, ASME B16.10, 8" LG** in the **Long Description (Size)** edit box and **8** in the **Length** edit boxes in the **Piping Component Properties** area.

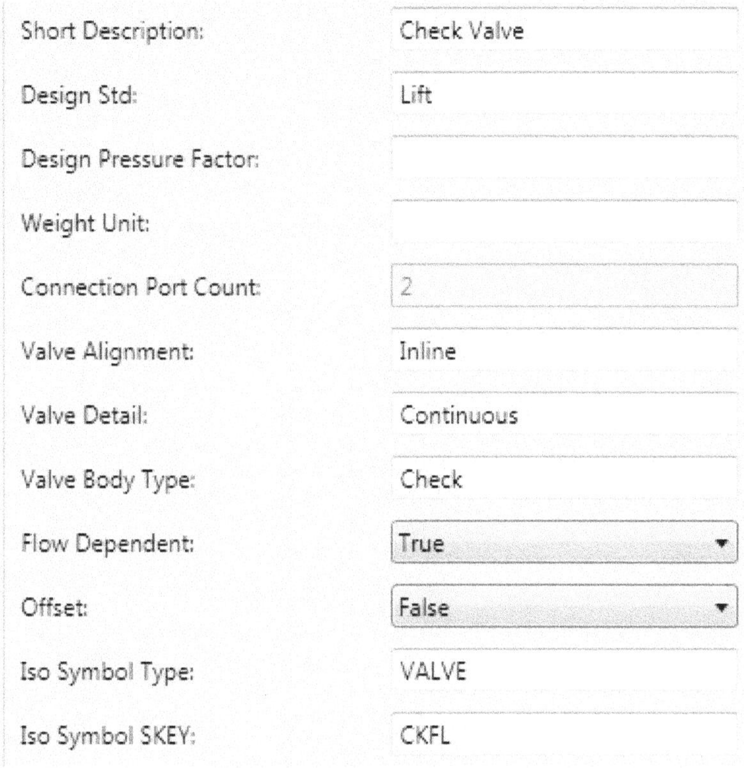

Short Description:	Check Valve
Design Std:	Lift
Design Pressure Factor:	
Weight Unit:	
Connection Port Count:	2
Valve Alignment:	Inline
Valve Detail:	Continuous
Valve Body Type:	Check
Flow Dependent:	True ▾
Offset:	False ▾
Iso Symbol Type:	VALVE
Iso Symbol SKEY:	CKFL

*Figure 5-39 Partial view of the **Piping Component Properties** (Family)*

Next you need to specify the size parameters of the check valve for nominal diameter 2". Figure 5-40 displays the size parameters of a check valve.

17. Specify values in the **Size Parameters** area, as shown in Figure 5-41.

18. Select **8"** from the **Size** list and enter **8.625** in the **Matching Pipe OD** edit box in the **Connection Port Properties** area.

19. Enter **CHECK VALVE, LIFT, 8" ND, 150 LB, RTJ, ASME B16.10, 18" LG** in the **Long Description (Size)** edit box and **18** in the **Length** edit box in the **Piping Component Properties** area.

20. Now, specify values in the **Size Parameters** area, as shown in Figure 5-42.

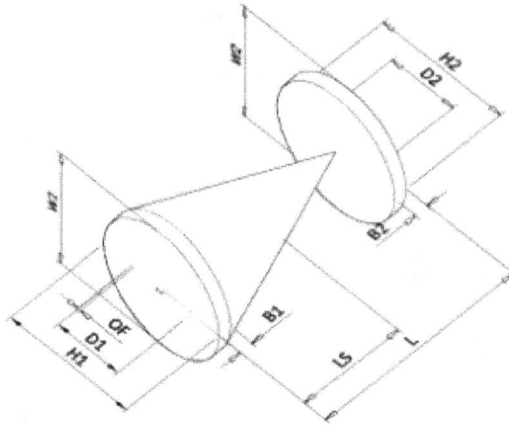

Figure 5-40 *The size parameters of a check valve*

Size Parameters	
These dimensions affect the actual size of the component in the 3D model.	
D1:	2.375
D2:	2.375
L:	8
LS:	0
H1:	5
H2:	5
W1:	0
W2:	0
OF:	-0.5
B1:	0.90
B2:	0.90

Figure 5-41 *Size parameters for 2" nominal diameter*

Size Parameters	
These dimensions affect the actual size of the component in the 3D model.	
D1:	8.625
D2:	8.625
L:	18
LS:	0
H1:	12.5
H2:	12.5
W1:	0
W2:	0
OF:	-0.5
B1:	1.5
B2:	1.5

Figure 5-42 *Size parameters for 8" nominal diameter*

21. Choose the **Save to Catalog** button located below the **Size Parameters** area; the part will be saved to the catalog.

Saving the Catalog File

1. Choose **File > Save** from the main menu; the catalog will be saved.

Tutorial 3

In this tutorial, you need to download the *c05tut03.dwg* file from *http://www.cadcim.com*. The path of the file is as follows: *Textbooks>CAD/CAM>AutoCAD Plant 3D>AutoCAD Plant 3D 2014 for Designers*. The file contains the 3D model of a valve body and actuators, as shown in Figure 5-43. You will add these models to the catalog. The parameters of the parts are given next.   **(Expected time: 45 min)**

Hand Wheel Actuator Parameters

Long Description (Family):	Hand Wheel
Long Description:	CUSTOM HAND WHEEL, H=36", W=24"
Short Description:	Custom Hand Wheel
Operator Type:	Manual
Actuator Type:	Wheel
Actuator Height:	36
Actuator Width:	24

Diaphragm Actuator Parameters

Long Description (Family):	Diaphragm
Long Description (Size):	CUSTOM HAND WHEEL, H=30", W=16"
Short Description:	Custom Diaphragm
Operator Type:	Pnematic
Actuator Type:	Diaphragm
Actuator Height:	30
Actuator Width:	16

Valve Parameters

Long Desciption(Family):	Globe Valve, 150 LB, ASME B16.10
Operator Size:	Diaphragm, H=30", W=16"
Matching Pipe Outer Diameter:	4.5

Valve Port Properties

Port Diameter:	4"
Nominal Unit:	Inch
End Type:	BV
Flange Standard:	ASME B16.10
Gasket Standard:	ASME B16.10
Facing:	RF
Pressure Class:	150

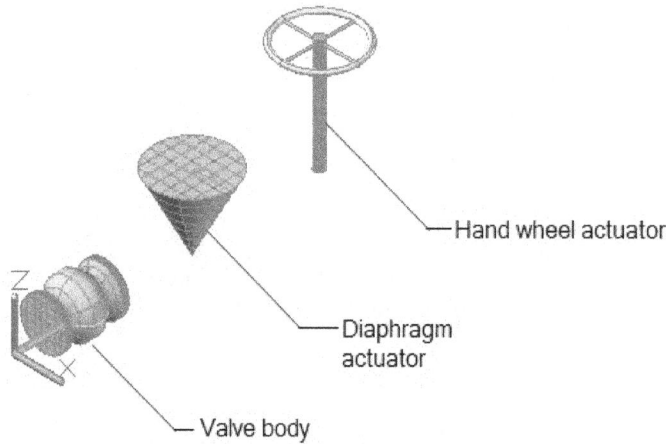

Figure 5-43 *AutoCAD blocks of valve body, diaphragm actuator, and hand wheel actuator*

The following steps are required to complete this tutorial:

a.   Convert the AutoCAD blocks to Plant 3D components.
b.   Add ports to the component.
c.   Open the catalog created in Tutorial 2.
d.   Add the newly created piping component.

Converting AutoCAD Blocks to Plant 3D Components

In this section, you will convert blocks of a valve body and actuators into AutoCAD Plant 3D components, refer to Figure 5-42.

1.   Download the *c05tut03.dwg* file from *http://www.cadcim.com*.

2.   Open the *c05tut03.dwg* file in AutoCAD Plant 3D 2014.

3.   Choose the **Create Block** tool from the **Block definition** panel in the **Insert** tab and create three separate blocks and name them as **Valve Body**, **Diaphragm**, and **Hand Wheel**. The insertion points of the these blocks are shown in Figure 5-44.

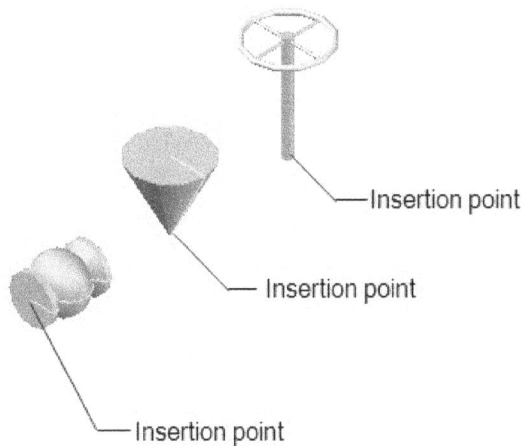

Figure 5-44 *Insertion points of the blocks*

4. Invoke the **PLANTPARTCONVERT** command and select the valve body block; the prompt **Select a port operation [Add/Delete/Move/eXit]** will be displayed at the Command prompt. Make sure that the **Ortho Mode** and **Dynamic Input** button are selected at the status bar.

5. Choose the **Add** option from the contextual menu; you are prompted to specify a port location.

6. Specify the port location on the block, as shown in Figure 5-45; you are prompted to specify the port direction. Move the cursor in the x-direction and left-click; the contextual menu is displayed. Choose the **Accept** option from the contextual menu; the port is created, as shown in Figure 5-45. Also, the **Select a port operation** prompt is displayed.

7. Choose the **Add** option from the contextual menu and create another port, as shown in Figure 5-46. Next, choose the **eXit** option to get out of the command.

8. Invoke the **PLANTPARTCONVERT** command and convert the hand wheel actuator and the diaphragm actuator into piping components. Note that you should not add ports to actuators.

9. Save the file at the location *C:\Documents\CADCIM*.

Figure 5-45 Specify the first port location

Figure 5-46 Specify the second port location

Opening the Catalog and Adding Actuators to It

1. Start **AutoCAD Plant 3D Spec Editor** and open the *c05tut02.pcat* file from the welcome window. Alternatively, Choose **File > Open Catalog** from the main menu; the **Open** dialog box is displayed. Browse to the location *C > Documents > CADCIM* and open the *c05tut02.pcat* file.

2. Choose the **Catalog Editor** tab and then choose the **Create New Component** button; the **Create New Component** dialog box is displayed.

3. Select the **Custom - AutoCAD DWG Block based graphics** radio button from the **Graphics** area.

4. Select **Actuators** from the **Component Category** drop-down list and specify other options, as shown in Figure 5-47.

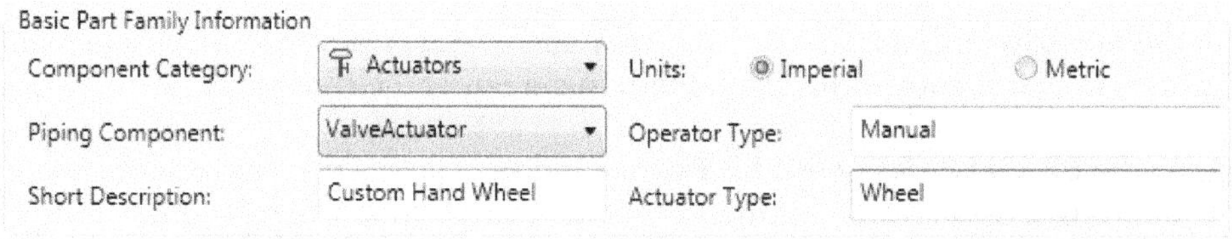

*Figure 5-47 The options to be specified in the **Basic Part Family Information** area*

5. Enter **Custom Hand Wheel** in the **Short Description** edit box. Also, enter **Manual** and **Wheel** in the **Operator type** and **Actuator type** edit boxes, respectively.

6. Choose the **Create** button from the dialog box; the new entry is created in the catalog. Now you need to enter the long description (family) and size description.

7. Enter **Hand Wheel** in the **Long Description (Family)** edit box and then choose the **Sizes** tab; various options to specify the size parameters are displayed.

8. Choose the **Select Model** button from the preview window located beside the size list; the **Open** dialog box is displayed.

9. Browse to the location *C >Documents > CADCIM* and open the file *c05tut03.dwg*; the **Select Block Definition** dialog box is displayed.

10. Select **Hand Wheel** from the **Select Block Definition** dialog box and choose the **OK** button; the block is displayed in the preview window.

11. Enter the values in the **Piping Component Properties** area, as shown in Figure 5-48. Next, choose the **Save to Catalog** button; the actuator is saved to the catalog.

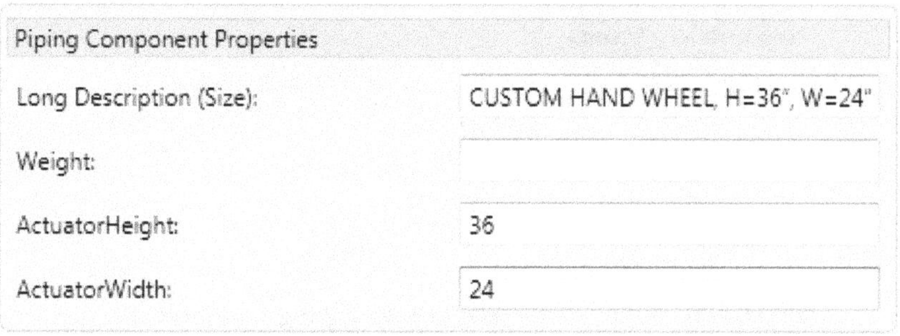

Piping Component Properties	
Long Description (Size):	CUSTOM HAND WHEEL, H=36", W=24"
Weight:	
ActuatorHeight:	36
ActuatorWidth:	24

*Figure 5-48 Values to be entered in the **Piping Component Properties** area*

Next, you need to add the cone actuator to the catalog.

12. Invoke the **Create New Component** dialog box and select the **Custom - AutoCAD DWG Block based graphics** radio button in the **General Properties** area.

13. Select **Actuators** from the **Component Category** drop-down list and then specify other options, as shown in Figure 5-49.

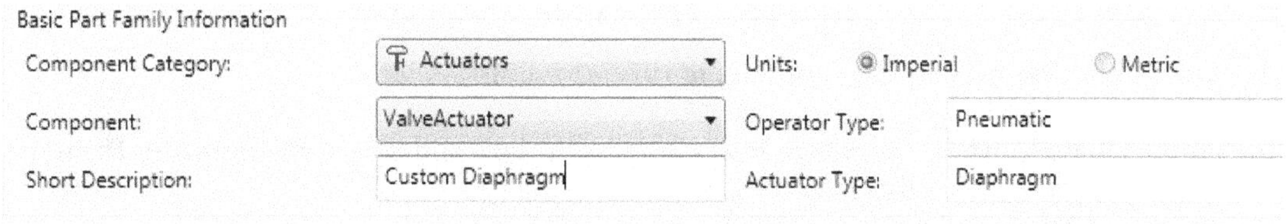

Basic Part Family Information				
Component Category:	Actuators	Units:	◉ Imperial	○ Metric
Component:	ValveActuator	Operator Type:	Pneumatic	
Short Description:	Custom Diaphragm	Actuator Type:	Diaphragm	

*Figure 5-49 The options to be specified in the **General Properties** area*

14. Enter **Custom Diaphragm** in the **Short Description** edit box. Also, enter **Pneumatic** and **Diaphragm** in the **Operator type** and **Actuator type** edit boxes, respectively.

15. Choose the **Create** button from the dialog box; the new entry is created in the catalog. Now you need to enter the long description (family) and size description.

16. Enter **Diaphragm** in the **Long Description (Family)** edit box and choose the **Sizes** tab; various options to specify the size parameters are displayed.

17. Choose the **Select Model** button from the preview window located beside the size list; the **Open** dialog box is displayed.

18. Browse to the location *C > Documents > CADCIM* and open the file *c05tut03.dwg*; the **Select Block Definition** dialog box is displayed.

19. Select **Diaphragm** from the **Select Block Definition** dialog box and choose the **OK** button; the block will be displayed in the preview window.

20. Enter the values in the **Piping Component Properties** area, as shown in Figure 5-50. Next, choose the **Save to Catalog** button; the actuator is saved to the catalog.

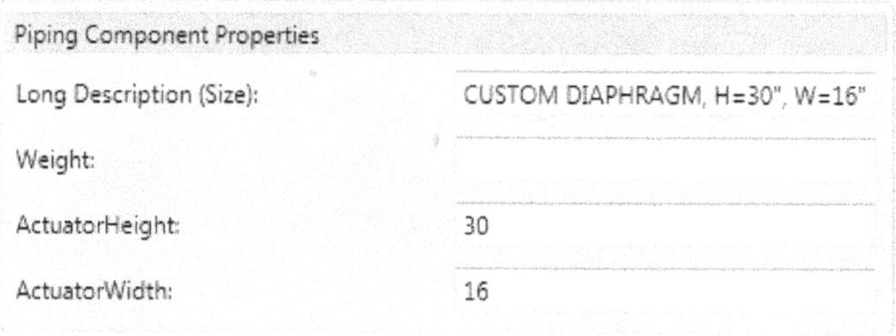

Piping Component Properties	
Long Description (Size):	CUSTOM DIAPHRAGM, H=30", W=16"
Weight:	
ActuatorHeight:	30
ActuatorWidth:	16

*Figure 5-50 Values to be entered in the **Piping Component Properties** area*

Adding a Valve to the Catalog

1. Choose the **Catalog Editor** tab and then choose the **Create New Component** button; the **Create New Component** dialog box is displayed.

2. Select the **Custom - AutoCAD DWG Block based graphics** radio button from the **Graphics** area.

3. Select **Valves** from the **Component Category** drop-down list.

4. Select **ValveBody** from the **Component** drop-down list.

5. Enter **Globe Valve** in the **Short Description** edit box and select **BV** from the **Primary End Type** drop-down list.

6. Select **2** from the **Number of Connection Ports** drop-down list and select the **Imperial** radio button from the **Units** area.

7. Specify **4"** in the **Size From** and **To** drop-down lists.

8. Choose the **Create** button; the **Create Component** dialog box is closed and the **Piping Component Editor** pane is displayed.

9. Specify the connection port properties in the **General Properties** tab, as shown in Figure 5-51.

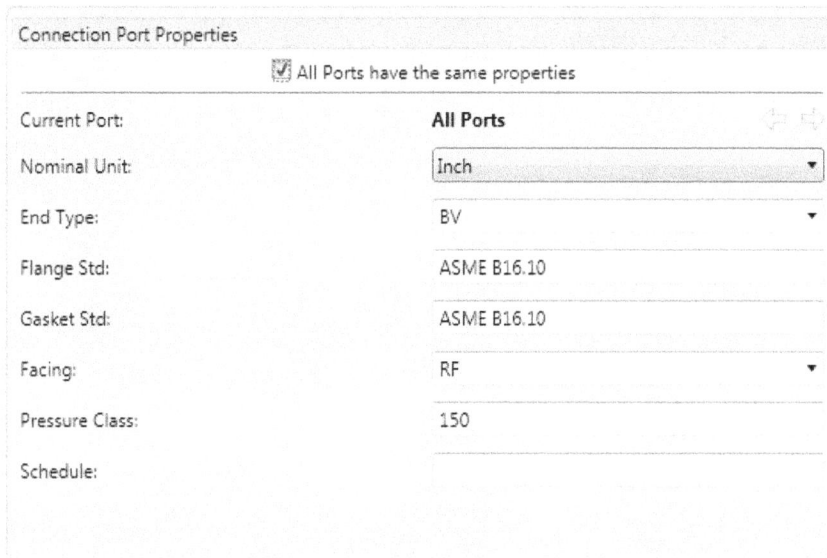

Figure 5-51 *The **Connection Port Properties** area (Family)*

10. Enter **Globe Valve, 150 LB, ASME B16.10** and **ASME B16.10** in the **Long description (Family)** and **Compatable Standard** edit boxes, respectively, in the **Piping Component Properties** area.

11. Select the **Sizes** tab and remove all other sizes except size **4"** from the **Size** list.

12. Choose the **Select Model** button from the preview window located beside the size list; the **Open** dialog box is displayed.

13. Browse to the location *C* > *Documents* > *CADCIM* and open the file *c05tut03.dwg*; the **Select Block Definition** dialog box is displayed.

14. Select **Valve body** from the **Select Block Definition** dialog box and choose the **OK** button; the block will be displayed in the preview window.

15. Enter **4.5** in the **Matching Pipe OD** edit box and select the **All Ports have the same properties** check box from the **Component Port Properties** area.

16. Enter **Globe Valve, 150 LB, ASME B16.10** in the **Long Description (Size)** edit box and **24** in the **Length** edit box.

17. Choose the **Save to Catalog** button; the component is saved to the catalog.

Assigning Operators to the Valve

1. Choose the **General Properties** tab and then choose the **Edit Operator Assignments** button which is located below the **Piping Component Properties** area; the **Valve Operator Mapping** dialog box is displayed.

2. In this dialog box, choose the **Add** button located below the **Operator Assignments** table; a new row is added to the table.

3. Select **Diaphragm, H=30", W=16"** from the drop-down list in the **Operator (Size)** column, refer to Figure 5-52.

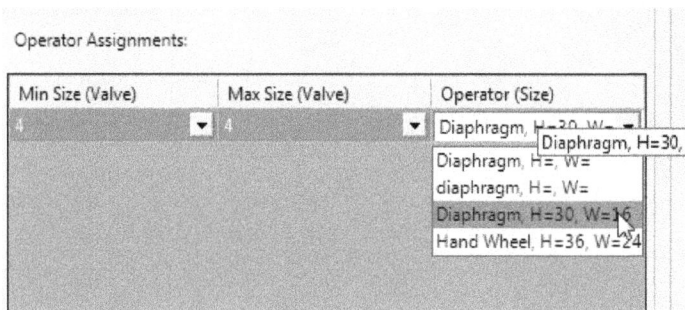

Figure 5-52 *Selecting an operator from the drop-down list in the Operator (Size) column*

4. Choose the **Add** button to add a new row to the table. Next, select **Hand Wheel, H=36", W=24"** from the **Operator (Size)** column; the diaphragm actuator is assigned to the gate valve.

5. Choose the **OK** button from the **Valve Operator Mapping** dialog box; the dialog box is closed.

6. Choose the **Save to Catalog** button from the **Piping Component Editor**; the valve is saved in the catalog.

Saving the Catalog File

1. Choose **File > Save** from the menu bar; the catalog is saved.

2. Choose **File > Exit** from the menu bar to close the application.

Self Evaluation Test

Answer the following questions and then compare them to those given at the end of this chapter:

1. The _____ displays the parts which are added to a spec file.

2. The _____ is used to modify the catalog content and customize it according to the requirement.

3. You can apply the specified properties to multiple ports by selecting the _____ check box.

4. The _____ button displays the component size properties in the form of a table.

5. The _____ determines the type of fitting to be used while joining a header pipe with a branch pipe.

6. To create a _____ for a branch table, first you need to identify all branch fittings available in a spec.

7. You can assign more than one branch fitting to a branch. (T/F)

Review Questions

Answer the following questions:

1. To create a new component using blocks, first you need to create a block and convert it into a AutoCAD Plant 3D component using the _____ command.

2. You need to set the part use priority if there are more than one part existing in the same group. (T/F)

3. Actuators are displayed in the spec sheet. (T/F)

4. You can create a new component by using two different methods. (T/F)

Answers to Self-Evaluation Test

1. Spec Sheet, 2. Catalog Editor, 3. **All Ports have the same properties, 4. Show Advanced Table, 5.** Branch table, **6.** legend (symbol), **7.** T

Chapter 6

Routing Pipes

After completing this chapter, you will be able to:

- *Select a Spec*
- *Route a pipe line*
- *Route a pipe using a P&ID*
- *Work with the compass*
- *Connect two pipes*
- *Route pipe at an offset*
- *Route pipes at a slope*
- *Create branches*
- *Create Autodesk connection points*

INTRODUCTION

In the previous chapter, you have learned to create equipment for a plant 3D model. Now, in this chapter, you will learn various methods to route pipes.

SELECTING A SPEC

Before you start routing a pipe, you need to select the piping spec. It defines specifications of the pipes, valves, fittings and other piping components to be used in the 3D piping model. You can select pipes with specific material standard, pressure class, pipe schedule, and so on from a spec file. In addition to pipe specifications, the piping spec also provides you the information about the type of joints, fittings, and fasteners to be used. You can also select valves and valve actuators from a spec.

The **Spec Selector** drop-down list available in the **Part Insertion** panel of the **Home** tab contains various specs, refer to Figure 6-1. Select a spec from the **Spec Selector** drop-down list; various components available in the selected spec are displayed in the Tool Palette which is located at right side of the drawing window. You can select a pipe or piping component from this Tool Palette.

After selecting the spec, you need to select the pipe size from the **Pipe Size Selector** drop-down list available in the **Part Insertion** panel, refer to Figure 6-2. The pipe will be routed using the selected pipe size.

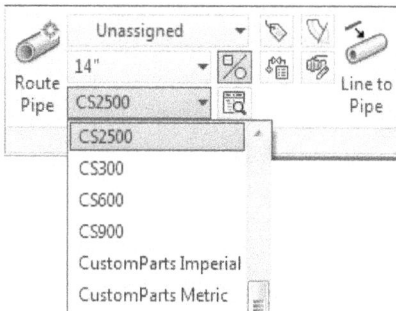

Figure 6-1 Selecting a spec from the
Spec Selector drop-down list

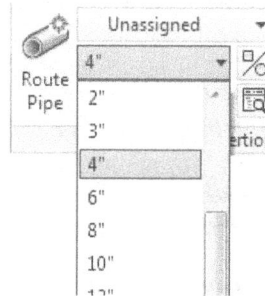

Figure 6-2 Selecting a pipe size from
the **Pipe Size Selector** drop-down list

WORKING WITH THE SPEC VIEWER

In AutoCAD Plant 3D, you can use the **Pipe Spec Viewer** to view the components present in the selected spec. To do so, choose the **Spec Viewer** button available in the **Part Insertion** panel; the **Pipe Spec Viewer** palette will be displayed, as shown in Figure 6-3. It contains the **Spec Sheet** rollout and the **Part Sizes** rollout. When you select a part from the **Spec Sheet**, the **Part Sizes** rollout displays various sizes available for the selected part. Also, the details of the component such as part type, end connection, material grade and rating are displayed in the **Part details** area. The **Pipe Spec Viewer** has many other options. The usage of these options is discussed next.

Adding a Part to the Tool Palette

To add a part to the Tool Palette, select a part from the **Spec Sheet** in the **Pipe Spec Viewer**; the **Part Size** pane displays various sizes available for the selected part. Select a part size and choose the **Add to Tool Palette** button; the selected part will be added to the Tool Palette.

Creating a New Tool Palette

You can also create a new Tool Palette. To do so, select a spec from the **Spec** drop-down list in the **Pipe Spec Viewer** and choose the **Create Tool Palette** button; a new Tool Palette is created and it will be loaded with the parts available in the selected spec.

Inserting a Part From the Spec Viewer

You can insert a part from the **Pipe Spec Viewer** into a model by using the **Insert in Model** button from the **Pipe Spec Viewer**. However, this option will be useful to you while placing components after routing a pipe.

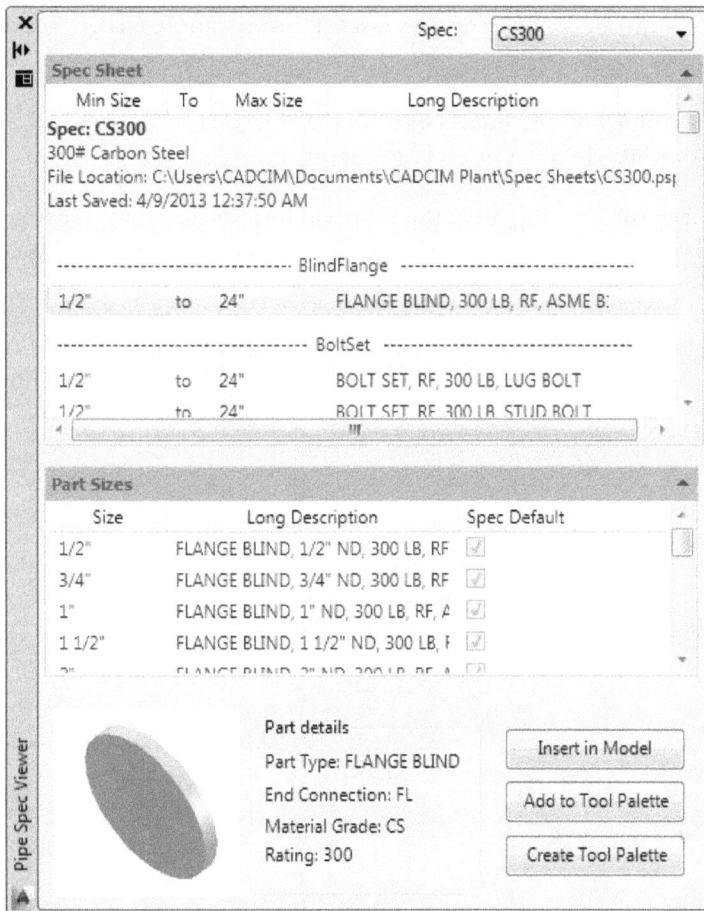

*Figure 6-3 The **Pipe Spec Viewer** palette*

ROUTING A PIPE

Piping is an important part of a plant 3D model. In AutoCAD Plant 3D, you can route pipes between the equipment and add inline components to them. The methods to route a pipe are discussed next.

Routing a Pipe With a New Line Number

You can route a pipe and assign it a new line number. To do so, select the **Route New line** option from the **Line Number Selector** drop-down list in the **Part Insertion** panel of the **Home** tab; the **Assign Tag** dialog box will be displayed, as shown in Figure 6-4. Next, follow the steps given below:

1. Specify a line number in the **Number** edit box.

2. Select a line size from the **Size** drop-down list.

3. Select a new spec from the **Spec** drop-down list, if you want to change the already selected spec.

4. Choose the **Assign** button; the **Assign Tag** dialog box will be closed and you will be prompted to specify the start point of the pipe.

5. Select a point in the drawing area; the pipe will be attached to the cursor and a compass will be displayed, refer to Figure 6-5. You can rotate the pipe at an incremental angle using the compass.

6. Move the cursor and select the end point of the pipe; the pipe will be created and you will be prompted to specify the next point. Press ENTER to finish the line.

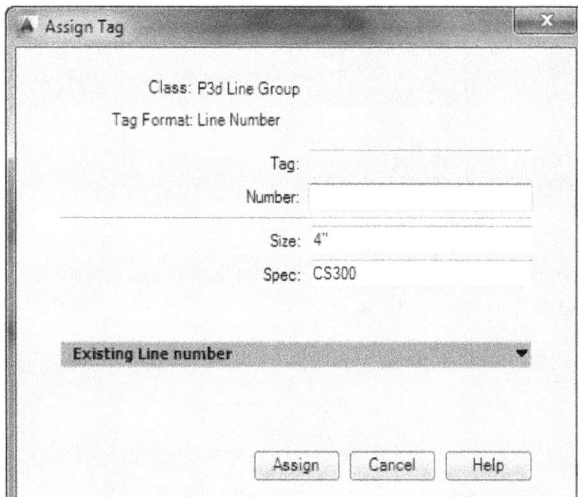

*Figure 6-4 The **Assign Tag** dialog box*

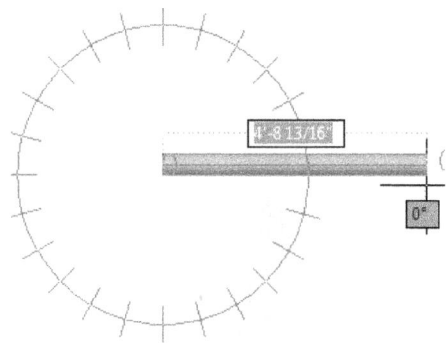

Figure 6-5 Creating a pipe

Note

*If the **Dynamic Input** mode is on, the dynamic edit boxes are displayed along with pipe. You can enter values for pipe length and orientation angle in these edit boxes.*

Setting the Route Line

While routing a pipe, you are actually drawing a line. The pipe is created with reference to this line. This reference line is known as the route line. By default, the drawn route line is the center line of the pipe. However, you can change the position of the route line. To do so, select an option from the **Set Routing Line** drop-down list in the **Elevation & Routing** panel of the **Home** tab; the position of the route line changes with respect to the selected option. The options in the **Set Routing Line** drop-down list are discussed next.

Top of Pipe (TOP)

On selecting this option, the route line will be set at the top of center point of the pipe.

Center of Pipe (COP)

On selecting this option, the route line will set at the center point of the pipe.

Bottom of Pipe (BOP)

On selecting this option, the route line will set at the bottom of the center point of the pipe.

Top Left

On selecting this option, the route line will set at the top left point of the pipe.

Top Right

On selecting this option, the route line will set at the top right point of the pipe.

Center Left

On selecting this option, the route line will set at the left of center point of the pipe.

Center Right

On selecting this option, the route line will set at the right of center point of the pipe.

Bottom Left

On selecting this option, the route line will set at the bottom left point of the pipe.

Bottom Right

On selecting this option, the route line will set at the bottom right point of the pipe.

Routing a Pipe From a Line

In AutoCAD Plant 3D, you can also route a pipe by converting a line into a pipe. To do so, first you need to draw a line, polyline, arc, or 3D polyline. Next, choose the **Line to Pipe** tool from the **Part Insertion** panel of the **Home** tab; you are prompted to select a line, polyline, or arc. Select a line entity and press ENTER; it will be converted into a pipe. Figure 6-6 shows a 3D polyline and Figure 6-7 shows a piping created from it. Notice that the elbows and tees are automatically placed while creating pipe. You can also create a pipe from an arc. Figure 6-8 shows an arc and Figure 6-9 shows a piping created from that arc.

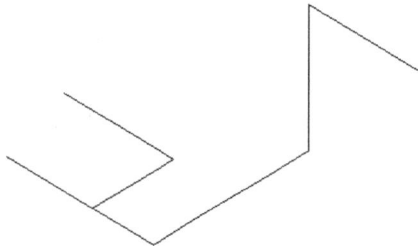

Figure 6-6 *A 3D Polyline*

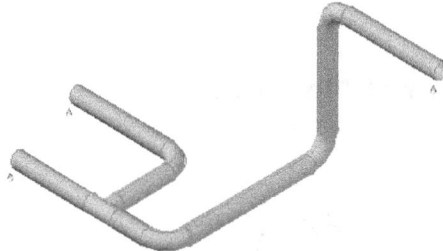

Figure 6-7 *Piping created from a 3D polyline*

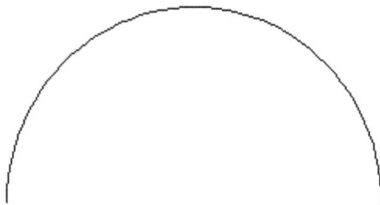

Figure 6-8 *An arc created using the **Arc** tool*

Figure 6-9 *Piping created from a arc*

The line entity gets deleted after being converted into a pipe. If you want to retain the line entity, press **E** after selecting the entity; a prompt will be displayed asking whether to erase the selected entity or not. Choose the **No** option to retain it.

Routing a Pipe using a P&ID

You can route a pipe in a Plant 3D model using a P&ID. Note that the P&ID and the Plant 3D files should be located in the same project. To route a pipe using a P&ID, choose the **P&ID Line List** button from the **Part Insertion** panel, refer to Figure 6-10; the **P&ID Line List** window will be displayed, as shown in Figure 6-11. Next, select the P&ID from the drop-down list available in this window; the Pipe Line Groups of the selected P&ID will be displayed in the tree view in the **P&ID Line List** window. Expand the Pipe Line Group from the tree and then select a line segment, refer to Figure 6-11. Next, choose the **Place** button; you will be prompted to specify the start point of the line. Select a point in the drawing area; you will be prompted to specify the end point of the pipe. Move the cursor upto a desired length and specify the end point. Next, press ENTER to exit the command.

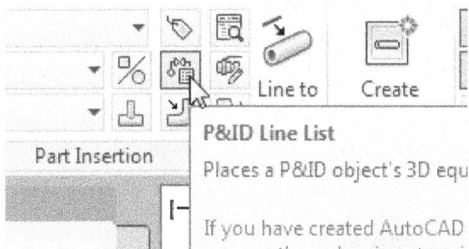

Figure 6-10 Choosing the **P&ID Line List** button

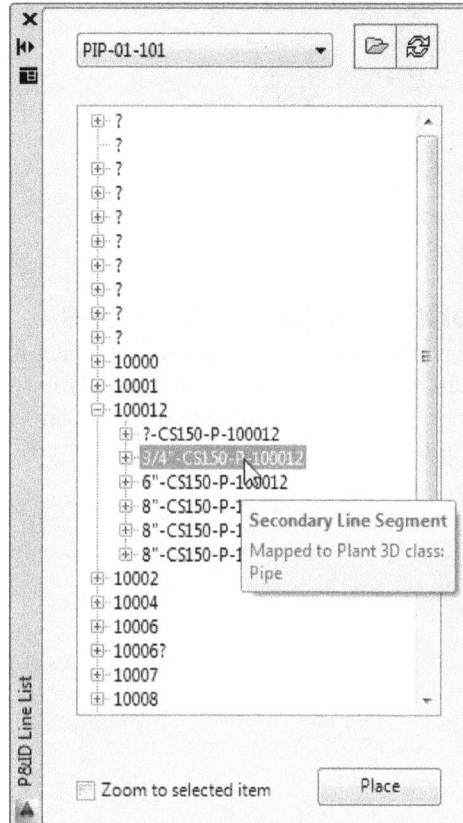

Figure 6-11 The **P&ID Line List** window

Routing a Pipe From an Equipment

In earlier chapters, you learned how to create equipment and add nozzles to them. Now, you will learn how to route a pipe directly from an equipment. To do so, click on an equipment in the model space; the equipment will be highlighted and a continuation grip (+ symbol) will be displayed on the nozzles located on it, refer to Figure 6-12. Click on the continuation grip and move the cursor away from the nozzle; a pipe is attached to the cursor and a compass is displayed on it, refer to Figure 6-13. Also, the size of the pipe is automatically set to match the nozzle size. Next, specify the end point of the pipe by clicking in the drawing area. Press ENTER to end the routing of the pipe.

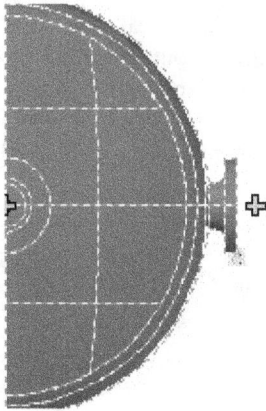

Figure 6-12 *The continuation grip displayed on the nozzle*

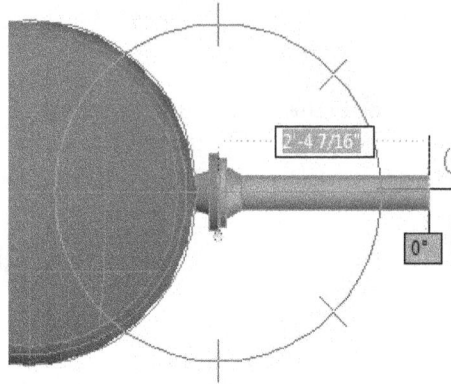

Figure 6-13 *Routing a pipe from an equipment*

Working with the Compass

Compass is an on-screen tool which makes it easy to route a pipe. Compass which is displayed automatically while routing a pipe helps you to rotate the pipe at a precise angle. You can change the default compass settings such as size, color, angle increments and so on by using the options in the **Compass** panel of the **Ribbon**, refer to Figure 6-14.

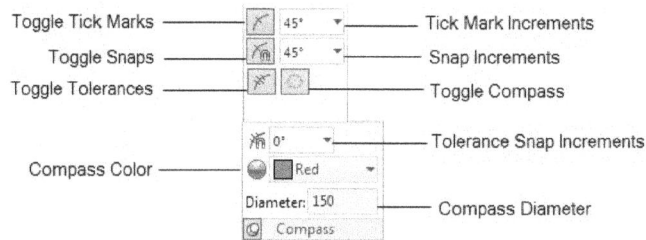

Figure 6-14 *The Compass panel*

The options in the **Compass** panel are discussed next.

Toggle Tick Marks

This toggle button controls the display of tick marks on the compass. By default, this toggle button is chosen in the **Compass** panel. As a result, tick marks are displayed on the compass. Figure 6-15 shows a compass with tick marks and Figure 6-16 shows a compass without tick marks.

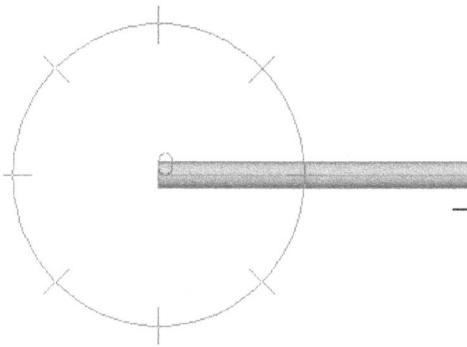

Figure 6-15 *Compass with tick marks*

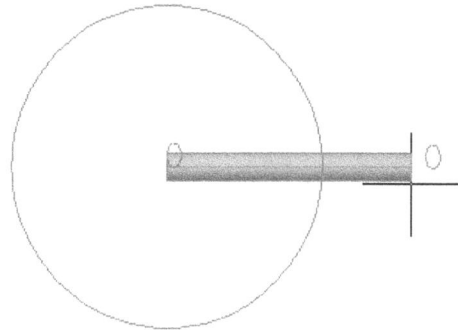

Figure 6-16 *Compass without tick marks*

Tick Mark Increments

This edit box is used to specify the angle between tick marks displayed on the Compass.

Snap Increments

This edit box is used to specify the angular increments through which cursor will snap. For example, when you specify the snap increment as 60-degree, the cursor snaps through 0, 60, 120, and 180 degrees. Figure 6-17 shows a compass with 46-degree snap increments.

Toggle Snaps

This toggle button controls whether the angle snap is turned on or off. On choosing this button, the cursor will snap through the angle specified in the **Snap Increments** edit box. Figure 6-17 shows the compass when the angle snap is turned on and Figure 6-18 shows the compass when the angle snap is turned off.

Figure 6-17 *Compass when the angle snap is turned on*

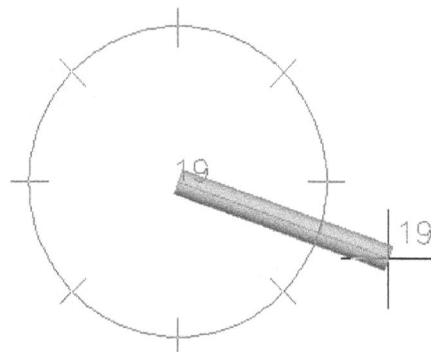

Figure 6-18 *Compass when the angle snap is turned off*

Tolerance Snap Increments

This edit box is used to set the tolerance angle when a pipe connects to another pipe or fitting.

Toggle Tolerances

On choosing this toggle button, the fitting tolerance is turned on.

Toggle Compass

This toggle button is used to turn the display of compass on or off while routing a pipe.

Compass Color

This drop-down list is used to specify the color of the compass.

Compass Diameter

This edit box is used to specify the diameter of the compass.

Connecting Two Open Ports of Pipes

You can connect ports of two different pipes. Before connecting these ports, you need to specify the **Object Snap** settings. To do so, right-click on the **Object Snap** button on the status bar and choose the **Settings** option from the shortcut menu displayed; the **Drafting Settings** dialog box will be displayed. Select the **Node** check box from this dialog box and choose the **OK** button; the cursor will snap to the center point of the pipe.

After specifying the **Object Snap** settings, choose the **Route Pipe** tool from the **Part Insertion** panel; you will be prompted to select the start point of the pipe. Snap to the center point of the first pipe and click to select the first port, as shown in Figure 6-19; you will be prompted to select the next point. Press and hold the SHIFT key and right-click to display the shortcut menu. Choose the **Node** option from the menu and then select the center point of the second pipe, as shown in Figure 6-20; the system generates multiple solutions for connecting the two open ports. Choose the **Next** option from the Command Prompt until the desired solution is displayed. To accept the desired solution, choose the **Accept** option; the two ports are connected by the selected solution. Figure 6-21 shows two open ports connected.

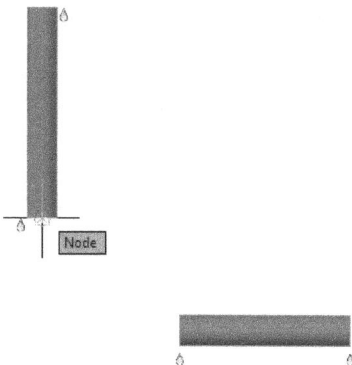

Figure 6-19 Selecting the first port

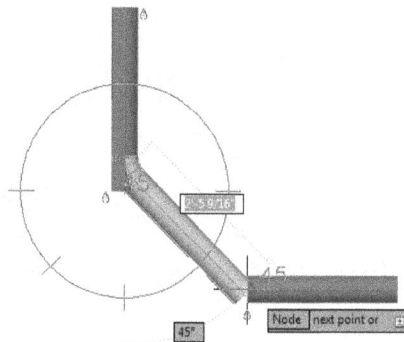

Figure 6-20 Selecting the second port

You can also connect two equipment by using the same procedure. Figure 6-22 through 6-24 show examples of two equipment connected using the automatically generated solutions.

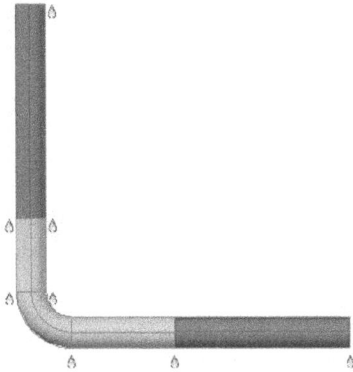

Figure 6-21 *Two pipes connected by using system generated solutions*

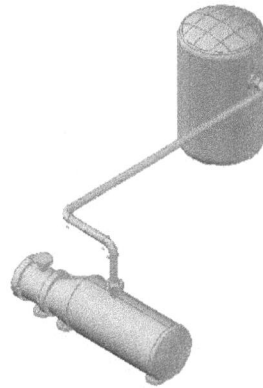

Figure 6-22 *Solution 1 for connecting two equipment*

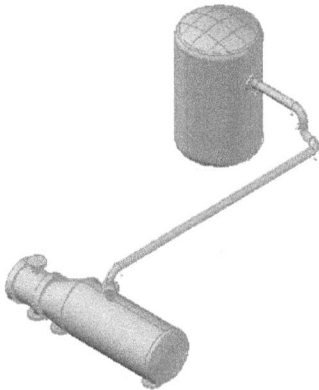

Figure 6-23 *Solution 2 for connecting two equipment*

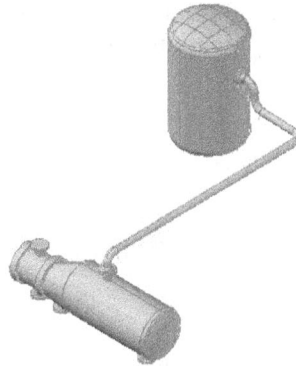

Figure 6-24 *Solution 3 for connecting two equipment*

Changing the Pipe Size while Routing

You can change the pipe size while routing. To do so, invoke the **Route Pipe** tool and then choose the **Size** option from the Command prompt; you will be prompted to specify the nominal size of the pipe. Also, the prompt displays a symbol "?" at the Command prompt. If you enter "?", the sizes available in the selected spec will be listed at the Command prompt. Next, select the desired size from the list displayed; the pipe size changes to the size specified. Figure 6-25 shows an example when a different pipe size is set for the continuation pipe. As a result, a reducer is automatically added between the pipes. Similarly, if you set a pipe size different from the nozzle size, a reducer is placed between the pipe and the nozzle, as shown in Figure 6-26.

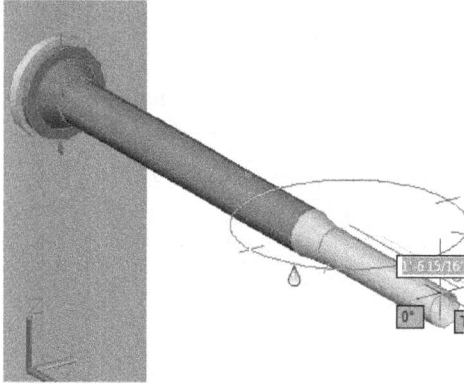

Figure 6-25 *Different pipe sizes connected automatically by a reducer*

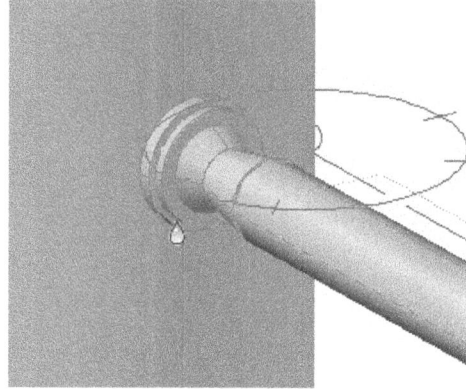

Figure 6-26 *A reducer placed between the nozzle and the pipe*

Changing the Orientation Plane while Routing a pipe

While routing a pipe, you can change the plane orientation by choosing the **Plane** option from the Command Prompt. To do so, invoke the **Route Pipe** tool and then enter **P** at the Command Prompt; the plane in which the pipe has been routed changes. Figure 6-27 through 6-29 show a pipe routed on different planes.

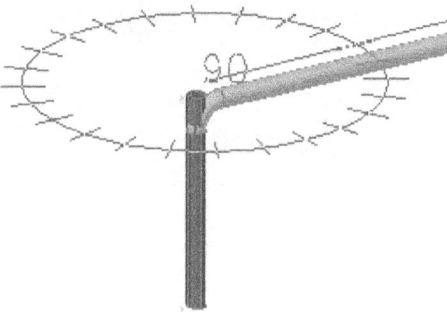

Figure 6-27 *Routing a pipe on XY plane*

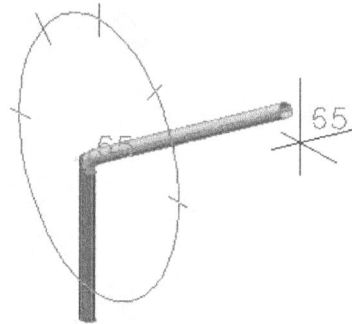

Figure 6-28 *Routing a pipe on XZ plane*

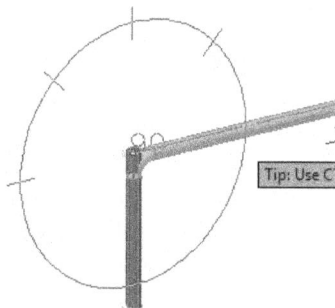

Figure 6-29 *Routing a pipe on YZ plane*

Creating a Cutback Elbow

When you rotate a pipe, you will notice that an elbow is automatically placed. You will also notice that the elbow is created only at 45 degree and 90 degree. This is because, by default, the cutback mode is turned off. As a result, you can only place elbows available in the spec (45 or 90 degree elbows). To turn the cutback mode on, choose the **Toggle Cutback Elbows** button from the **Part Insertion** panel, refer to Figure 6-30; the cutback mode gets turned on and you can now create an elbow whose angle is different from the angle of elbow available in the spec. Figures 6-31 and 6-32 show a pipe being rotated when the cutback mode is off and on, respectively.

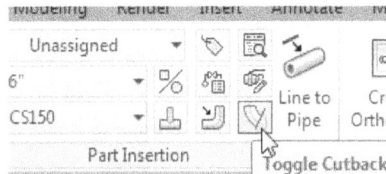

Figure 6-30 *Choosing the* **Toggle Cutback Elbows** *button*

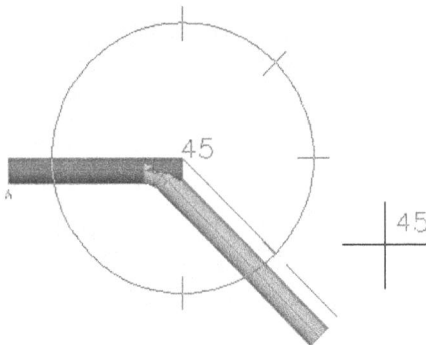

Figure 6-31 *Creating an elbow with the cutback mode turned off*

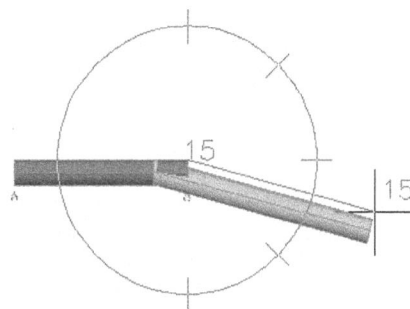

Figure 6-32 *Creating an elbow with the cutback mode turned on*

Creating a Roll Elbow

You can create a rolled elbow (twisted elbow) by specifying rotation angles in two planes which are perpendicular to each other. To do so, invoke the **Route Pipe** tool and specify the start and end point of the pipe. Next, choose the **Rollelbow** option from the Command Prompt; you will be prompted to specify an angle on the first plane. Rotate the pipe to the required angle and click to specify the angle on the first plane, refer to Figure 6-33; you will be prompted to specify an angle on the second plane which is perpendicular to the first plane. Rotate the pipe and click to specify the angle on the second plane, refer to 6-34. A rolled elbow is created after specifying angles on both the planes, refer to Figure 6-35.

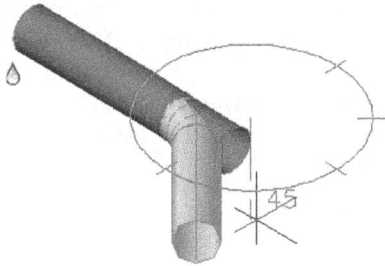

Figure 6-33 *Specifying an angle in the first plane*

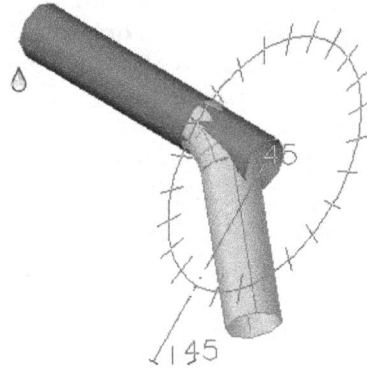

Figure 6-34 *Specifying an angle in the second plane (perpendicular plane)*

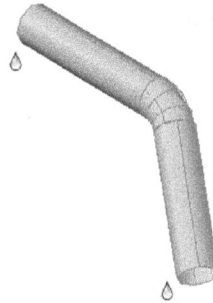

Figure 6-35 *Roll elbow created*

Creating Bends

You can create a pipe bend upto to an angle of 180 degrees. To do so, first you need to set the maximum bend angle to 180 degrees by entering the **PLANTMAXBENDANGLE** command at the Command Prompt. Next, start routing a pipe and choose the **Toggle Pipe Bends** button from the **Part Insertion** panel, refer to Figure 6-36. Now, you can create a pipe bend upto an angle of 180 degrees. Figure 6-37 shows bends created at different angles.

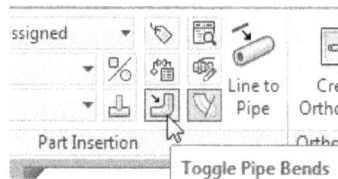

Figure 6-36 *Choosing the **Toggle Pipe Bends** button*

Figure 6-37 *Pipe bends created at different angles*

Changing the Elevation while Routing a Pipe

To change the elevation while routing a pipe, enter the desired value in the **Set Elevation** edit box located in the **Elevation & Routing** panel; the cursor elevates to the specified height. Next, you can route the pipe at the specified elevation. You can also enter an elevation value relative to the current elevation. To do so, invoke the **Elevation** option and then enter **R** at the Command Prompt; the relative option will be invoked. Next, enter an elevation value at the Command Prompt. Figure 6-38 shows a pipe before specifying the elevation and Figure 6-39 shows a pipe after specifying the elevation.

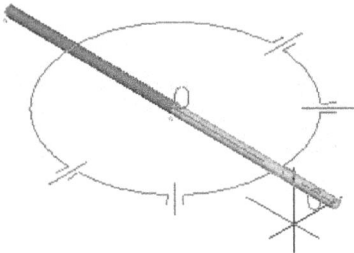

Figure 6-38 *Routing a pipe before specifying the elevation*

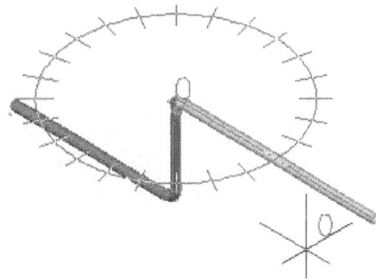

Figure 6-39 *Routing a pipe after specifying the elevation*

Routing a Pipe at an Offset

You can route a pipe at an offset distance to the routing line. To do so, first select the justification point of the pipe. By default, the justification point is at the center of the pipe. To change it, select the required justification point option from the **Set Routing Line** drop-down list in the **Elevation & Routing** panel. Next, click on the down arrow in the **Elevation & Routing** panel; it gets expanded and displays the **Horizontal Offset** and **Vertical Offset** edit boxes, as shown in Figure 6-40. Enter a required value in the **Horizontal Offset** edit box and then choose the **Route Pipe** tool. Next, specify the start point and the end point of the pipe; you will notice that the pipe is being created at an offset distance from the specified line, as shown in Figure 6-41. Continue routing the pipe. Figure 6-42 shows a pipe created at an offset. Similarly, you can also route a pipe at an elevation by entering a required value in the **Vertical Offset** edit box.

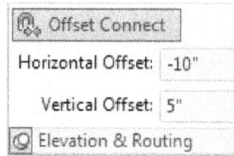

Figure 6-40 *Expanded portion of the **Elevation** & **Routing** panel*

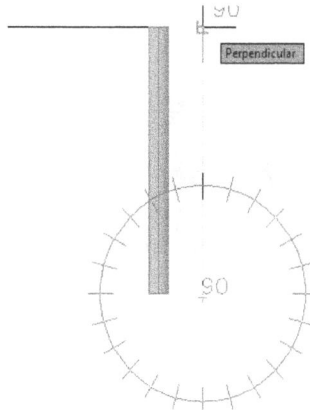

Figure 6-41 *Routing a pipe at an offset*

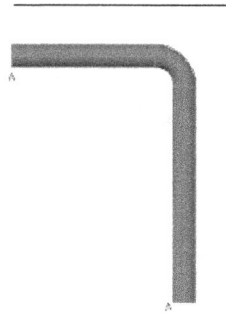

Figure 6-42 *Pipe created at an offset*

Offset Connect

This option is very useful while routing an offset pipe. The **Offset Connect** button is chosen by default in the **Elevation & Routing** panel. As a result, you can connect an offset pipe to a piping component. Figure 6-43 shows an offset pipe snapped to a piping component and Figure 6-44 shows the offset pipe connected to it. Figure 6-45 shows a case when the offset connect mode is turned off.

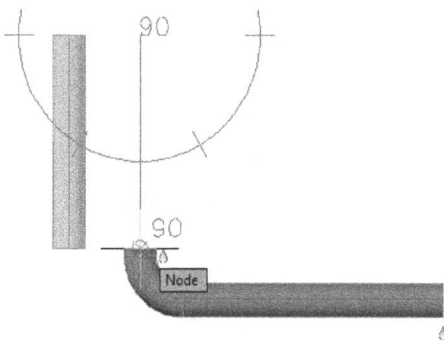

Figure 6-43 *An offset pipe snapped to a piping component*

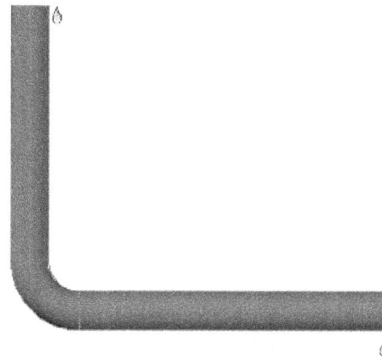

Figure 6-44 *Offset pipe connected to the snapped component*

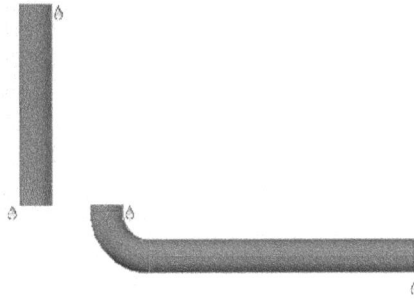

Figure 6-45 *Offset pipe not connected to the snapped component*

Routing a Pipe at a Slope

In a piping layout, there are many cases where you need to create a sloped pipe. For example, you need a sloped pipe for lifting a fluid to a higher gradient. To create such a pipe, first you need to enter a value in the **Slope rise** edit box and then in the **Slope run** edit box. These two edit boxes are located in the **Slope** panel of the **Home** tab. On doing so, the system automatically calculates the slope angle and displays it in the **Slope** edit box. The slope angle is calculated by the specified vertical distance (rise) and the horizontal distance (run). Next, choose the **Toggle Slope** button, if not chosen already. The **Toggle Slope** button controls whether the pipe is to be created at a slope or not. Next, choose the **Route pipe** tool and start routing the pipe; you will notice that the pipe has been created at a slope, as shown in Figure 6-46. An example of a sloped pipe joining two horizontal pipes located at different elevations is shown in Figure 6-47.

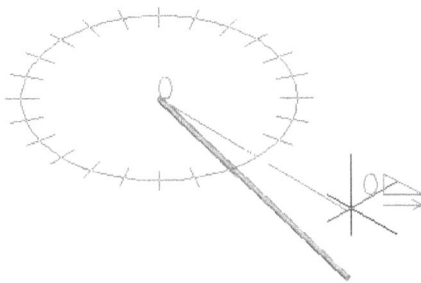

Figure 6-46 *Pipe created at a slope*

Figure 6-47 *Sloped pipe joining two pipes located at different gradients*

Converting a Straight Pipe to a Sloped pipe

You can convert a straight pipe to a sloped pipe. To do so, select a horizontal straight pipe from the drawing area and right-click on it; a shortcut menu will be displayed. Choose the **Pipe Slope Editing** option from the shortcut menu; the **Edit Slope** dialog box will be displayed, as shown in Figure 6-48. Choose the **Start point** button located on the left of the dialog box; you are prompted to select the start point on the highlighted component. Click on the pipe to specify the start point of the slope. Next, choose the **End point** button located on the right and then, specify the end point on the pipe, refer to Figure 6-49. Next, you need to specify the slope calculation method. There are three methods listed in the **Calculation** drop-down list to calculate the slope, Start elevation, End elevation, and Slope. Select the **Slope** option from the **Calculation** drop-down list and specify values in the **Start Elevation** and **End elevation** edit boxes; the slope angle is automatically calculated. Next, choose the **OK** button; the pipe will be sloped, as shown in Figure 6-50.

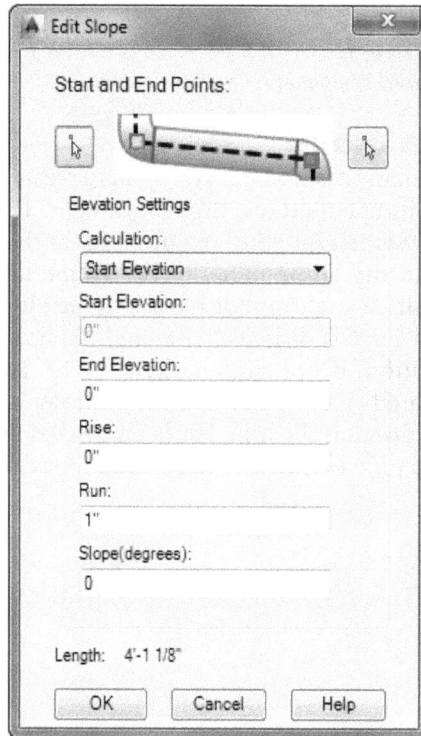

Figure 6-48   The **Edit Slope** *dialog box*

Figure 6-49   *Specifying the start point and end point*

Figure 6-50   *Slope applied to a straight pipe*

Note

It is recommended that you create a sloped pipe instead of converting a straight pipe into a sloped pipe.

CREATING BRANCHES

You can create and connect branches to a header pipe. The methods to create branches are discussed next.

Creating a Tee Branch

You can create a tee branch using the continuation grip displayed at the center of the pipe. To do so, first select the pipe from the drawing area; a continuation grip is displayed at the midpoint of the pipe, as shown in Figure 6-51. Click on the continuation grip located at the midpoint of the pipe. You will notice that a tee fitting is placed at the midpoint and also a branch pipe is connected to it. You can rotate this branch pipe to a required angle using the compass. Next, drag the branch pipe upto a required distance and press ENTER; the branch will be created, as shown in Figure 6-52.

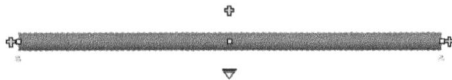

Figure 6-51 *Continuation grip displayed on selecting a pipe*

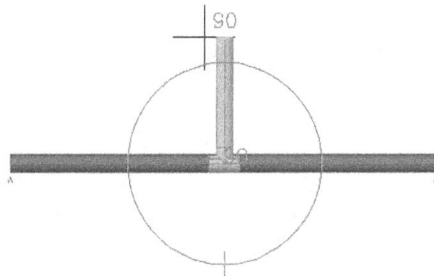

Figure 6-52 *Creating a tee branch*

You can also create a tee branch by manually placing a tee from the tool palette. To do so, scroll to the **Tee** area in the **Dynamic Pipe Spec** tool palette and select a tee fitting; the tee will be attached to the cursor. By default, the left port of the tee is attached to the cursor. You can change the cursor port attachment by pressing the CTRL key, refer to Figure 6-53. Next, select a point on the pipe in the drawing area; the tee will be placed at that point, as shown in Figure 6-54. You can rotate the tee using the compass. After placing the tee, press ESC and click on the tee; a continuation grip is displayed on it. Click on the continuation grip; a branch pipe is created, as shown in Figure 6-55.

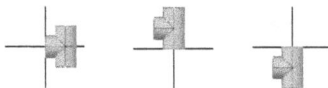

Figure 6-53 *Changing the cursor port attachment*

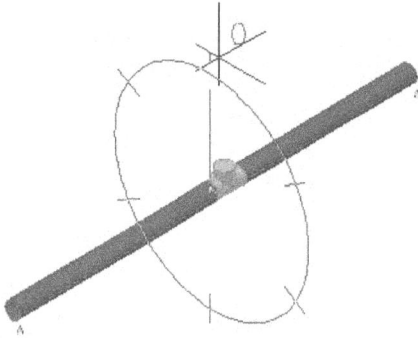

Figure 6-54 Tee placed on the pipe

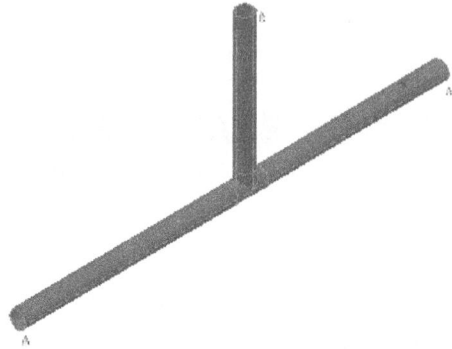

Figure 6-55 Branch pipe created from the tee

Creating an O-let Branch

To create an O-let branch, first choose the **Spec Viewer** button from the **Part Insertion** panel; the **Pipe Spec Viewer** will be displayed. Next, select a part from the **Olet** category in the **Spec Sheet** rollout; the sizes available for the selected olet are displayed in the **Part Size** rollout. Next, select the size from the **Part Size** rollout and choose the **Insert In Model** button; the olet will be attached to the cursor. Select a point on the pipe and place the olet, as shown in Figure 6-56. Next, press ESC and click on the olet; the continuation grip is displayed. Select it and create a branch, as shown in Figure 6-57.

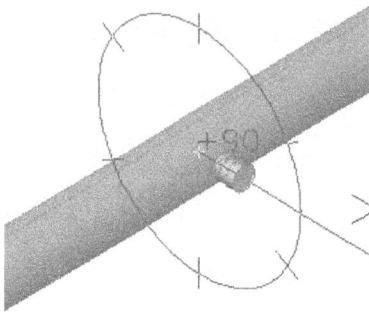

Figure 6-56 Tee placed on the pipe

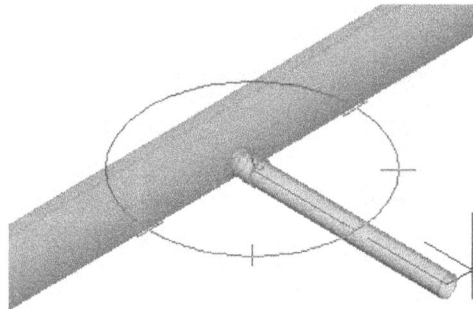

Figure 6-57 Creating an olet branch

Creating a Stub-In Branch

A Stub-In is a type of branch directly welded into the side of a header pipe. In this type of branch, you do not use any fitting. This is the most common and least expensive method of welding a full size pipe or reducing branch to a header pipe. To create a branch on an existing pipe using stub-in, first you need to invoke the **Route Pipe** tool. Next, choose the **Toggle Stub-In** button from the **Part Insertion** panel and specify the start point of the stub-in branch on the pipe, refer to Figure 6-58. Next, specify the end point of the branch pipe; the stub-in branch will be created, as shown in Figure 6-59.

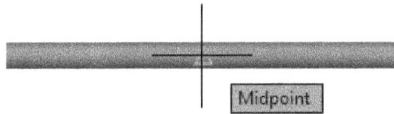

Figure 6-58 *Specifying the start point of the Stub-in branch pipe*

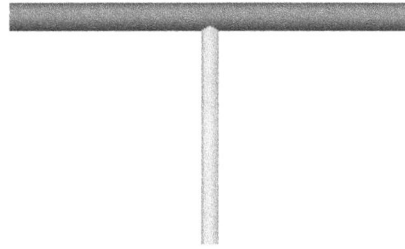

Figure 6-59 *Specify the end point of the Stub-in branch pipe*

Creating a Stub-In Branch at an Offset from the Center of the Header Pipe

You can create a stub-in branch at an offset from the center of the header pipe. To do so, first invoke the **Route Pipe** tool and then choose the **Toggle Stub-In** button from the **Part Insertion** panel. Next, select the pipe size from the **Pipe Size Selector** drop-down list and specify the start point on the header pipe. Next, choose the **connectionoffseT** option from the Command prompt; the **Offset connection to [Toptangent/Bottomtangent/Distance]** prompt will be displayed. Choose the appropriate option from the command prompt. For example, choose the **Toptangent** option if you want to create a stub-in branch at the top of the pipe; the stub-in branch will be created at the top of the header pipe, as shown in Figure 6-60.

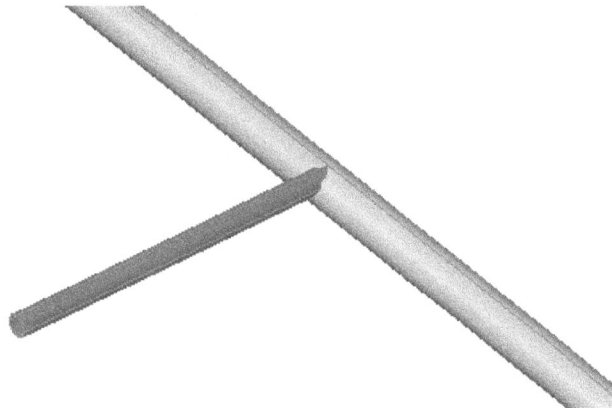

Figure 6-60 *A stub-in branch created on the top of the header pipe*

To create a stub-in branch at a specific offset distance from the center of the pipe, choose the **Distance** option and specify the offset distance value at the Command prompt; the stub-in branch will be created at the specified offset distance from the center of the pipe.

Creating a Branch from an Elbow

You can create a branch from an elbow. To do so, click on an elbow connected to a pipe in the drawing area. On doing so, two continuation grips are displayed on the elbow, as shown in

Figure 6-61. Click on one of the grips and create a pipe. Press ENTER after routing the pipe; a branch will be created, as shown in Figure 6-62.

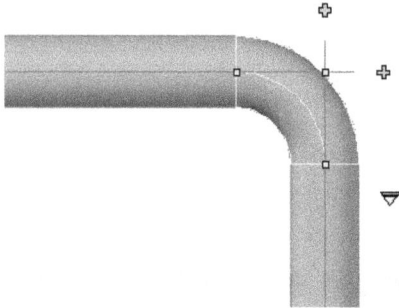

Figure 6-61 *Continuation grips displayed on an elbow*

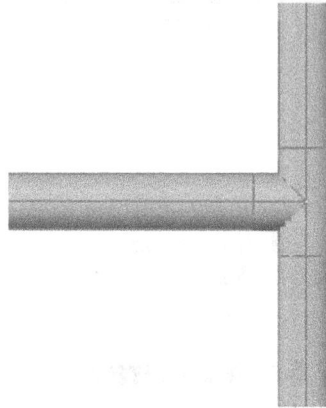

Figure 6-62 *Branch created from an elbow*

Creating a Stub-in Branch at a Precise Location

You can create a branch at a precise location on the header pipe. To do so, first invoke the **Route pipe** tool from the **Part Insertion** panel and make sure that the **Toggle Stub-In** button is chosen. Next, press and hold the SHIFT key and right-click to invoke the shortcut menu. Choose the **From** option from the shortcut menu displayed; you are prompted to specify a point from which the distance will be calculated. Next, move the cursor over the header pipe; the Osnap glyph will be displayed on the pipe, as shown in Figure 6-63. Next, enter the value of distance from the end point of the header pipe at the Command Prompt; the start point of the pipe will be specified and a pipe is connected to the cursor. Move the cursor and specify the end point of the branch pipe, and then press ENTER; the branch will be created at a precise location, as shown in Figure 6-64.

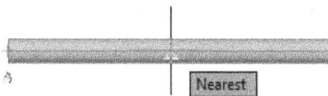

Figure 6-63 *Osnap glyph displayed on the pipe*

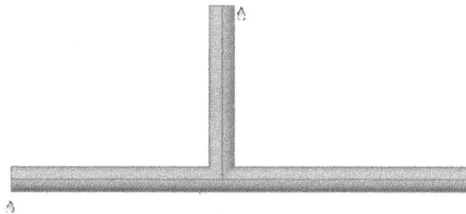

Figure 6-64 *Branch created at a precise location*

Adding a Reinforcing Pad to a Stub-In Branch

You can add a reinforcing pad to a stub-in branch. To do so, right-click on the stub-in branch and choose the **Add Reinforcing Pad** option from the shortcut menu displayed; a reinforcing pad will be added to the stub-in branch, as shown in Figure 6-65.

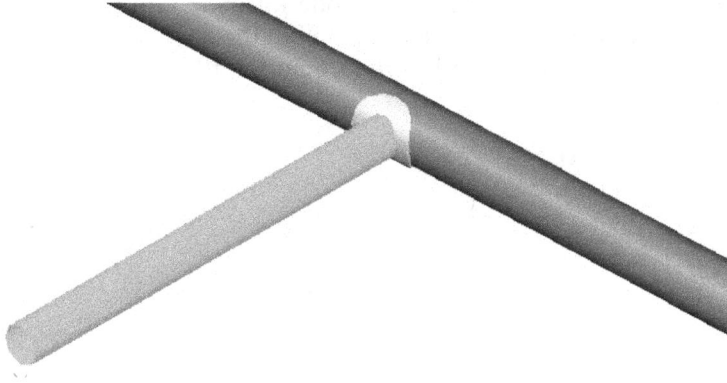

Figure 6-65 *A reinforcing pad added to a stub-in branch*

CREATING A WELD CONNECTION

Weld connections are automatically created when two piping components are connected to each other. You can also create weld connections manually at precise locations. You can break a long pipe into small pipes of a specific length by creating a weld connection. To create a welding connection, right-click on the pipe to display a shortcut menu. Next, choose the **Add Weld to Pipe** option from the shortcut menu displayed; you are prompted to specify the point location on the pipe. Specify a point on the pipe; the weld connection will be created, as shown in Figure 6-66.

You can also create a weld connection at a precise distance from the end point of the pipe. First, make sure that the **Dynamic input** mode is turned on. Next, enter a distance value in the dynamic edit box, as shown in Figure 6-67; the weld will be created at the specified distance from the end point.

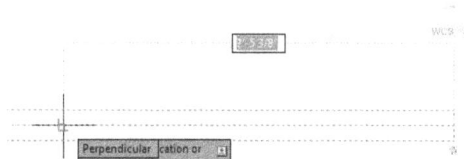

Figure 6-66 *Weld connection created on a pipe*

Figure 6-67 *Specifying distance in the dynamic edit box*

CREATING AUTODESK CONNECTION POINT

Insert

The Autodesk Connection point contains the port information of the piping components. Using the Autodesk connection point, you can connect an AutoCAD Plant 3D pipe to piping components created in other Autodesk applications such as AutoCAD MEP, Civil 3D, and so on. You can also connect a pipe to Xref objects using the Autodesk connect points. To create an Autodesk Connection point, choose the **Insert** button from the **Autodesk Connection Point** panel in the **Insert** tab; you will be prompted to select an insertion point. Select an insertion point on the pipe and place the Autodesk connection point on it, as shown in Figure 6-68.

Figure 6-68 *The Autodesk Connection point created on an open port*

Editing an Autodesk Connection Point

You can edit the properties of an Autodesk connection point. To do so, choose the **Edit** button from the **Autodesk Connection Point** panel; you will be prompted to select an Autodesk connection point. Select a point from the drawing area; the **Autodesk Connection Point Editor** dialog box will be displayed, as shown in Figure 6-69. Edit the property values in the dialog box and choose the **OK** button; the dialog box will be closed and the properties of the Autodesk connection point will be changed.

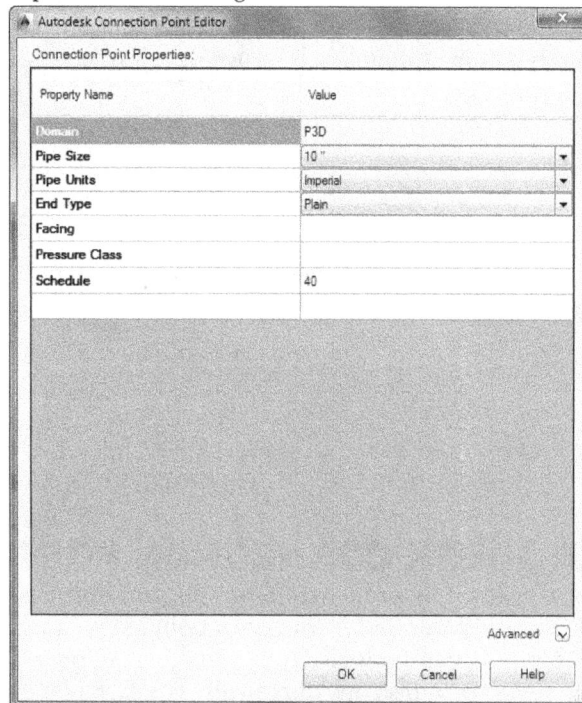

Figure 6-69 *The **Autodesk Connection Point Editor** dialog box*

Routing a Pipe from an Autodesk Connection Point

You can also route a pipe from the Autodesk connection point. To do so, choose the **Route Pipe from Point** button from the **Autodesk Connection Point** panel. Next, select an Autodesk connection point and start routing a pipe.

TUTORIALS

Tutorial 1

In this tutorial, you will open the **Piping Model.dwg** file created in Tutorial 2 of Chapter 4. Next, you will connect the equipment by routing pipes, as shown in Figure 6-70. Figure 6-71 shows the list of pipes used in the model. The dimensions of the model are given in Figures 6-72 through 6-74.

(Expected time: 45 min)

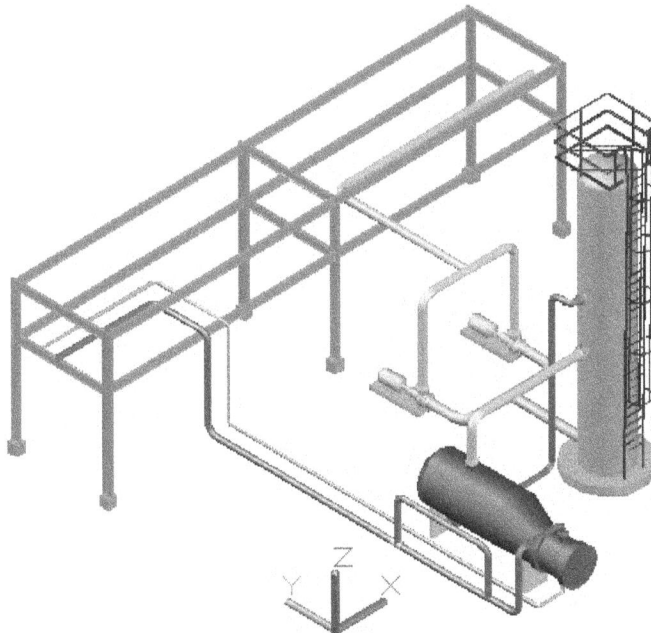

Figure 6-70 Model for Tutorial 1

The following steps are required to complete this tutorial:

a. Open the **Piping Model** file from the **Project Manager**.
b. Select the required spec and select the pipe size.
c. Insert the **Pipe_rack** file into the model space.
d. Place reducers at the inlet and outlet nozzles of the centrifugal pumps.
e. Connect the centrifugal pumps to the vertical vessel.
f. Connect the reboiler to the vertical vessel.
g. Route pipes from the other nozzles of the reboiler.

Pipe List			
Pipe Number	Spec	Size	Equipment connected
1004	CS150	10"	Centrifugal Pumps and Vertical Vessel
1012	CS150	8"	Centrifugal Pumps
1007	CS150	8"	Vertical Vessel and Reboiler
2002	CS150	3"	Reboiler
2000	CS150	6"	Reboiler

Figure 6-71 The Pipe list

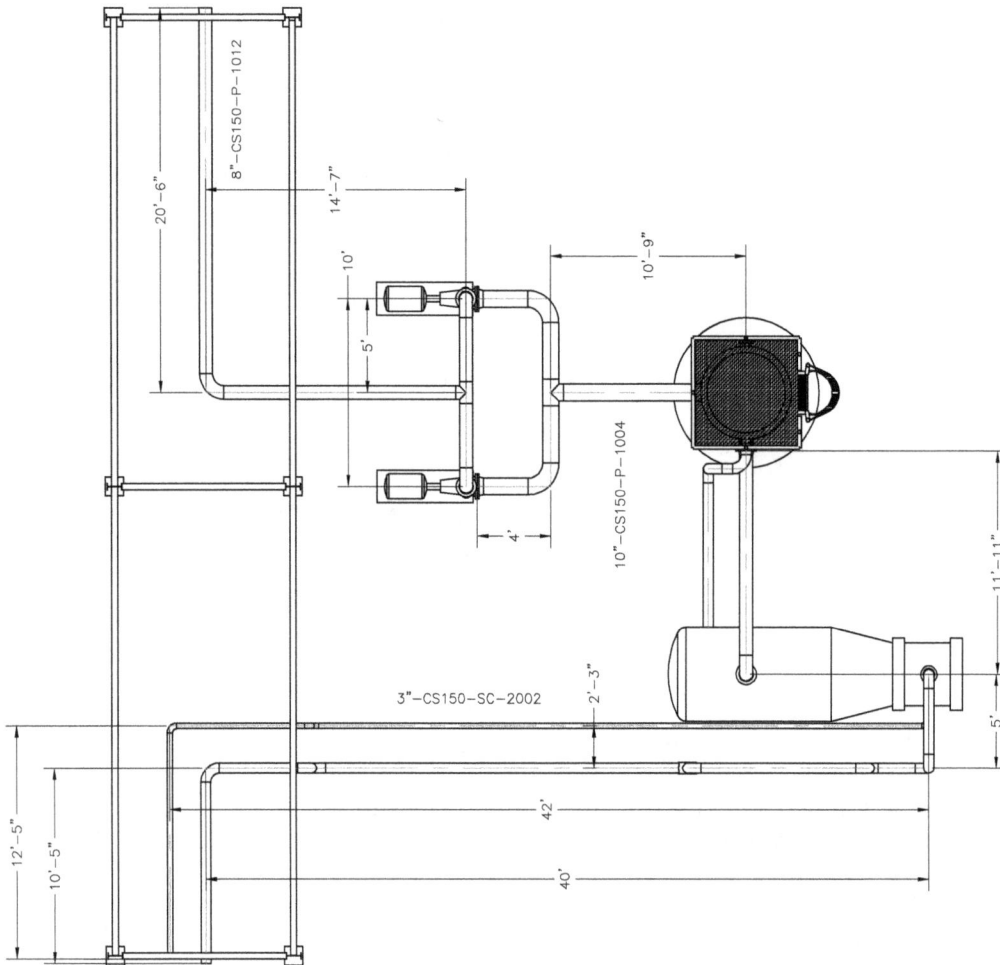

Figure 6-72 Top view of the model

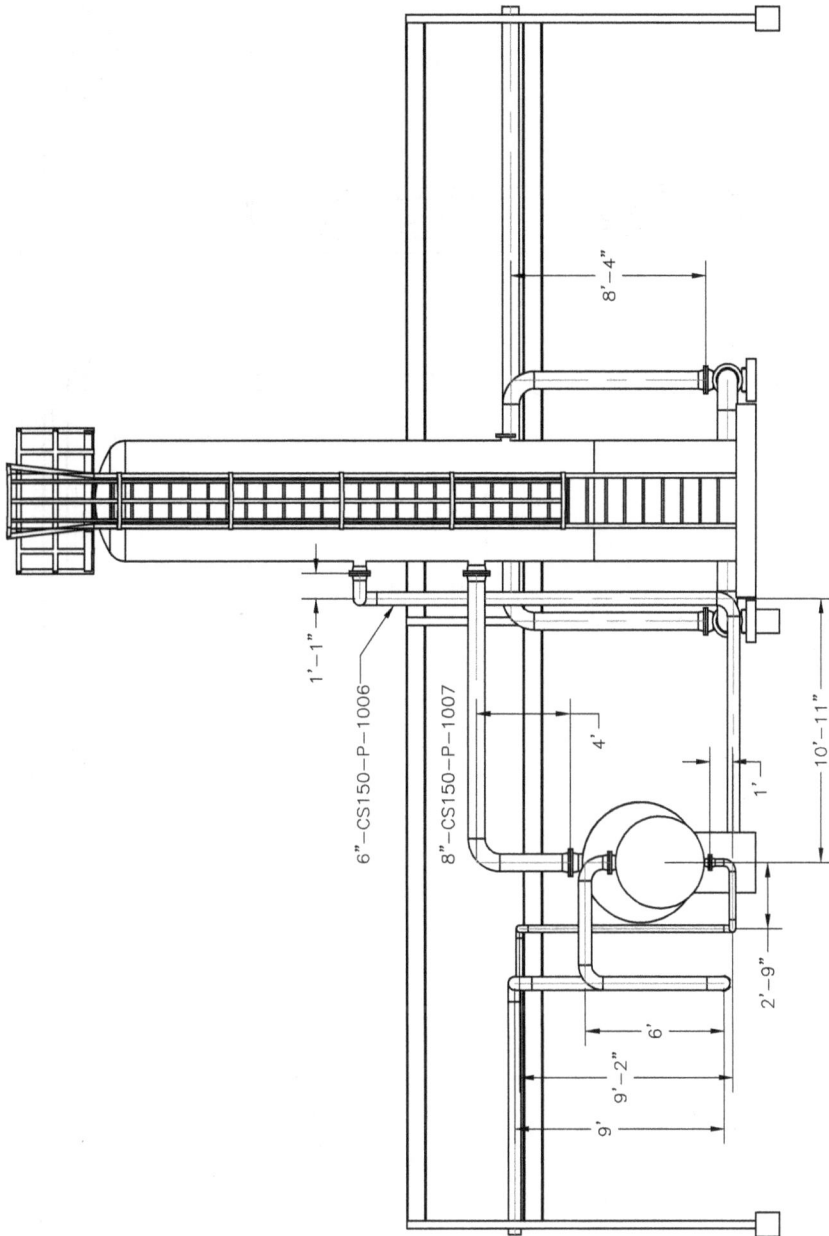

Figure 6-73 *Front view of the model*

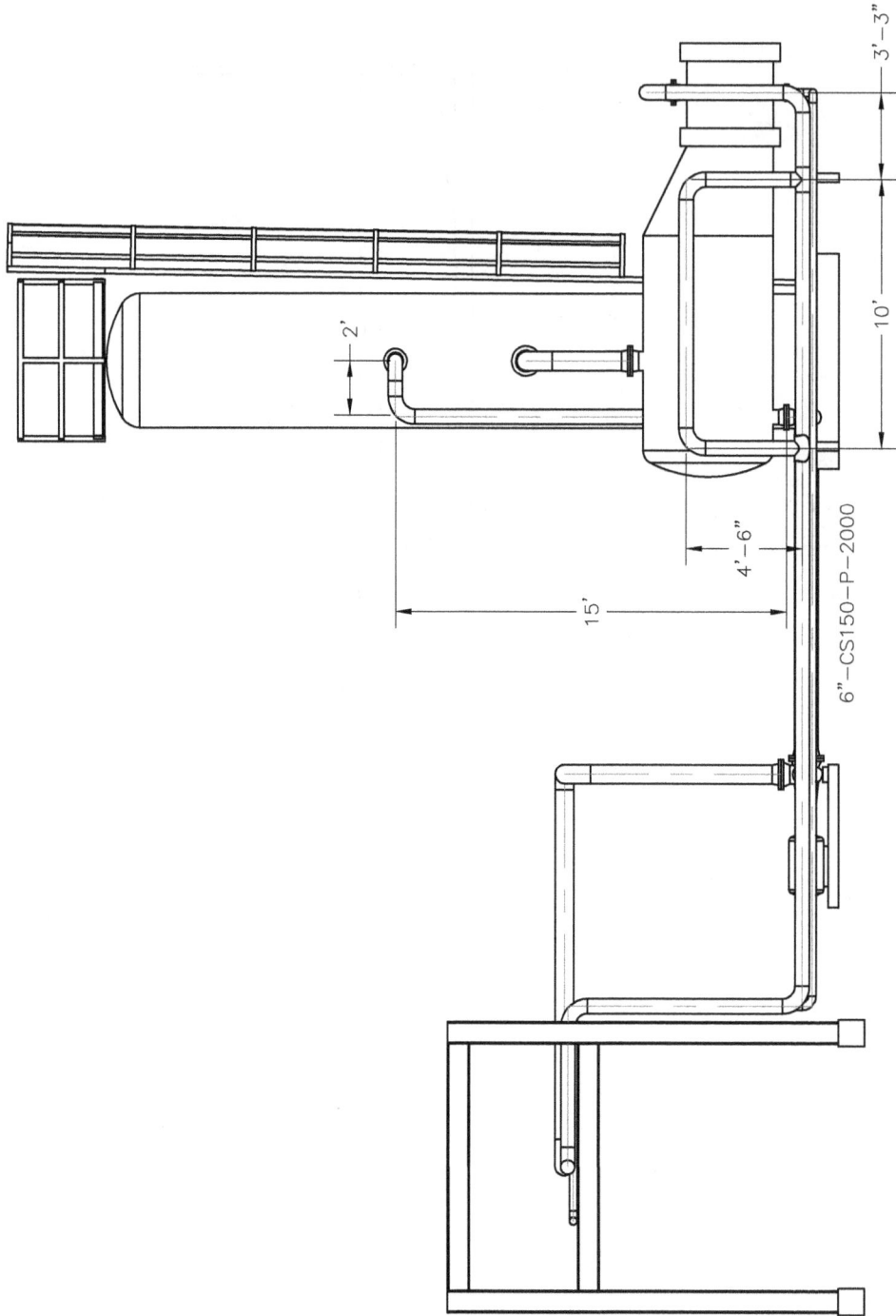

Figure 6-74 *Right view of the model*

Opening a File

1. Choose **Start > All Programs (or Programs) > Autodesk > AutoCAD Plant 3D 2014 > AutoCAD Plant 3D 2014**; AutoCAD Plant 3D gets started and the welcome screen is displayed.

2. Select the **CADCIM** from the **Current Project** drop-down list in the **Project Manager**.

3. Expand the **Plant 3D Drawings** node in the **Project Manager** and double-click on **Piping Model**; the selected drawing file is opened.

Selecting the Spec

In this section, you will select a spec.

1. Select *CS150* from the **Spec Selector** drop-down list in the **Part Insertion** panel; the **Dynamic Pipe Spec** Tool Palette on the left side of the window is loaded with *CS150* piping components.

Inserting the Pipe Rack into the Model

Next, you need to insert the pipe rack created in Tutorial 1 of Chapter 3.

1. Choose the **Insert** tool from the **Block** panel in the **Insert** tab; the **Insert** dialog box is displayed.

2. Choose the **Browse** button next to the **Name** display box in the **Insert** dialog box and then browse to the location *C: > Documents > CADCIM >Plant 3D Models*. Next, double-click on the **Pipe_rack.dwg** file; the filename is displayed in the **Name** edit box of the **Insert** dialog box.

3. Clear the **Specify On-Screen** check box in the **Insertion point** area, if selected. Next, enter **0, 25', 0** in the **X, Y, Z** edit boxes, respectively.

4. Choose the **OK** button from the **Insert** dialog box; the pipe rack is placed in the drawing area at the specified location, as shown in Figure 6-75. Next, change the view orientation to **SW Isometric**.

Connecting the Centrifugal Pumps and Vertical Vessel

In this section, you will connect the pumps and the vertical vessel. First, you need to hide the support at the bottom of the vertical vessel.

1. Select the support of the vertical vessel from the drawing area and then choose the **Hide Selected** button from the **Visibility** panel in the **Home** tab; the support will be hidden.

2. Select the **Route New line** option from the **Line Number Selector** drop-down list in the **Part Insertion** panel; the **Assign Tag** dialog box is displayed.

3.   Enter **1004** in the **Number** edit box in the dialog box.

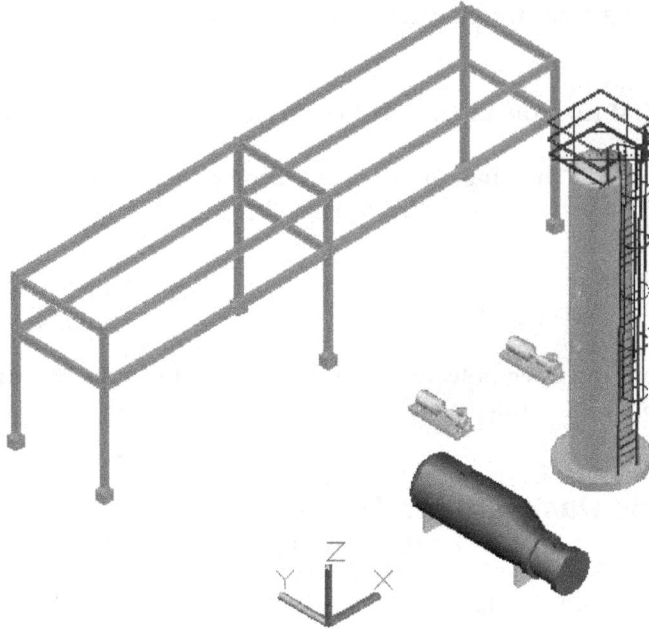

Figure 6-75 Model after placing the pipe rack

4.   Specify the line size as **10"** in the **Size** drop-down list.

5.   Choose the **Assign** button; the **Assign Tag** dialog box will be closed and you are prompted to specify the start point of the pipe.

6.   Press and hold the SHIFT key and right-click to display a shortcut menu. Choose the **Node** option from the shortcut menu displayed and then select the center point of the inlet nozzle of the left-side pump, as shown in Figure 6-76; you are prompted to specify the next point.

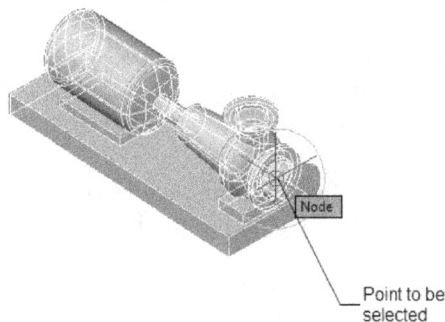

Figure 6-76 Selecting the center point of the nozzle

Note

*You can switch to the **Wireframe** view for easy selection of the end point. To do so, click on the down arrow of the **View Styles** drop-down list in the **View** panel and then select the **Wireframe** option.*

7. Move the cursor in the direction which is in-line with the nozzle. Next, enter **4'** at the Command Prompt to specify the end point; you are prompted to specify the next point of the pipe.

8. Press and hold the SHIFT key and right-click to display a shortcut menu. Choose the **Node** option from the shortcut menu and then select the center point of the inlet nozzle of the right-side pump; the inlet nozzles are connected to the pump, as shown in Figure 6-77.

9. Rotate the model using the **Orbit** tool such that the bottom nozzle of the vertical vessel is visible.

10. Make sure that the **Toggle Cutback Elbow** button is chosen in the **Part Insertion** panel. This creates an elbow whose angle is different from the angle of elbow available in the spec.

Figure 6-77 Pumps after connecting the inlet nozzles

11. Click on the vertical vessel to display the plus(+) symbol on the bottom nozzle.

12. Click on the plus(+) symbol displayed and then press and hold the SHIFT key and right-click to display a shortcut menu. Choose the **Midpoint** option from the shortcut menu and select the mid point of the pipe joining the two pumps, as shown in Figure 6-78; solutions are generated to connect the vessel.

13. Choose the **Next** option from the Command prompt until the required solution is displayed, as shown in Figure 6-79.

14. Next, choose the **Accept** option from the Command prompt to accept the solution.

Next, you need to connect the outlet nozzles of the centrifugal pumps.

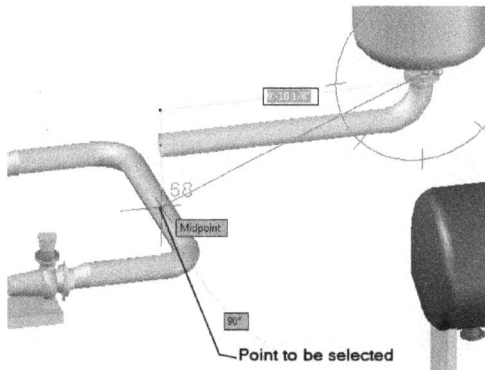

Figure 6-78 *Select the center point of the pipe*

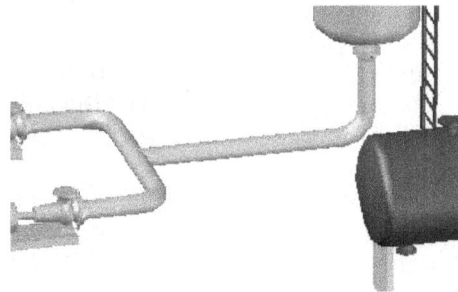

Figure 6-79 *Required solution for connecting the pipes*

Routing Pipes from the Outlet Nozzles of the Pumps

1. Select the **Route New line** option from **Line Number Selector** drop-down list in the **Part Insertion** panel; the **Assign Tag** dialog box is displayed.

2. Enter **1012** in the **Number** edit box.

3. Select the line size **8"** from the **Size** drop-down list.

4. Choose the **Assign** button; the **Assign Tag** dialog box will be closed and you are prompted to specify the start point of the pipe.

5. Press and hold the SHIFT key and right-click to display a shortcut menu and choose the **Node** option from it.

6. Next, select the center point of the nozzle on the left pump, as shown in Figure 6-80; the pipe is connected to it and you are prompted to specify the next point.

7. Move the cursor upward and enter **100"** at the Command prompt; a vertical pipe is created.

8. Next, press and hold the SHIFT key and right-click to display a shortcut menu and then choose the **Node** option from it. Next, select the outlet nozzle of the right-side pump; the outlet nozzles are connected, as shown in Figure 6-81.

9. Click on the horizontal pipe of the newly created pipe connection; a plus(+) symbol is displayed on it.

10. Select the plus(+) symbol and then move the cursor toward piperack.

11. Change the orientation plane, as shown in Figure 6-82 by entering the **Plane** option at the Command prompt.

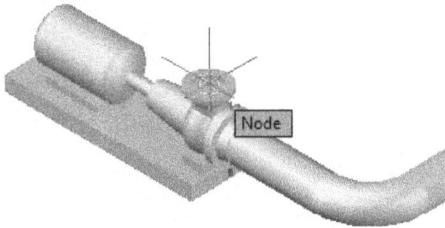

Figure 6-80 *Selecting the center point outlet nozzle of the left pump*

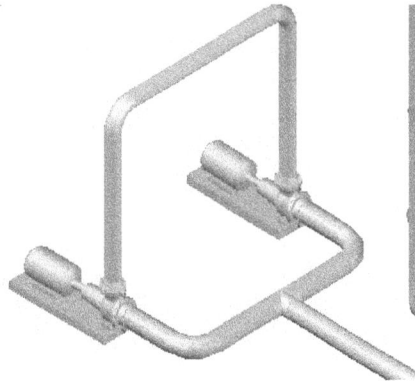

Figure 6-81 *Pumps after connecting the outlet nozzles*

12. Next, move the pipe toward the pipe rack, type **14'** at the Command Prompt, and then press ENTER to create a pipe passing toward the pipe rack, as shown Figure 6-82.

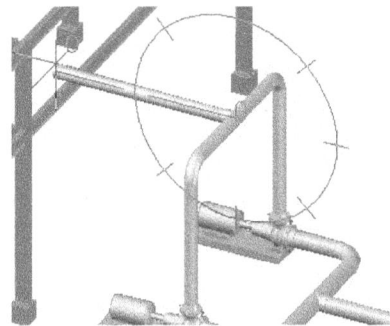

Figure 6-82 *Pipe extending from the right elbow*

13. Next, change the orientation plane and move the cursor toward right, as shown in Figure 6-83.

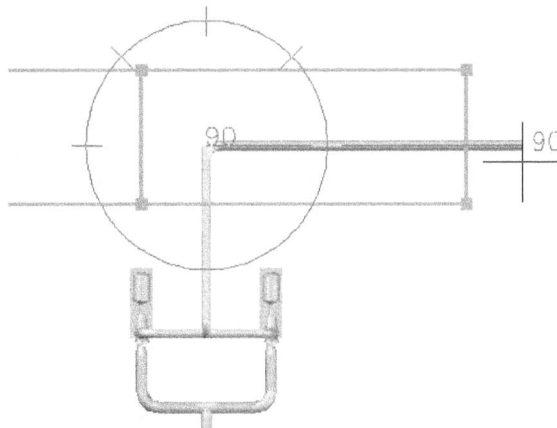

Figure 6-83 *Pipe passing toward the pipe rack*

14. Type **20'** at the Command prompt and then press ENTER; a pipe is created upto the right end of the pipe rack. Next, press ESC to exit the **Route Pipe** tool.

Connecting the Vertical Vessel and the Reboiler

In this section, you will connect the vertical vessel and the reboiler using the **Route Pipe** tool.

1. Select the **Route New line** option from the **Line Number Selector** drop-down list in the **Part Insertion** panel; the **Assign Tag** dialog box is displayed.

2. Enter **1007** in the **Number** edit box.

3. Specify the line size as **8"** from the **Size** drop-down list.

4. Choose the **Assign** button; the **Assign Tag** dialog box is closed and you are prompted to specify the start point of the pipe.

5. Specify the start point of the pipe on the nozzle of the vertical vessel, as shown in Figure 6-84; you are prompted to specify the end point.

6. Press and hold the SHIFT key and right-click to display a shortcut menu and then choose the **Node** option.

7. Select the center point of the nozzle on the reboiler, as shown in Figure 6-85; the pipe connect is created between the two selected nozzles, as shown in Figure 6-86.

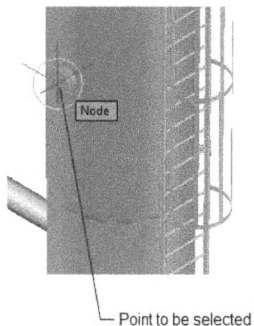

Figure 6-84 Point to be selected to specify the start point of the pipe

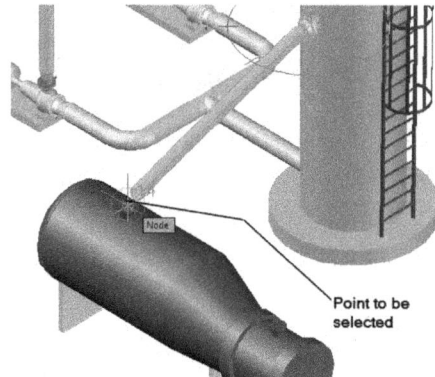

Figure 6-85 Point to be selected to specify the end point of the pipe

8. Next, choose the **Accept** option from the Command prompt to accept the solution.

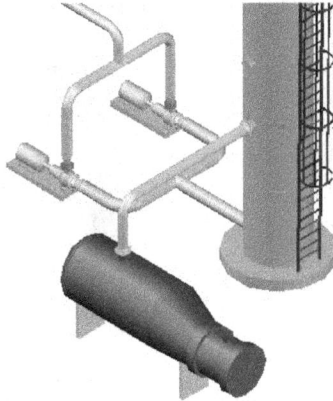

Figure 6-86 Pipe connection created between
the selected nozzles

Next, you need to connect the nozzle located at the bottom of the reboiler to the nozzle on the vertical vessel.

9. Rotate the model using the **Orbit** tool such that the nozzle at the bottom of the reboiler is visible.

10. Select the **Route New line** option from **Line Number Selector** drop-down list in the **Part Insertion** panel; the **Assign Tag** dialog box is displayed.

11. Enter **1006** in the **Number** edit box.

12. Specify the line size as **6"** from the **Size** drop-down list.

13. Choose the **Assign** button; the **Assign Tag** dialog box is closed and you are prompted to specify the start point of the pipe.

14. Select the center point of the nozzle located at the bottom of the reboiler, refer to Figure 6-87.

15. Next, move the cursor toward the vertical vessel, type **3'** at the Command prompt, and then press ENTER.

16. Next, invoke the **Object Snap** shortcut menu and choose the **Node** option from it.

17. Select the center point of the nozzle, as shown in Figure 6-88; solutions are generated to create the connection between the two selected nozzles.

18. Choose the **Next** option from the Command prompt until the required solution is displayed, as shown in Figure 6-89.

19. Choose the **Accept** option to accept the required solution.

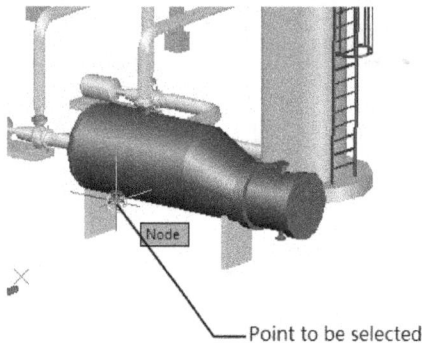

Figure 6-87 Point to be selected

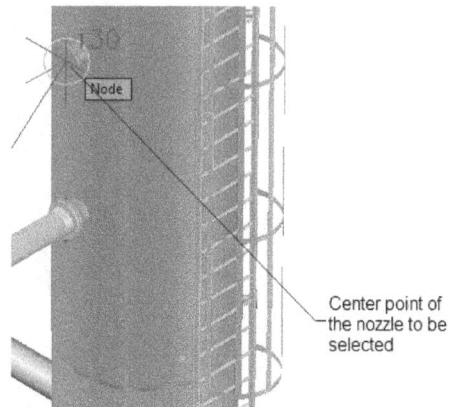

Figure 6-88 Point to be selected

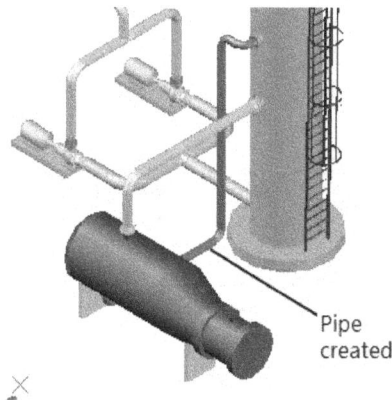

Figure 6-89 The solution to be selected to create the pipe connection

Routing Pipes from the Other Nozzles of the Reboiler

Next, you need to route a pipe from the bottom nozzle of the reboiler.

1. Select the **Route New line** option from the **Line Number Selector** drop-down list in the **Part Insertion** panel; the **Assign Tag** dialog box is displayed.

2. Enter **2002** in the **Number** edit box.

3. Select the line size **3"** from the **Size** drop-down list.

4. Choose the **Assign** button; the **Assign Tag** dialog box is closed and you are prompted to specify the start point of the pipe.

5. Select the bottom nozzle of the reboiler, as shown in Figure 6-90. Next, move the cursor downward.

6. Type **1'** at the Command prompt and the press ENTER.

7. Move the cursor toward left, enter **3'** at the Command prompt, and then press ENTER.

8. Change the orientation plane and move the cursor toward the pipe rack. Enter **34'** at the Command prompt and then press ENTER.

9. Change the orientation plane, if required, and move the cursor upward. Next, enter **122"** at the Command prompt and press ENTER.

10. Move the cursor into the pipe rack, enter **8'** at the Command prompt, and then press ENTER.

11. Change the orientation plane, if required and move the cursor toward left upto the end of the pipe rack.

12. Next, click to specify the end point. Figure 6-91 shows the pipe routed from the bottom nozzle of the reboiler.

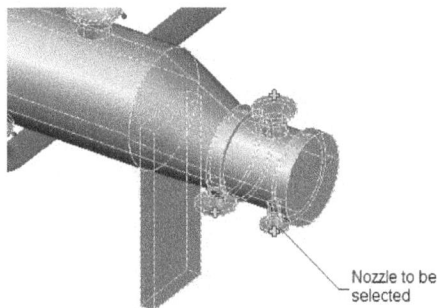

Figure 6-90 Nozzle to be selected

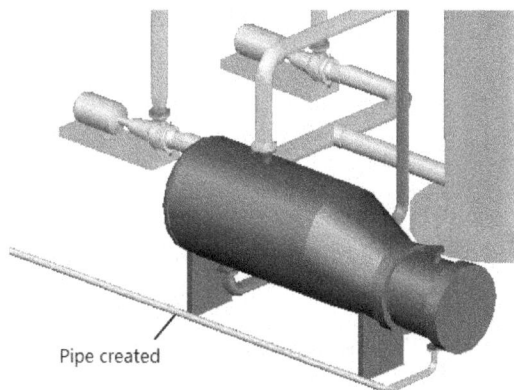

Figure 6-91 Pipe routed from the bottom nozzle of the reboiler

Routing Pipe from Another Nozzle of the Reboiler

Next, you need to create other pipes connecting the reboiler.

1. Select the **Route New line** option from the **Line Number Selector** drop-down list in the **Part Insertion** panel; the **Assign Tag** dialog box is displayed.

2. Enter **2000** in the **Number** edit box.

3. Select the line size **6"** from the **Size** drop-down list.

4. Choose the **Assign** button; the **Assign Tag** dialog box is closed and you are prompted to specify the start point of the pipe.

5. Specify the start point on the nozzle of the reboiler, as shown in Figure 6-92.

6. Next, move the cursor toward left, type **5'** at Command prompt and press ENTER.

7. Next, move the cursor downward, type **6'** at the Command prompt and press ENTER.

8. Next, choose the **Plane** option from the command prompt to change the orientation plane, as shown in Figure 6-93.

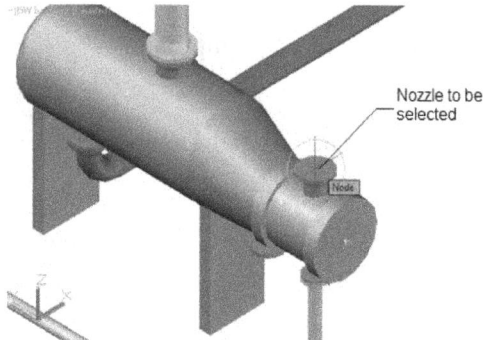

Figure 6-92 Point to selected

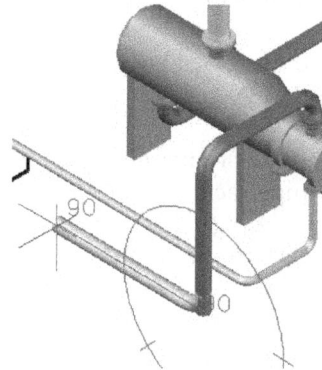

Figure 6-93 Orientation plane to be set

9. Move the cursor toward the pipe rack, enter **34'** at the Command prompt, and press ENTER.

10. Change the orientation plane, if required, and move the cursor upward. Next, enter **120"** at the Command prompt and press ENTER.

11. Move the cursor into the pipe rack, enter **6'** at the Command prompt, and then press ENTER.

12. Change the orientation plane, if required and move the cursor toward left upto the end of the pipe rack. Next, click to specify the end point.

    Next, you need to create a tee branch on the previously created pipe.

13. Select **Tee, BV 40(CS150)** from the Tool Palette, refer to Figure 6-94; the tee is attached to the cursor and you are prompted to specify the insertion point.

    The insertion point is to be specified at a precise distance from the end point of the pipe.

14. Invoke the **Object Snap** shortcut menu (Press SHIFT and right-click) and choose the **From** option. Next, select the end point of the previously created pipe, as shown in Figure 6-95.

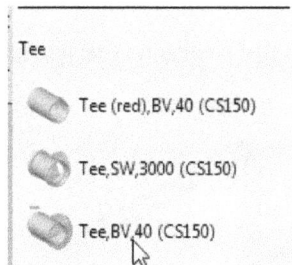

Figure 6-94 *Selecting* **Tee,BV** *40(CS150) from the Tool Palette*

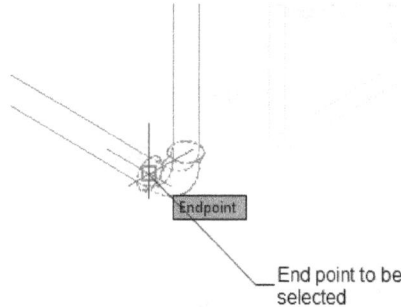

Figure 6-95 *Selecting the end point of the pipe*

15. Next, move the cursor over the pipe and enter **4'** at the Command Prompt. Next, press ENTER; the tee is placed at the specified point, as shown in Figure 6-96. Also, you are prompted to specify the rotation angle.

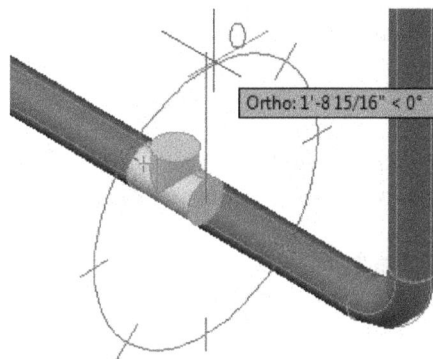

Figure 6-96 *Tee placed at the specified point*

16. Press ENTER to accept the default value. Press ESC to exit the tool.

17. Click on the tee to display a plus(+) symbol on it.

18. Click on the plus(+) symbol and move the cursor upward.

19. Enter **4'** at the Command prompt and press ENTER.

20. Change the orientation plane and move the cursor toward the pipe rack.

21. Enter **10'** at the Command prompt and press ENTER.

22. Move the cursor downward and select the **STub-in** option from the Command prompt.

23. Select the center point of the horizontal pipe, as shown in Figure 6-97; two solutions are created.

24. Choose the **Next** option from the Command prompt until the required solution is displayed, as shown in Figure 6-98.

25. Choose the **Accept** option from the Command prompt to accept the solution.

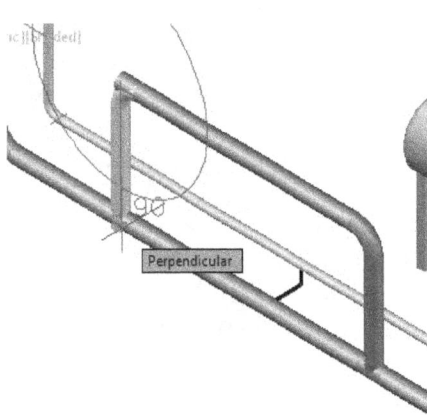

Figure 6-97 Point to be selected on the horizontal pipe

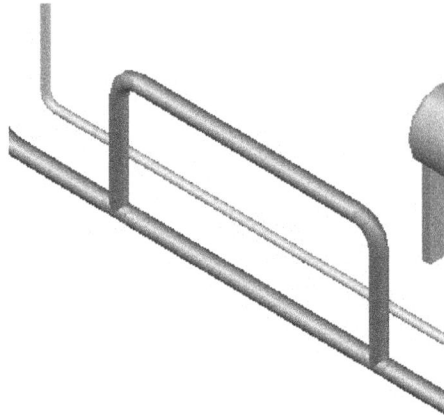

Figure 6-98 Required solution

Note

Sometimes the required solution is displayed automatically when you select the center point of the horizontal pipe. In that case, you may skip the 24th and 25th steps.

Adding the Reinforcing Pad

1. Select the vertical pipe, as shown in Figure 6-99.

2. Right-click on the selected pipe and choose the **Add Reinforcing Pad** option from the shortcut menu displayed; the reinforcing pad is added to the stub-in pipe, as shown in Figure 6-100.

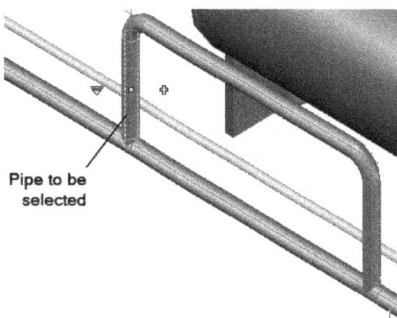

Figure 6-99 Pipe to be selected

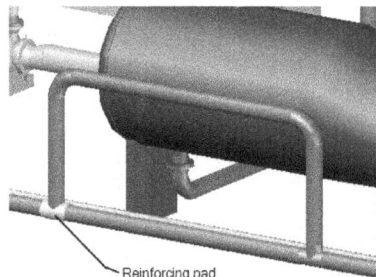

Figure 6-100 Reinforcing pad created

The piping system after adding all pipes is displayed, as shown in Figure 6-101.

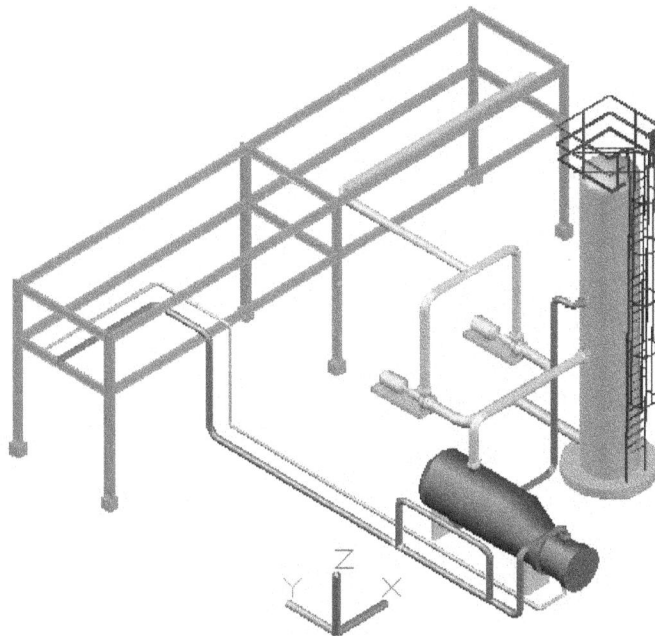

Figure 6-101 Model for Tutorial 1

3. Choose the **Save** button from the **Application** menu and then the **Close** button to close the file.

Self-Evaluation Test

Answer the following questions and then compare them to those given at the end of this chapter:

1. The _____ edit box is used to specify the angular increments through which cursor will snap.

2. The **Pipe Spec Viewer** contains the _____ rollout and the _____ rollout.

3. You can insert a part from the **Pipe Spec Viewer** by using the _____button.

4. You can route a pipe with a new line number assigned to it using the _____option.

5. You can change the position of the route line using the_____ drop-down list.

6. The _____tool is used to convert a line into a pipe.

7. The _____ button is used to control the display of tick marks on the compass.

8. Before creating a 3D piping model, you need to select a piping spec. (T/F)

9. Solutions are generated for connecting two open ports. (T/F)

10. Compass helps you to rotate the pipe at a precise angle. (T/F)

Review Questions

Answer the following questions:

1. You can select the pipe size from the _____ drop-down list in the **Part Insertion** panel.

2. Using the _____ button, you can create a new tool palette with parts available in the selected spec.

3. You can change the compass settings by using the options in the _____ panel.

4. You can turn on/off the display of tick marks on the compass by using the _____ button.

5. With the _____ mode turned on, you can create elbow at the angles other than 90 or 45 degrees.

6. You can create a rolled elbow (twisted elbow) by specifying rotation angles in two planes which are _____ to each other.

7. You can convert a straight pipe into a sloped pipe by invoking the _____ dialog box.

8. On selecting a pipe, the _____ is displayed at the mid point of the pipe.

9. Using the _____ , you can connect an AutoCAD Plant 3D pipe to piping components created in other Autodesk applications such as AutoCAD MEP, Civil 3D, and so on.

10. You can change the size of the pipe while routing. (T/F)

Answers to Self-Evaluation Test
1. Snap Increments, **2.** Spec Sheet, Part Sizes, **3.** Insert in Model, **4.** Route New line, **5.** Set Routing Line, **6.** Line to Pipe, **7.** Toggle tick marks, **8.** T, **9.** T, **10.** T.

Chapter 7

Adding Valves, Fittings, and Pipe Supports

Learning Objectives

After completing this chapter, you will be able to:
- *Place valves and fittings*
- *Map P&ID objects to Plant 3D objects*
- *Add supports to pipes*
- *Insulate a pipe*
- *Modify pipe components*
- *Validate a 3D model*

ADDING VALVES AND FITTINGS

In AutoCAD Plant 3D, you can add valves and fittings to a pipe by using different methods. All these methods are discussed next.

Adding Valve and Fittings to a pipe using the Spec Sheet

To add a valve or a fitting onto a pipe from a spec sheet, invoke the **Spec Viewer** by choosing the **Spec Viewer** button from the **Part Insertion** panel in the **Home** tab. Next, double-click on the required valve in **Spec Sheet** present in **Pipe Spec Viewer**; the valve will be attached to the cursor. You can press the CTRL key to change the port attached to the cursor. Next, you need to specify the insertion point of the valve, refer to Figure 7-1. Make sure that the **Object Snap** is turned on for easy selection of the point. Next, you need to specify the rotation of the valve by using the compass or just press ENTER to specify the zero degree rotation, refer to Figure 7-2.

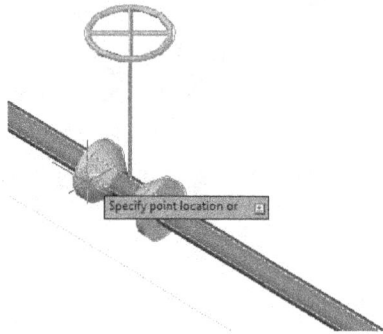

Figure 7-1 Specifying the insertion point of the valve

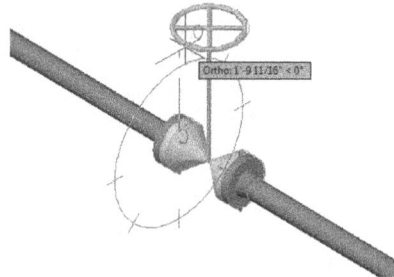

Figure 7-2 Specifying the rotation of the valve

You can also place the valves at a precise distance from a point. To do so, press and hold the SHIFT key and then right-click in the drawing area to display a shortcut menu. Next, choose the **From** option from the shortcut menu displayed, refer to Figure 7-3; you are prompted to specify a point from which the distance will be calculated. Press and hold the SHIFT key and right-click to invoke the shortcut menu. Choose the **Node** option from the shortcut menu and then select the end point of the pipe. Next, move the cursor over the pipe, refer to Figure 7-4 and enter a distance at the Command prompt; the insertion point of the valve will be specified. Next, you need to specify the rotation angle of the valve.

You can also place a valve or a fitting from the Tool Palette located at the right-side of the drawing area. To do so, choose the **Dynamic Pipe Spec** tab from the Tool Palettes and select the required valve or fitting; the selected valve or fitting will be attached to the cursor. Next, place the valve or fitting in the model.

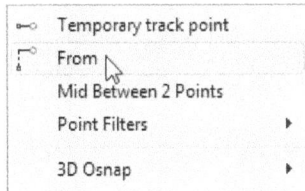

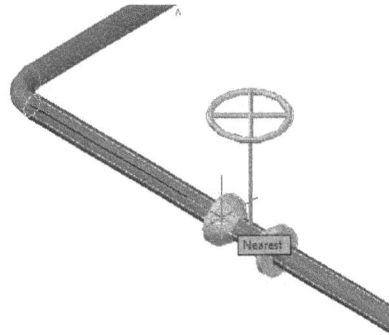

*Figure 7-3 Choosing the **From** option from the shortcut menu displayed*

Figure 7-4 Placing the valve at a precise distance from the end point of the pipe

Adding Valves and Fittings Using a P&ID

To place a valve using a P&ID, choose the **P&ID Line List** button from the **Part Insertion** panel to display the **P&ID Line List** window, refer to Figure 7-5. In this window, select the P&ID from the drop-down list located at the top; all the line numbers present in the selected P&ID are displayed. The components present in the P&ID are grouped under different line numbers. Next, expand the respective pipe line number from the tree view and select the valve available under the pipe line number. Next, choose the **Place** button from the **P&ID Line List** window; the selected component will be attached to the cursor and you will be prompted to specify the insertion point. Specify the insertion point on the pipe.

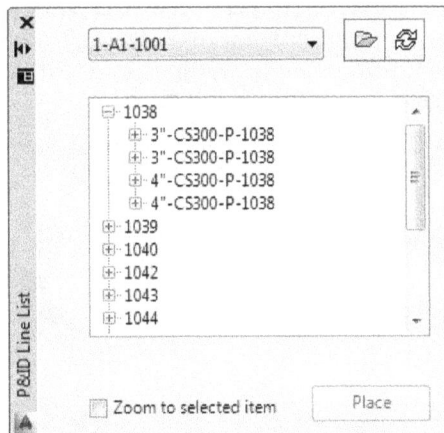

*Figure 7-5 The **P&ID Line List** window*

Note that the component to be placed from a P&ID to plant 3D model should be mapped with the corresponding Plant 3D component. You will learn more about mapping P&ID components with 3D model later in this chapter.

If the selected component is not available in the spec, the **Select Size and Spec** dialog box will be displayed, as shown in Figure 7-6. In this dialog box, select the spec and size from the

Spec and **Size** list boxes, respectively. Next, select the **Always substitute the selected size value for** check box and choose the **Select** button to place the valve.

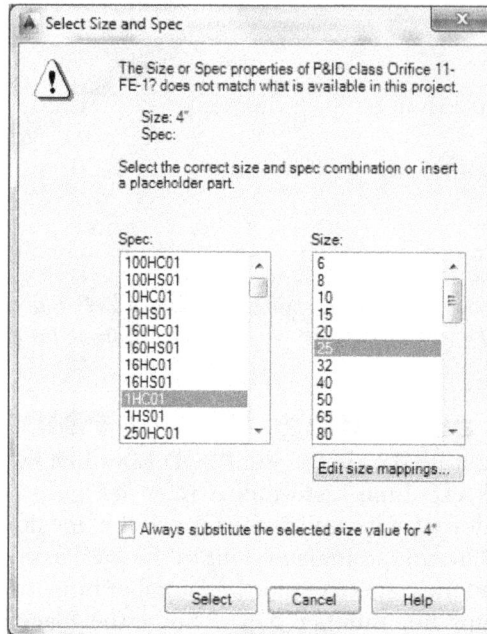

*Figure 7-6 The **Select Size and Spec** dialog box*

Placing Valves and Fittings while Routing a Pipe

You can also place a valve or a fitting while routing a pipe. To do so, invoke the **Route Pipe** tool. Next, choose the **pipeFitting** option from the Command prompt; the **Part Placement** dialog box will be displayed, as shown in Figure 7-7.

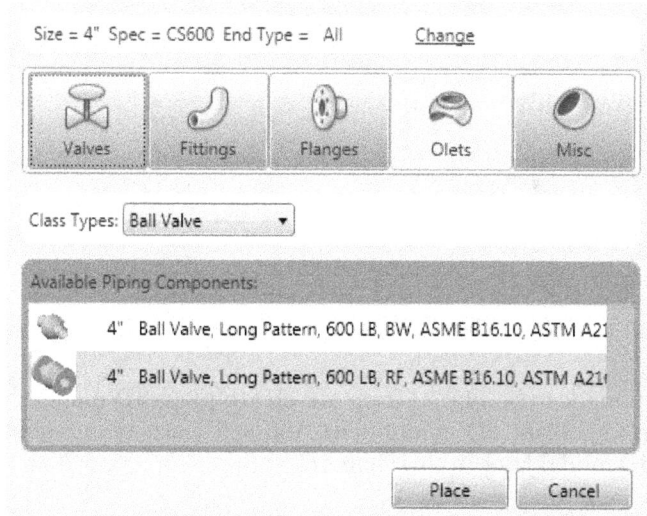

*Figure 7-7 The **Part Placement** dialog box*

To place a valve, choose the **Valve** button from the **Part Placement** dialog box. Next, select an option from the **Class Type** drop-down list; the available valves will be displayed. Select the required valve from the **Available Pipe Components** area and choose the **Place** button; the selected valve will be attached to the end of the pipe, as shown in Figure 7-8. Next, click to place the valve; the compass will be displayed and you are prompted to specify the valve rotation angle. Specify the rotation angle and then press ENTER. Similarly, you can place other pipe components such as fittings, flanges, caps, and so on.

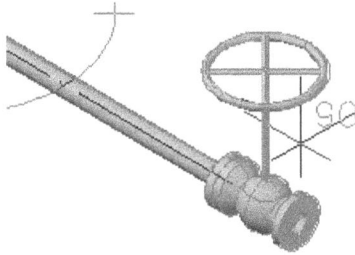

Figure 7-8 *Valve attached to the end of the pipe*

Placing Custom Parts

You can place a custom part or instrument into a 3D model. To do so, choose the **Custom Part** button from the **Part Insertion** panel; the **Custom Parts Builder** window will be displayed, as shown in Figure 7-9. In this window, choose the **Plant 3D Shape** button from the **Graphics** area and select the part type from the **Part Type** drop-down list. Next, choose the **Shape Browser** button from the **Graphics** area; the **Plant 3D Shape Browser** dialog box will be displayed, as shown in Figure 7-10. Select the required shape from the **Graphics** area and choose the **OK** button; the preview of the selected shape will be displayed in the **Custom Parts Builder** window. Select the **Permanent** option from the **Custom Part type** drop-down list in the **Part Properties** area. Also, specify the size and other properties in the **Custom Parts Builder** window. Next, choose the **Insert in Model** button from the **Custom Parts Builder** window and place the part in the model.

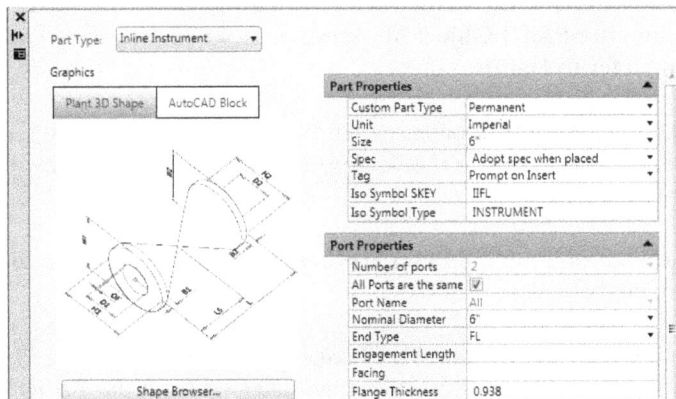

Figure 7-9 *Partial view of the **Custom Parts Builder** window*

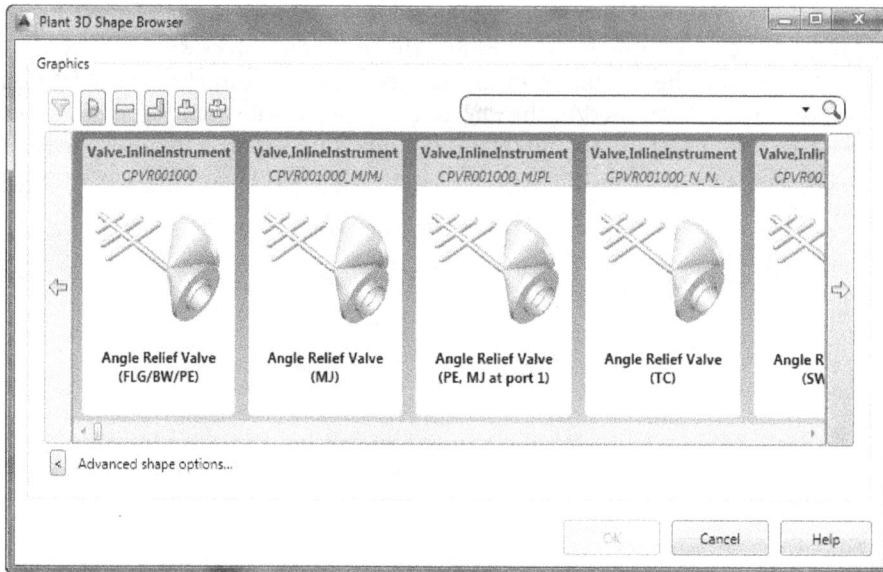

*Figure 7-10 The **Plant 3D Shape Browser** dialog box*

Mapping a P&ID object onto Plant 3D object

Before placing an object from a P&ID into a Plant 3D model, it has to be mapped to the corresponding Plant 3D object. This is the prerequisite to place a P&ID object. Most of the P&ID objects are mapped to the Plant 3D object, by default. However, you need to map a newly created P&ID object to the Plant 3D object. For example, when you place a control valve in P&ID, you need to select the required valve body and actuator. This makes the control valve a new object type in the P&ID and you need to map it to a Plant 3D object. To map a P&ID object onto a Plant 3D object, follow the steps given next.

1. Choose the **Project Setup** tool from the **Project** drop-down in the **Project** panel of the **Home** tab from the **Ribbon**; the **Project Setup** dialog box will be displayed.

2. In this dialog box, expand the **Plant 3D DWG Settings** node and select the **P&ID object Mapping** option; the **P&ID Object Mapping** pane will be displayed at the right side of the dialog box, refer to Figure 7-11.

3. In the **P&ID Object Mapping** pane, expand the **P&ID Classes** tree and then select the required class; the properties of the corresponding Plant 3D object will be displayed in the **Plant 3D Class** area.

4. Choose the **Add** button from the **Plant 3D Class** area; the **Select Plant 3D Class Mapping** dialog box will be displayed.

5. In this dialog box, expand the **Plant 3D Classes** tree and then select the Plant 3D class. Note that the class should be same as that you have selected from the **P&ID Classes** tree. For example, if you have selected **Control Valve** from **Engineering Items > Instrumentation**

> Inline Instruments node in the **P&ID Classes** tree, you need to select the **Valve** class from **Pipe and Equipment > Pipe run Components** in the **Plant 3D Classes** tree.

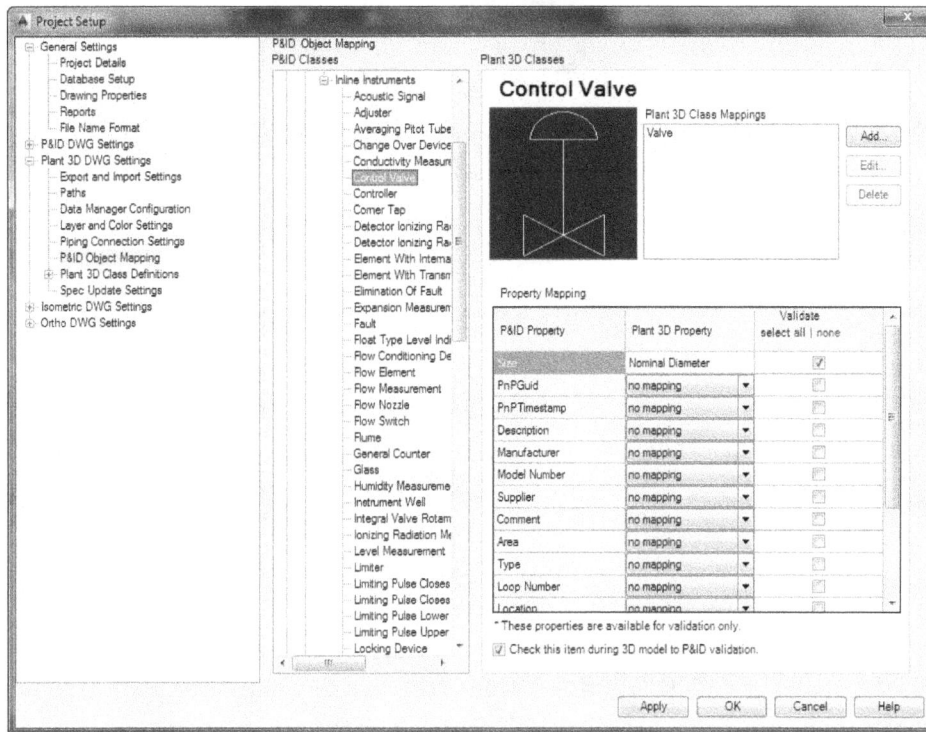

*Figure 7-11 The **Project Setup** dialog box*

6. Next, select the required check box from the **Map to one or more specific subtypes of this class** list box in the **Properties** area, refer to Figure 7-12. If you want to select all the sub types, then select the **Map to all subtypes of this class** check box.

7. Choose the **OK** button; the **Select Plant 3D Class Mapping** dialog box will be closed and the **Project Setup** dialog box will be displayed.

   Next, you need to set the properties that are to be mapped.

8. In the **Project Setup** dialog box, select appropriate values from the drop-down lists located next to the properties in the **Property Mapping** table of the **Plant 3D Classes** area. For example, you need to select the **Actuator Type** option from the drop-down list next to the **Actuator Type** property.

9 Next, select the required check boxes in the **Validate** Column in the **Property Mapping** table. The corresponding properties will be checked while running the validation.

10. Choose the **OK** button to close the **Project Setup** dialog box.

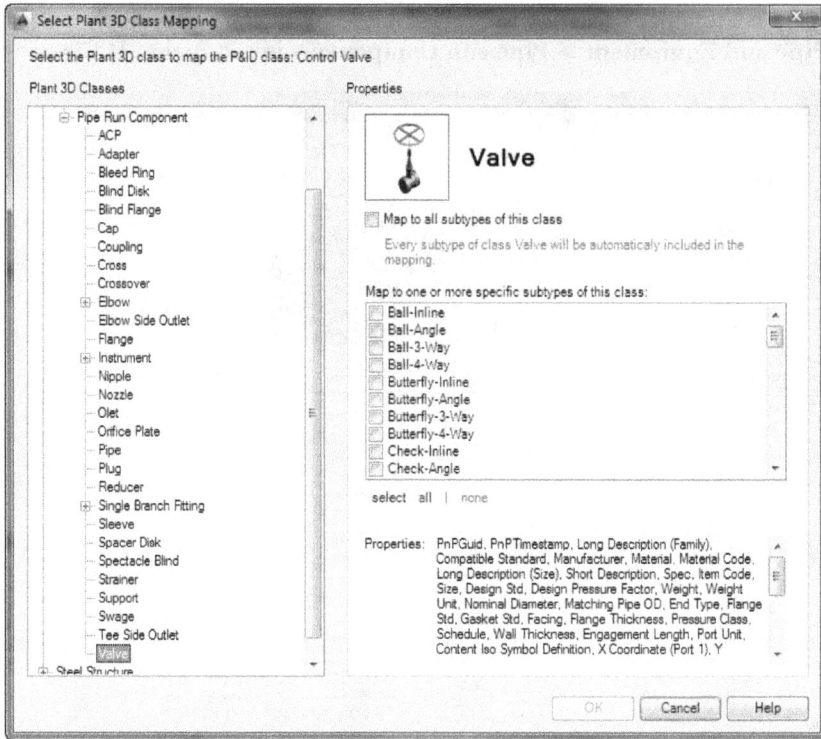

Figure 7-12 The Select Plant 3D Class Mapping dialog box

PIPE SUPPORTS

Create

To add a piping support, choose the **Create** button from the **Pipe Supports** panel in the **Home** tab; the **Add Pipe Support** dialog box will be displayed, as shown in Figure 7-13.

Next, select a pipe support from the **Graphics** area. You can filter the display of pipe supports by using the filter buttons present at the top left of the dialog box. For example, if you want to display only the base supports, choose the **Base Supports** button. You can further filter base supports by using the search bar present at the top right of the dialog box. To do so, click on the down arrow present in the search bar; a drop-down list will be displayed with four options: **Bolted Supports**, **Clamped Supports**, **Rolled Supports**, and **Spring Supports**. Select the required option from the drop-down list; the pipe supports will filter as per the option selected in the drop-down list. After selecting the required pipe support, choose the **OK** button; the selected pipe support will attach to the cursor and you will be prompted to select an insertion point. Select a point on the pipe where you want to place the support. Make sure that the **Object Snap** is turned on for easy selection of the point. Next, press ENTER to exit the tool. Figure 7-14 shows a Side Clamped Stanchion placed as a pipe support.

You can also add a pipe from the **Pipe Support Spec** tab. To do so, choose the **Pipe Support Spec** tab from the Tool Palettes present at the right-side of the drawing window; the pipe supports will be displayed, as shown in Figure 7-15. Scroll through the Tool Palette and select the required pipe support from it.

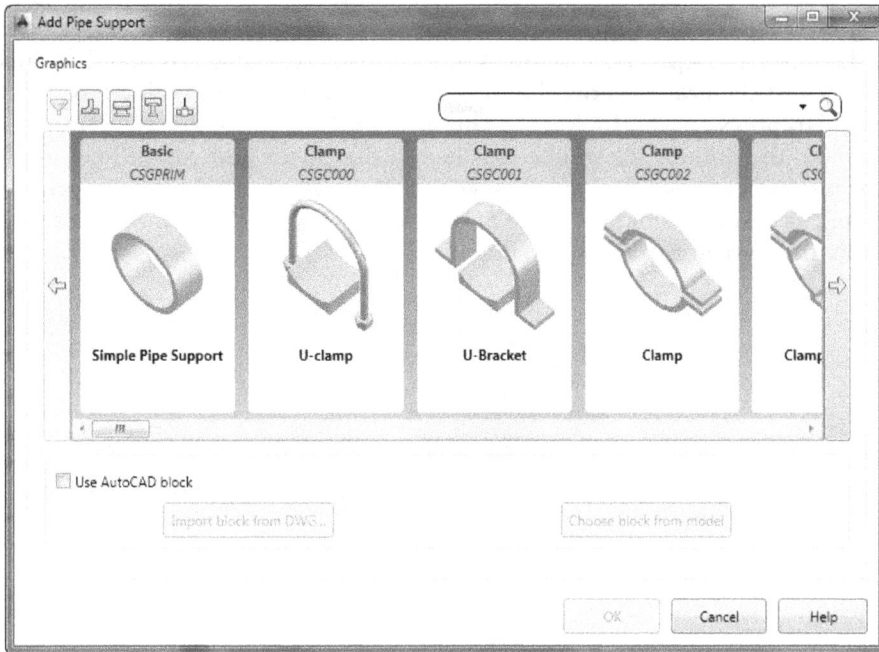

Figure 7-13 The **Add Pipe Support** *dialog box*

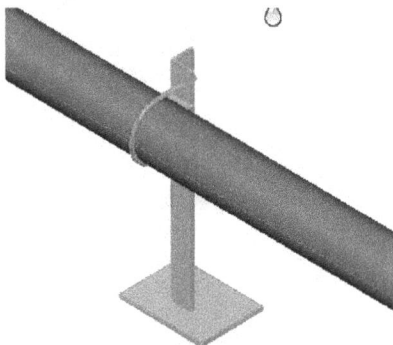

Figure 7-14 A Side Clamped Stanchion placed as a pipe support

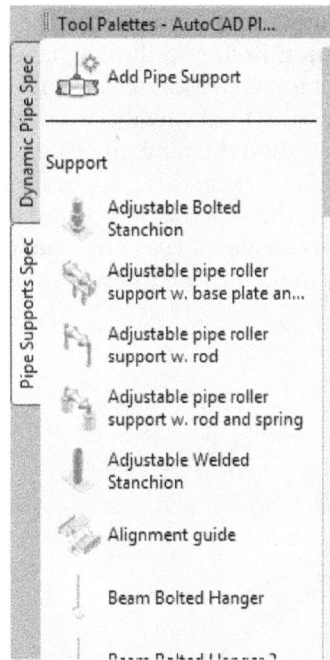

Figure 7-15 Pipe Supports displayed in the Tool Palettes

The methods to add and modifying the pipe supports are discussed next.

Adding a Dummy Leg

To add a dummy leg to a pipe, invoke the **Add Pipe Support** dialog box and then choose the **Dummy legs and General Supports** button from the top left of the dialog box. Next, scroll through the **Graphics** area and select the **Dummy Leg** from it. Choose the **OK** button from the dialog box; the dummy leg will be attached to the cursor and you will be prompted to specify the insertion point. Select an elbow connection point from the drawing area, refer to Figure 7-16. For easy selection, press the SHIFT key and right-click. Now, choose the **Node** option from the shortcut menu displayed.

Figure 7-16 *A Dummy Leg placed at the elbow*

Adding a Hanger and Connecting it to a Structural Member

To add a hanger to a pipe, invoke the **Add Pipe Support** dialog box and then choose the **Hangers** button from it to display hangers. Next, select a hanger and choose **OK**. Place the hanger at the required location and press ENTER. After placing the hanger, you can connect it to an existing structural member located at the top of the pipe. To do so, select the hanger that you have placed on the pipe; the Change Support Elevation grip will be displayed, as shown in Figure 7-17. Click on this grip, drag, and then snap to the point on the structural member located above the pipe, as shown in Figure 7-18; the hanger will connect to the structural member. You can also specify an elevation value in the Dynamic Input boxes. To do so, press the TAB key and enter a required value in the dynamic input. Next, press ENTER.

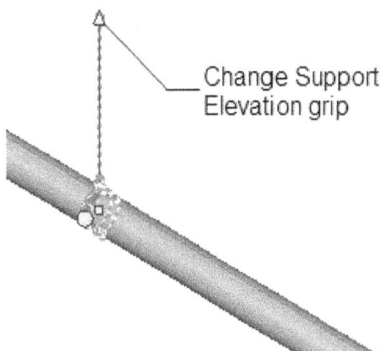

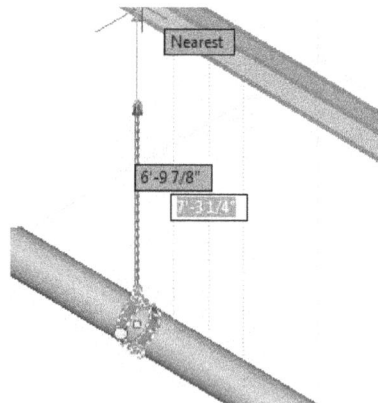

Figure 7-17 *The Change Support Elevation grip*

Figure 7-18 *Dragging the Change Support Elevation grip*

Modifying the Pipe Supports

To modify the pipe supports, right-click on the pipe support in the drawing area; a shortcut menu will be displayed. Choose the **Properties** option from the shortcut menu displayed; the **Properties** palette will be displayed. In the **Properties** palette, scroll down to the **Dimensions** area and edit the dimensions. The dimension details will be shown in the **Preview** window of the **Part Geometry** area, as shown in Figure 7-19. Click in the **Preview** window in order to expand the view. After modifying the dimensions, press ENTER; the pipe support will be modified in the drawing area. Figure 7-20 shows a spring hanger with default dimensions and Figure 7-21 shows the hanger after modifying the dimensions.

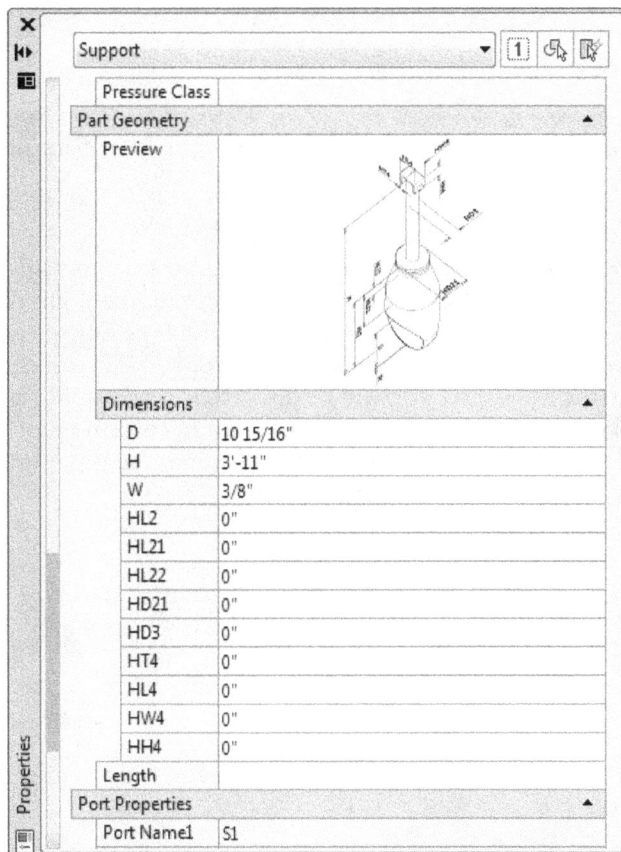

Figure 7-19 The **Part Geometry** area of the **Properties** palette

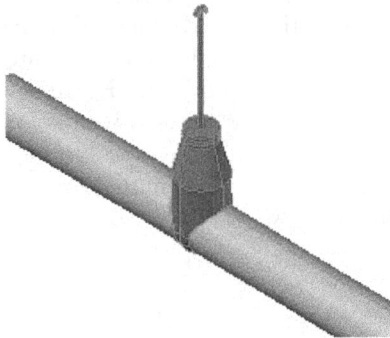

***Figure* 7-20** *A spring hanger with default dimensions values*

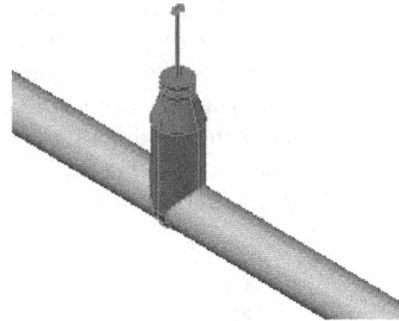

***Figure* 7-21** *The spring hanger after modifying dimension values*

Copying and Moving a Pipe Support

In AutoCAD Plant 3D, you can copy and move an existing pipe support from one location to another and also maintain the contact with the point of support. To do so, choose the **Toggle Lock of Support mode** button from the **Pipe Supports** panel in the **Home** tab; the pipe support will be locked to the point of support. Next, select the pipe support to be copied or moved; the Move Part grip will be displayed, as shown in Figure 7-22. Select the Move Part grip and move the pipe support. If you want to copy the pipe support, enter **P** at Command prompt and then copy it to the desired location; the pipe support will be copied. Figure 7-23 shows an original part and copied part.

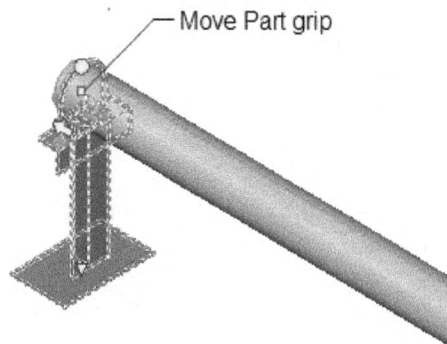

***Figure* 7-22** *The Move Part grip displayed on the pipe support*

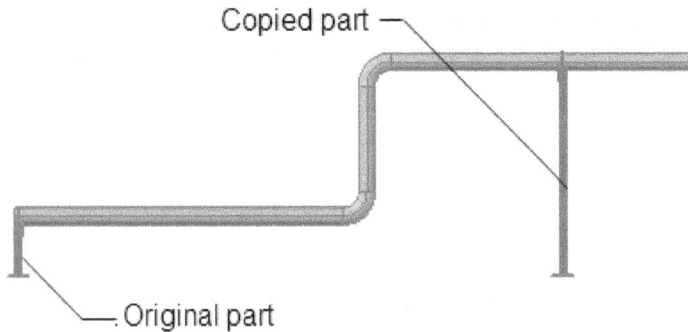

Figure 7-23 *The original pipe support and the copied one*

The **Toggle Lock Point of Support mode** button is much more useful when you are placing a pipe support for a sloped pipe. If you move a pipe support with the **Toggle Lock Point of Support mode** button chosen, height of the pipe support gets updated automatically, refer to Figure 7-24.

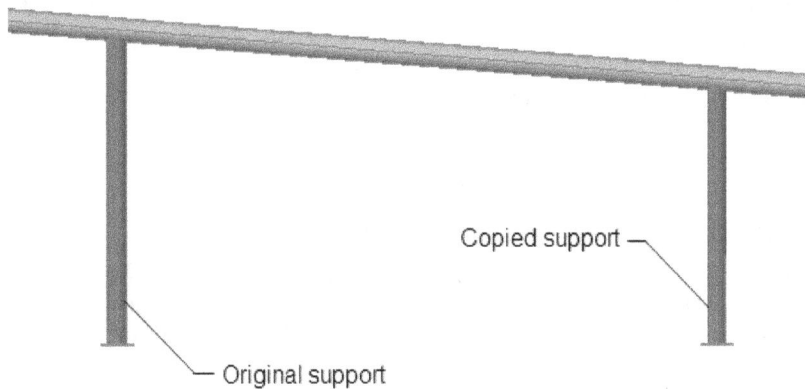

Figure 7-24 *The original pipe support and the copied one*

Connecting Two Pipe Supports

To connect two pipe supports, choose the **Support on Supports** button from the **Pipe Supports** panel in the **Home** tab; you will be prompted to select pipe supports to connect together. Select two or more pipe supports to be connected and then press ENTER; the selected pipe supports will be connected to each other.

Converting Solids to Pipe Supports

To convert a solid model into a pipe support, you need to create a solid model by using tools that are available in the **Modeling** tab. To do so, create a solid model supporting a pipe, as shown in Figure 7-25. Next, choose the **Convert Supports** tool from the **Pipe Supports** panel; you will be prompted to select the solid model. Select the solid model and press ENTER; you will be prompted to specify an insertion point on the pipe or pipe component. Use object snap options to select a point on the pipe or pipe component. After selecting a point on the pipe, the solid will be converted into a pipe support.

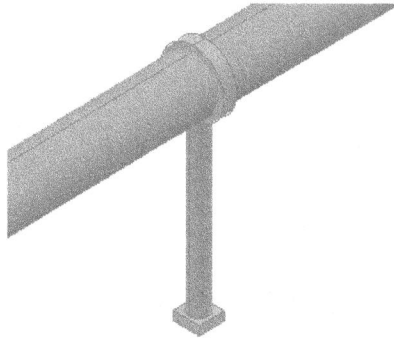

Figure 7-25 *A solid converted into a pipe support*

Attaching Objects to a Pipe Support

To attach objects to a pipe support, choose the **Attach Supports** tool from the **Pipe Supports** panel; you will be prompted to select a pipe support. Select the pipe support to which the object will be attached; you will be prompted to select objects that are to be added to the pipe support. Select the objects from the drawing area, and then press ENTER; the selected object will be attached to the pipe support and become its integral part.

Detaching Objects From the Support

To detach previously added objects from a support, choose the **Detach Supports** tool; you will be prompted to select the support from which the objects will be detached. Select a pipe support with attached objects; the attached objects will be detached from the pipe support.

INSULATING A PIPE

You can insulate a pipe using the **Properties** palette. To do so, invoke the **Properties** palette of the pipe by double-clicking on it. Next, scroll to the **Process Line** area in the **Plant 3D** rollout in the **Properties** palette and select an option from the **Insulation Type** drop-down list, refer to Figure 7-26. Select the insulation thickness from the **Insulation thickness** drop-down list which is located above the **Insulation Type** drop-down list. After applying the insulation, you need to choose the **Toggle Insulation Display** button from the **Visibility** panel to turn on the insulation display.

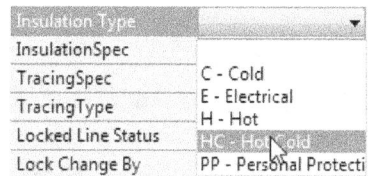

Figure 7-26 *Selecting the option from the* ***Insulation Type*** *drop-down list*

MODIFYING THE PIPE COMPONENTS USING GRIPS

You can modify pipes and its components by using the grips, which are displayed on selecting the pipe components. The various methods to modify pipes using grips are discussed next.

Substituting a Pipe Component

To substitute a pipe component with a new pipe component, select it from the drawing area; the Substitute Part grip will be displayed, as shown in Figure 7-27. Click on the Substitute Part grip; a flyout will be displayed with a list of substitute components. Choose the required substitute component from the flyout; the existing component will be replaced with the newly selected component.

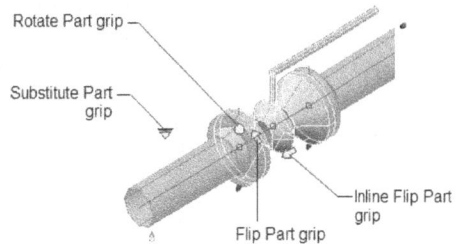

Figure 7-27 *The Substitute Part grip displayed on the selected component*

Rotating a Pipe Component

To change the angular orientation of the pipe component, select the pipe component from the drawing area; the Rotate Part grip will be displayed, refer to Figure 7-27. Click on the Rotate Part grip; the compass will be invoked. Next, use the compass to rotate the pipe component and then press ENTER; the component will be reoriented.

Flipping a Pipe Component

To flip a pipe component, select it from the drawing area to display the Flip Part grip, refer to Figure 7-27. Click on the Flip Part grip; the pipe component will be flipped.

Flipping a Component Inline with the Pipe

To flip a pipe component inline with the pipe, select; the Inline Flip Part grip will be displayed, refer to Figure 7-27. Click on this grip to flip the component in line with the pipe, refer to Figure 7-28 and 7-29.

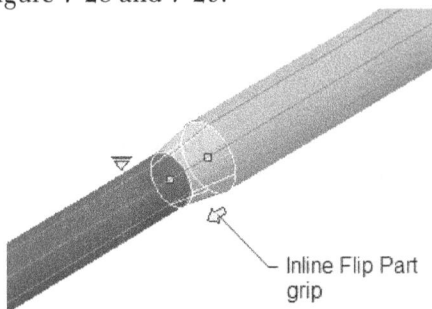

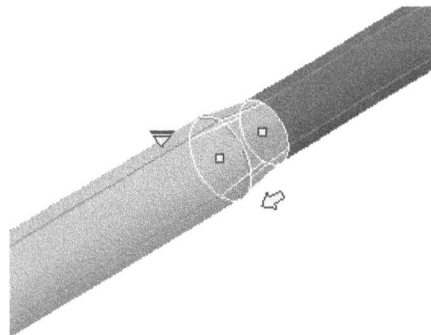

Figure 7-28 *The Inline Flip Part grip displayed on the pipe component*

Figure 7-29 *The component after flipping it in the inline direction*

Changing the Elevation of the Pipe

To change the elevation of a pipe, select it from the drawing area to display the Change Elevation grip, as shown in Figure 7-30. Click on this grip and move the cursor to the required elevation. You can also enter an elevation value in the dynamic input box. Figure 7-31 shows the pipe with the changed elevation.

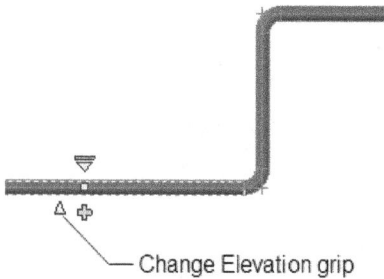

Change Elevation grip

Figure 7-30 The Change Elevation grip displayed on the pipe

Figure 7-31 The pipe with the changed elevation

Changing the Valve Operator

After placing the valve in a pipe, you may need to change the valve operator. To do so, select the valve from the drawing area. Next, right-click and choose the **Properties** option from the shortcut menu displayed; the **Properties** palette will be displayed. In the **Properties** palette, expand the **Part Properties** rollout and then expand the **Valve Operator** sub-rollout. Next, select the required operator type from the **Operator** drop-down list, as shown in Figure 7-32; the selected valve operator will replace the existing one.

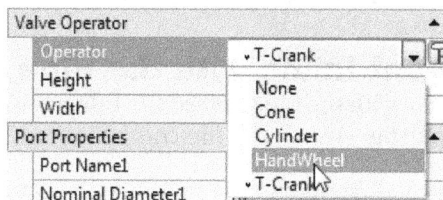

*Figure 7-32 Selecting an option from the **Operator** drop-down list*

You can also select an operator by invoking the **Override Valve Operator** dialog box. To do so, choose the button next to the **Operator** drop-down list; the **Override Valve Operator** dialog box will be displayed, as shown in Figure 7-33.

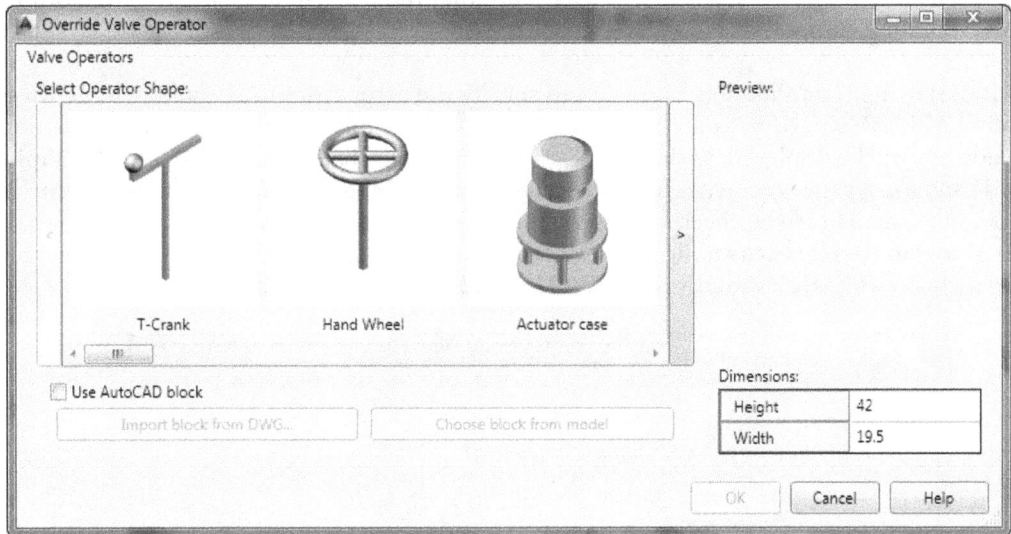

*Figure 7-33 The **Override Valve Operator** dialog box*

Select an operator from the **Select Operator Shape** area and then specify the dimensions in the **Dimensions** area. Next, choose the **OK** button; the new operator will replace the old one. You can also use a block to override an existing operator. To do so, select the **Use AutoCAD block** check box present below the **Select Operator Shape** area. Next, choose the **Import block from DWG** or the **Choose block from model** button to import a block or select it from an existing model. Figure 7-34 shows a gate valve with the hand wheel operator and Figure 7-35 shows the valve after changing the operator to T-Crank.

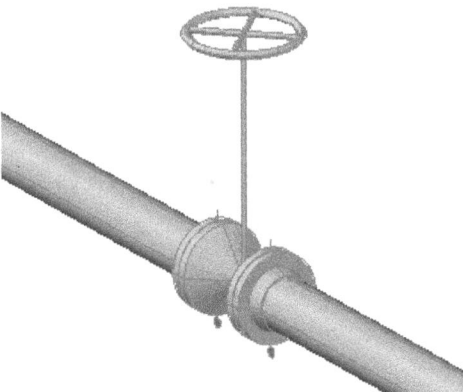

Figure 7-34 Gate valve with a Hand Wheel operator

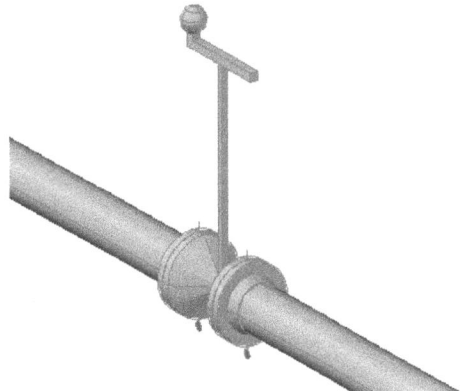

Figure 7-35 Valve with a T-Crank operator

VALIDATING THE 3D MODEL

In Chapter 2, you learned to validate a P&ID. Similarly, you need to validate the 3D model and correct the errors in it. In addition, you also need to validate the 3D model against the P&ID. It means that you need to make sure that the parts in 3D model match with the corresponding components in P&ID.

Before starting the validation, you need to specify the error types to be validated. To do so, enter VALIDATECONFIG command at the Command prompt; the **P&ID Validation Settings** dialog box will be displayed, as shown in Figure 7-36. In this dialog box, expand the **3D Piping** node and specify the error types to be validated in the 3D model. Next, you need to specify the types of mismatches to be checked between the 3D model and the P&ID. To do so, expand the **3D Model to P&ID checks** node and select the check boxes of the mismatches to be checked. Next, choose the **OK** button from the **P&ID Validation Settings** dialog box.

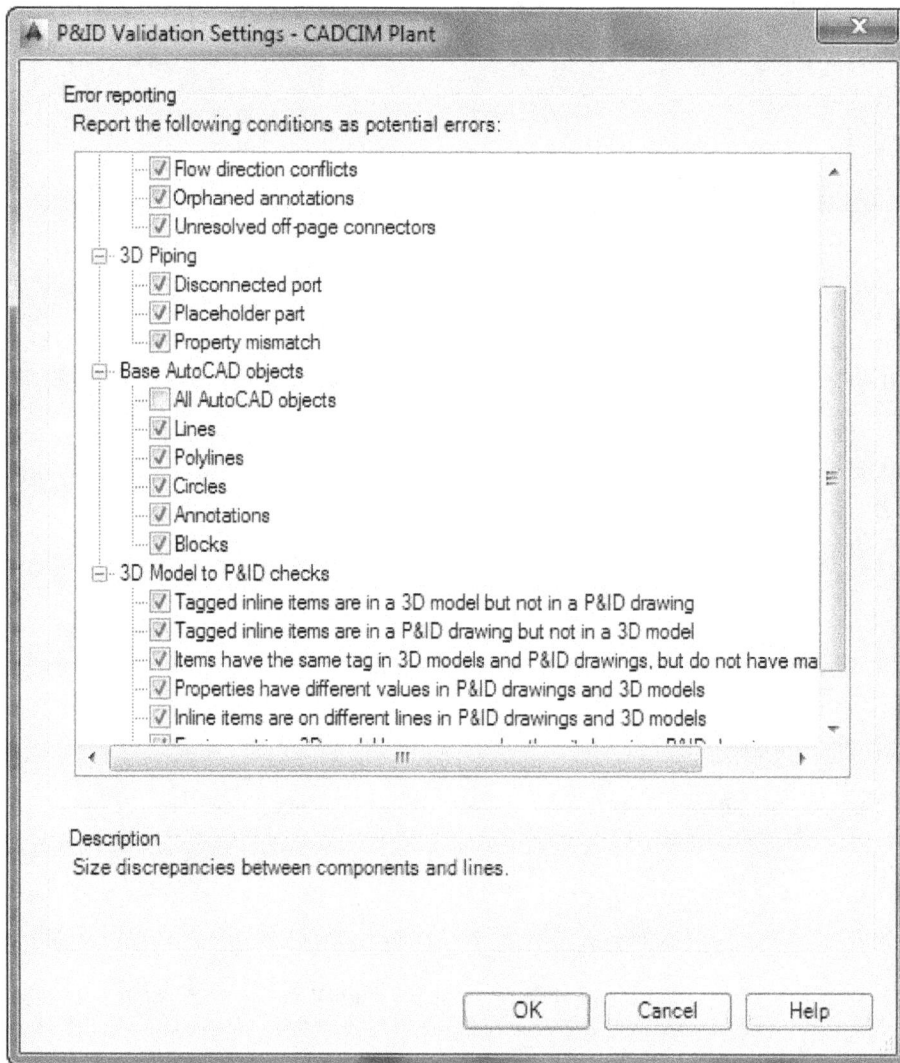

*Figure 7-36 The **P&ID Validation Settings** dialog box*

After specifying the validation settings, expand the **Plant 3D Drawings** node in the **Project Manager**. Next, right-click on the drawing file to be validated and choose the **Validate** option from the shortcut menu displayed; the **Validation Progress** message box will be displayed and the drawing will be validated. The **Validation Summary** window will be displayed after validation is completed. Select an error from **Validation Summary** window; the error type and the action to be taken are displayed in the **Details** area. Ignore or correct the error. To ignore the error, select the **Don't display errors marked as ignored** check box; the error will be ignored.

TUTORIAL

Tutorial 1

In this tutorial, you will open the **Piping_Model** file from the **CADCIM** project located in the **Project Manager** and place valves and fittings in the model space. You can also download this model from www.cadcim.com by following the path: *Home > Textbooks > CAD/CAM > AutoCAD Plant 3D > AutoCAD Plant 3D 2014*. Figures 7-37 through 7-40 show the locations of the valves and valve list. **(Expected time: 45 min)**

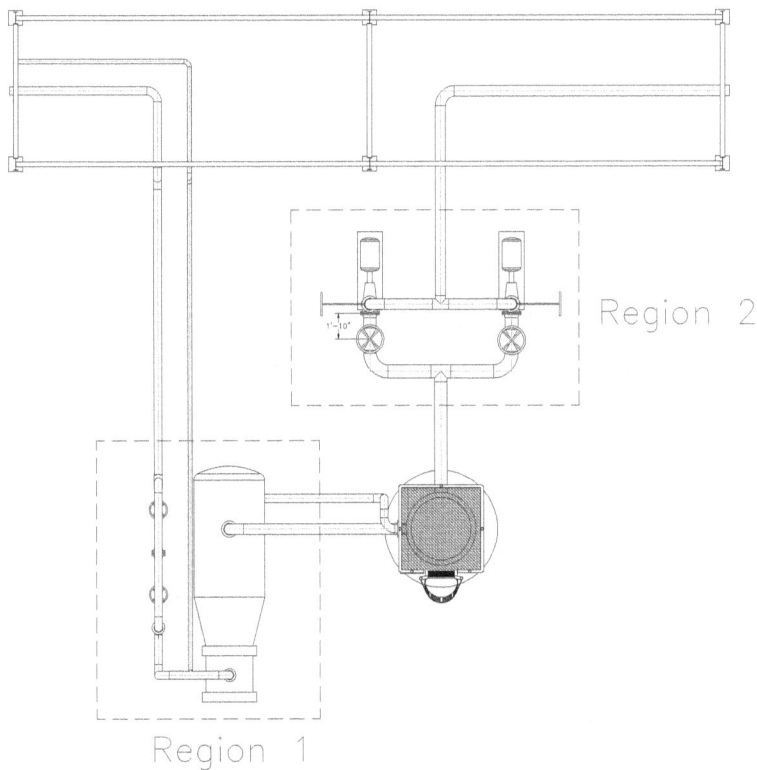

Figure 7-37 Top view of the model

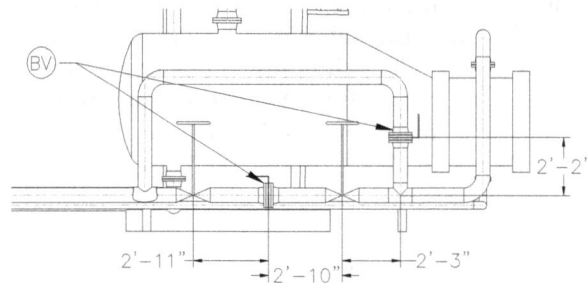

Figure 7-38 *Valves in Region 1*

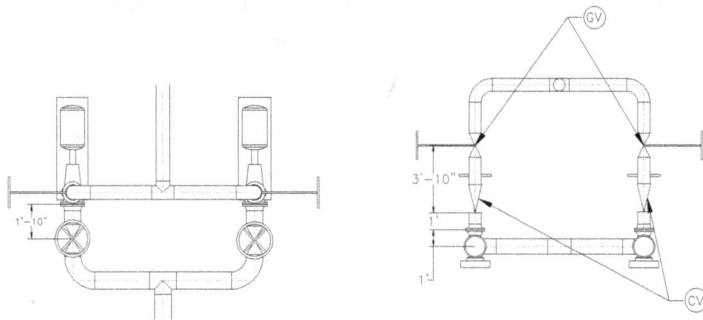

Figure 7-39 *Valves in Region 2*

Valve List			
	Valve Type	Line Number	Actuator Type
BV	Butterfly Valve	2000	Hand Lever
	Butterfly Valve	2000	Hand Lever
CV	Check Valve	1012	
	Check Valve	1012	
GV	Gate Valve	1012	Hand Wheel
	Gate Valve	1012	Hand Wheel
	Gate Valve	1004	Hand Wheel
	Gate Valve	1004	Hand Wheel
	Gate Valve	2000	Hand Wheel
	Gate Valve	2000	Hand Wheel

Figure 7-40 *The Valve list*

The following steps are required to complete this tutorial:

a.   Open the **Piping_Model** from the **Project Manager**.
b.   Select the required spec and select the pipe size.

c. Place valves at the required locations.
d. Save the file.

Opening the File

1. Choose **Start > All Programs (or Programs) > Autodesk > AutoCAD Plant 3D 2014 > AutoCAD Plant 3D 2014**; AutoCAD Plant 3D gets started and the welcome screen is displayed.

2. Select **CADCIM** from the **Current Project** drop-down list in the **Project Manager**.

3. Expand the **Plant 3D Drawings** node in the **Project** area and double-click on **Piping_Model**; the selected drawing file is opened.

Selecting the Spec and the Pipe Size

In this section, you will select a spec and specify the pipe size.

1. Select *CS150* from the **Spec Selector** drop-down list in the **Part Insertion** panel; the **Dynamic Tool Palette** on the left side of the window is loaded with *CS150* piping components.

Placing Valves Using the P&ID Line List

In this section, you will place valves in the Plant 3D model with reference to **P&ID1** and **P&ID2** files of the **CADCIM Plant** project. Before that, you need to map the new P&ID objects that you have placed in P&ID.

1. Choose the **Project Setup** tool from the **Project** drop-down in the **Project** panel; the **Project Setup** dialog box is displayed.

2. In this dialog box, expand the **Plant 3D DWG Settings** node and select the **P&ID Object Mapping** option; the **P&ID Object Mapping** pane is displayed at the right side of the dialog box.

3. In the **P&ID Object Mapping** pane, select **Control Valve** from **Engineering Items > Instrumentation > Inline Instruments** node in the **P&ID Class** tree; the properties of the corresponding Plant 3D object are displayed in the **Plant 3D Classes** area.

4. Choose the **Add** button from the **Plant 3D Class** area; the **Select Plant 3D Class Mapping** dialog box is displayed.

5. In this dialog box, select the **Valve** class from **Piping and Equipment > Pipe Run Components** in the **Plant 3D Classes** tree.

6. Select the **Map to all subtypes of this class** check box and choose the **OK** button to close the **Select Plant 3D Class Mapping** dialog box.

7. In the **Project Setup** dialog box, select *Actuator Type—wait.

7. In the **Project Setup** dialog box, select ***Actuator Type** option from the drop-down list next to the **Actuator Type** property in the **Property Mapping** table. Similarly, select the **Valve Body Type** option for the body type.

8. Select the **Body Type** and **Actuator Type** check boxes in the **Validate** Column in the **Property Mapping** table. Make sure that the **Check this Item during 3D model to P&ID validation** check box is selected.

9. Choose the **OK** button to close the **Project Setup** dialog box.

    Next, you need to place a valve using a P&ID.

10. Choose the **P&ID Line List** button from the **Part Insertion** panel in the **Home** tab; the **P&ID Line List** window is displayed.

11. Select **P&ID1** from the drop-down list located at the top of the window; all the line numbers in this P&ID are listed in the form of tree.

12. Select the **Control Valve 02-CV-1002** from **2000 > 6"-CS150-P-2000** in the **P&ID Line list** tree. Figure 7-41 shows the corresponding pipe in the P&ID.

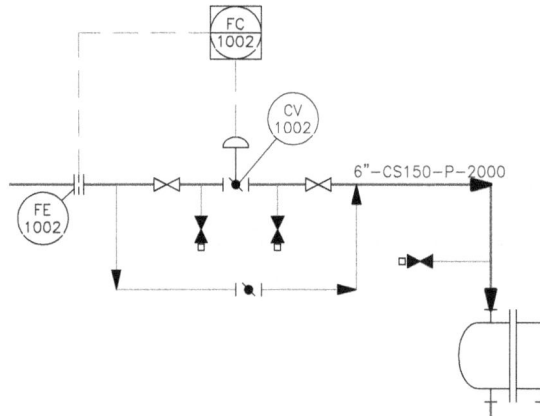

Figure 7-41 The selected pipe in the P&ID

13. Choose the **Place** button from the **P&ID Line List** window; the **Select Size and Spec** dialog box is displayed.

14. In this dialog box, select **CS150** from the **Spec** list box and then select **6"** from the **Size** list box.

15. Choose the **Select** button from the **Select Size and Spec** dialog box; you are prompted to specify the insertion point.

16. Select a point on the pipe connecting the reboiler, refer Figure 7-42; the control valve is placed at the specified point and you are prompted to specify the rotation angle.

17. Type **0** at the Command prompt and press ENTER. Figure 7-43 shows the control valve after specify the rotation angle.

    You may notice that the valve body in the P&ID is different from that you have placed. Next, you need to change the valve body.

18. Click on the valve body and select the **Substitute Part** grip displayed on it, refer to Figure 7-44; a flyout is displayed showing the list of parts that can be substituted.

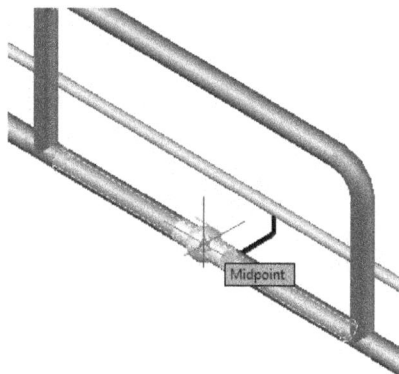

Figure 7-42 Selecting the midpoint of the pipe connecting the reboiler

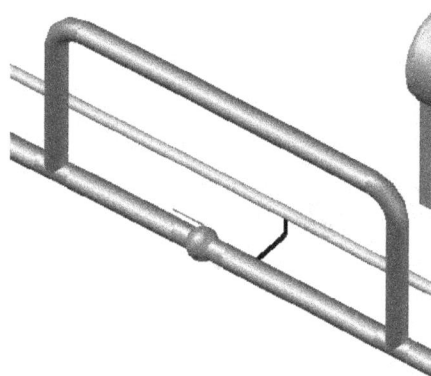

Figure 7-43 Control Valve after specifying the rotation angle

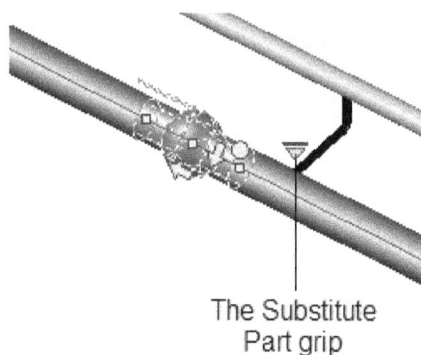

Figure 7-44 The Substitute Part grip displayed on the valve body

19. Choose the **6" Butterfly Valve** from the flyout, refer to Figure 7-45; the valve body of the control valve is changed to butterfly valve, as shown in Figure 7-46.

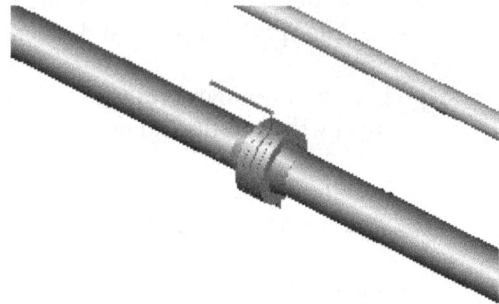

✓   6" BALL VALVE, LONG PATTERN, 6" ND, 150 LB, BW,
    6" BALL VALVE, LONG PATTERN, 6" ND, 150 LB, RF, .
    6" BUTTERFLY VALVE, OFFSET, 6" ND, 150 LB, LUG, F
    6" BUTTERFLY VALVE, OFFSET, 6" ND, 150 LB, WFR, I
    6" CHECK VALVE, SWING, 6" ND, 150 LB, BW, ASME

Figure 7-45 *Selecting the butterfly valve from the flyout*

Figure 7-46 *The control valve after changing the valve body to butterfly valve*

Placing Gate Valves

Next, you need to place the gate valves on the same pipe, refer to Figure 7-41.

1. Invoke the **P&ID Line list** window and select anyone of the **Gate Valve** from **2000 > 6"-CS150-P-2000** in the **P&ID Line list** tree.

2. Place a gate valve at the location as shown in Figure 7-47.

3. Similarly, select another gate valve from **2000 > 6"-CS150-P-2000** in the **P&ID Line list** tree and then place it into the model, as shown in Figure 7-47.

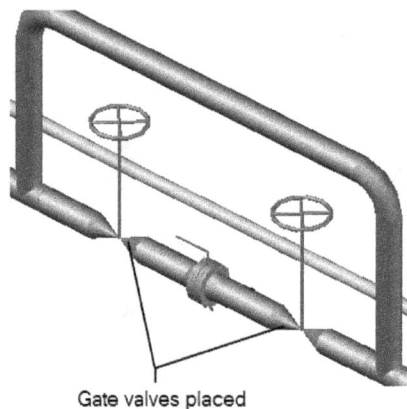

Gate valves placed

Figure 7-47 *Locations of the gate valves*

4. Next, select the **Butterfly Valve** from the **P&ID Line list**, as shown in Figure 7-48 and choose the **Place** button; the **Select 3D Class** dialog box appears. Select **Valve (Butterfly-Inline)** from the dialog box and then place it at the location shown in Figure 7-49.

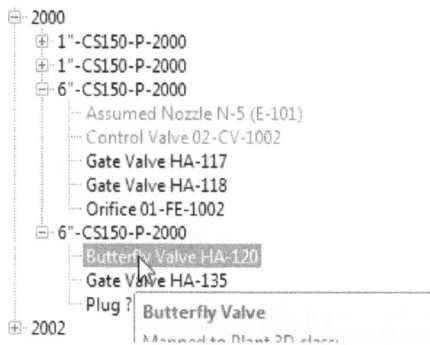

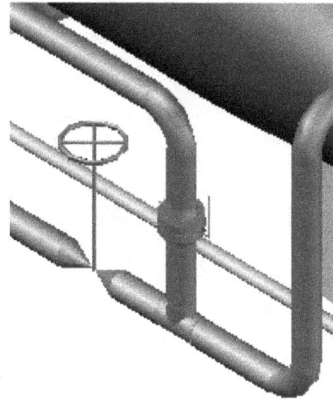

Figure 7-48 *Selecting the Butterfly Valve from the P&ID Line list tree*

Figure 7-49 *Location of the globe valve*

Placing Valves on the Pipes Connecting the Pumps

1. Select gate valves from the **P&ID Line list**, refer to Figure 7-50, and place them on the pipes connecting the pumps, as shown in Figure 7-51.

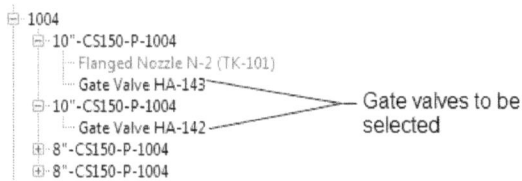

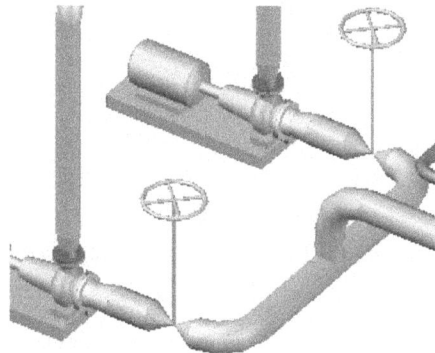

Figure 7-50 *Gate valves to be selected from the P&ID Line list*

Figure 7-51 *Locations of the gate valves*

2. Select the check valves from the **P&ID Line list**, refer to Figure 7-52 and place them on the pipes connecting the pumps, as shown in Figure 7-53.

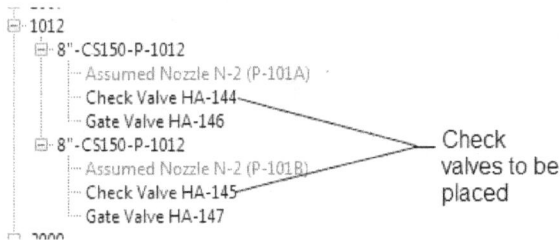

Figure 7-52 Check valves to be selected from the P&ID Line list

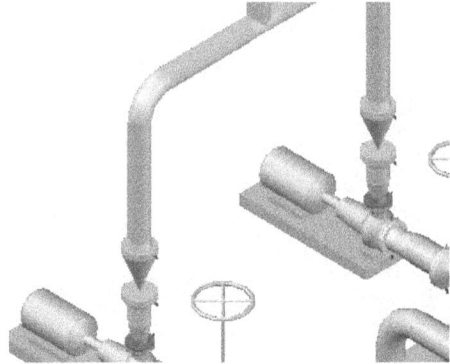

Figure 7-53 Location of the check valves

Note

You need to flip the direction of check valves if they are not placed in the required direction, as shown in Figure 7-53. To do so, select a check valve to display the Inline Flip Part grip on it, refer to Figure 7-54. Next, use this grip to flip the direction of the valve. Similarly, flip the direction of the other valve.

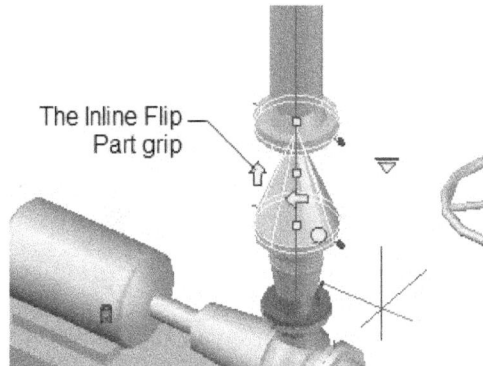

Figure 7-54 The Inline Flip Part grip displayed

3.   Similarly, select the gate valves from the **P&ID Line list** and place them on the pipes connecting the pumps, refer to Figure 7-55 and 7-56.

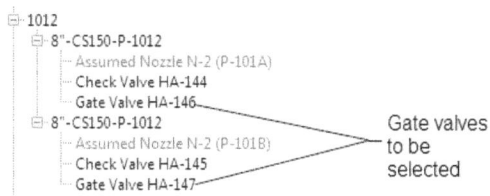

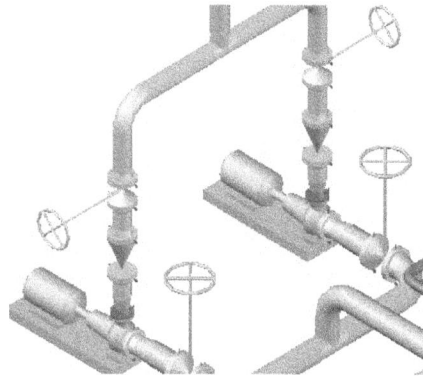

Figure 7-55 *Gate Valves to be selected from the **P&ID line list***

Figure 7-56 *Locations of the gate valves*

4. Choose the **Save** button from the **Application** menu and then the **Close** button to close the file.

Self-Evaluation Test

Answer the following questions and then compare them to those given at the end of this chapter:

1. If the selected component from **P&ID Line list** is not available in the spec, the _____ dialog box will be displayed.

2. The _____ dialog box is used to change the valve operator.

3. To modify a pipe support, you need to invoke its _____ palette.

4. You can maintain the contact between the pipe support and point of support with the _____ button turned on.

5. The _____ tool is used to connect two pipe supports.

6. To insulate a pipe, invoke its _____ palette and select the insulation type from the _____ drop-down list.

7. To substitute a pipe component with a new one, right-click on it and choose the _____ grip.

8. You can modify pipes and their components by using the grips, which are displayed on selecting the components. (T/F)

9. Validating can also be done between a P&ID and a 3D model. (T/F)

10. You can also place a valve or a fitting while routing a pipe. (T/F)

Review Questions

Answer the following questions:

1. Before placing an object from a P&ID into a Plant 3D model, it has to be mapped to the corresponding Plant 3D object. (T/F)

2. You can insulate a pipe using its **Properties** palette. (T/F)

3. To route a pipe using a P&ID, the Plant 3D model and the P&ID should be placed in the same project. (T/F)

4. You can place a custom part into a Plant 3D model using the _____ dialog box.

5. To place a placeholder part, choose the _____ option from the **Custom Part type** drop-down list in the **Custom Part Builder** window.

6. The _____ tool is used to convert a solid model into a pipe support.

7. The _____ tool is used to attach objects to a pipe support.

Answers to Self-Evaluation Test
1. Select Size and Spec, 2. Override Valve Operator, 3. Properties, 4. Toggle Lock of Support mode, 5. Support on Supports, 6. Properties, Insulation Type, 7. Substitute Part, 8. T, **9.** T, **10.** T.

Chapter 8

Creating Isometric Drawings

Learning Objectives

After completing this chapter, you will be able to:
- *Understand various types of isometric drawings*
- *Create a quick isometric drawing*
- *Create a production isometric drawing*
- *View isometric results*
- *Create iso messages*
- *Export a Part Component file*
- *Create an iso from a Part Component file*
- *Lock and unlock lines*
- *Configure the isometric drawing settings*

ISOMETRIC DRAWINGS

Isometric drawings are created to make it easy to visualize the 3D pipe lines, which is a challenging task in an orthographic drawing. Also, the isometric drawings are used for the fabrication of pipes. In this chapter, you will learn how to create and modify an isometric drawing. Also, you will learn to add iso messages and annotations such as insulation, flow arrows, and so on to the isometric drawings.

ISOMETRIC DRAWING TYPES

The most commonly used isometric drawing types are Check, Stress, Final, and Spool. These types are discussed next.

Check Isometric

The Check isometric drawing is created to check whether all the essential components are represented in the model or not. It also ensures that the Final isometric drawing is created without errors. You can also compare the details in the Check isometric drawing with P&ID.

Stress Isometric

The Stress isometric drawing presents the geometric data which is precise and relevant to the stress analysis of pipes. It is generally used to analyze the stress in pipelines. The pipelines that require stress analysis include high pressure lines, high temperature lines, lines with large pipe size and so on.

Final Isometric

The Final isometric drawing is the final product document created from the 3D piping model. It is generally produced at the last stage of a project. The final isometric drawing contains a bill of material (BOM) and is used for carrying out the fabrication and construction process.

Spool Drawings

A Spool drawing is the final isometric drawing separated in individual sections called spools. It is created for shop fabrication.

In this chapter, you will learn to create Quick Isometric drawings (Check Isometric) and the Production (Final) Isometric drawing.

CREATING A QUICK ISOMETRIC DRAWING

A Quick isometric drawing is created to check the piping before creating a Production isometric drawing. You can create a Quick isometric drawing by selecting pipes from the drawing area or from the line list of the **Project Manager**. To create a Quick isometric drawing, choose the **Quick Iso** tool from the **Iso Creation** panel in the **Isos** tab; you will be prompted to select components to create an isometric drawing. Select a component from the drawing area and press ENTER; the **Create Quick Iso** dialog box will be displayed, as shown in Figure 8-1. Next, choose the **Create** button to create a quick isometric drawing.

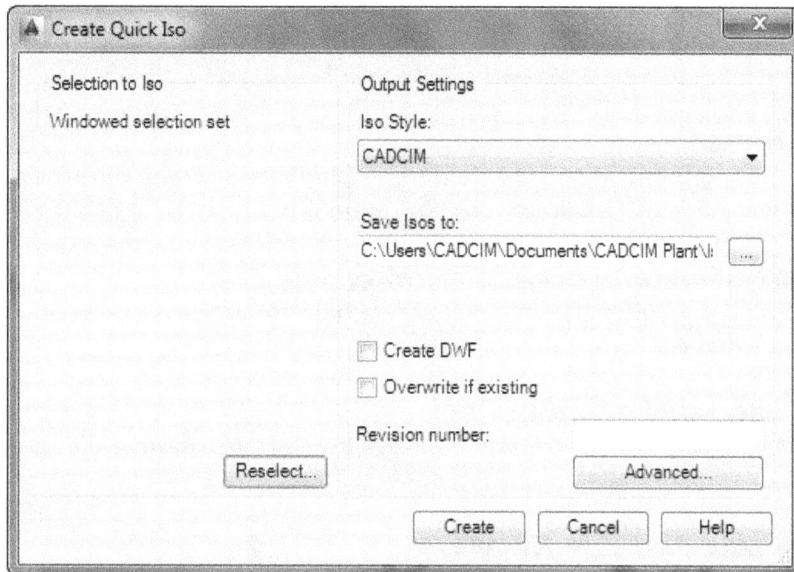

Figure 8-1 *The* **Create Quick Iso** *dialog box*

If you want to create a Quick isometric drawing of a particular pipe line, invoke the **Quick Iso** tool and choose the **Line number** option from the Command prompt; the **Create Quick Iso** dialog box will be displayed, as shown in Figure 8-2. The options in this dialog box are discussed next.

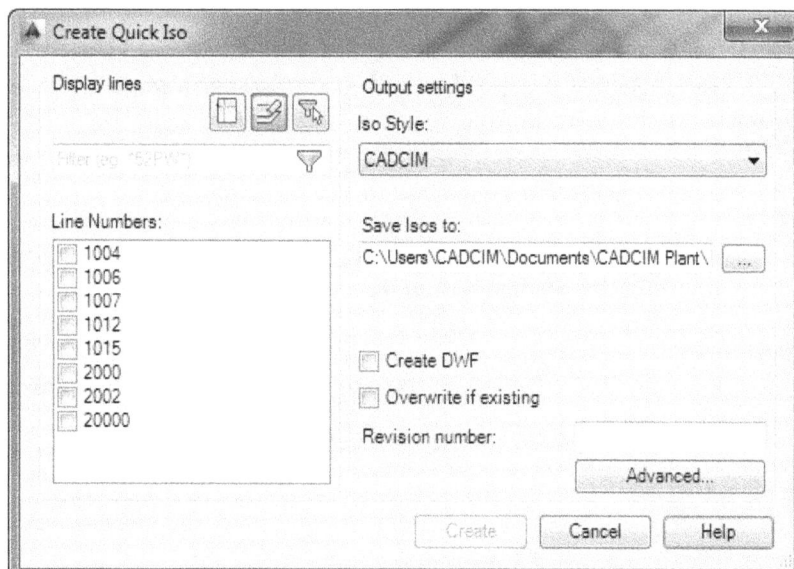

Figure 8-2 *The* **Create Quick Iso** *dialog box displayed on choosing the* **Line number** *option*

Selection to Iso

This area is displayed only when you select components from the drawing area. The **Reselect** button in this area is used to re-select the components from which the Iso is to be created.

Display lines

This area is displayed only when you choose the **Line number** option from the command prompt after invoking the **Quick Iso** tool. The options in this area are discussed next.

Show Line Numbers in current drawing only

On choosing this button, the line numbers of only the current drawing are displayed in the **Line Numbers** list box present in the **Display lines** area.

Show all Line Numbers in Project

On choosing this button, the line numbers from all the drawing files present in the project are displayed in the **Line Numbers** list box.

Show only selected Line Numbers

On choosing this button, only the selected lines are displayed in the **Line Numbers** list box. You can also enter the required line numbers in the **Filter** edit box present in the **Display lines** area to filter the line number list.

Output settings

This area is used to specify the settings of the isometric output file. The options in this area are discussed next.

Iso Style

This drop-down list is used to specify the style of the isometric drawing. You can select predefined isometric styles from it. You can also modify these styles by using the **Project Setup** dialog box.

Save Isos to

This display box displays the path of the output file. You can change the default location of the output file by choosing the **Browse** button next to it.

Create DWF

If you select this check box, you can create a Design Web Format (DWF) file from the output file.

Overwrite if Existing

This check box is used to overwrite the existing isometric file. On clearing this check box, the file will be renamed.

Revision number

This edit box is used to enter the revision number. The entered number will be displayed in the title block of the isometric drawing.

Advanced

On choosing this button, the **Advanced Iso Creation Options** dialog box will be displayed. You can use this dialog box to specify advanced settings while creating an isometric drawing.

After specifying the options in the **Create Quick Iso** dialog box, choose the **Create** button from it; the Quick isometric drawing will be created in the background and a message will be displayed at the status bar, as shown in Figure 8-3.

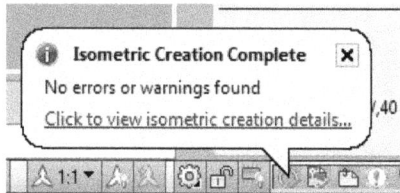

Figure 8-3 Message displayed at the status bar

CREATING A PRODUCTION ISOMETRIC DRAWING

To create a Production isometric drawing, choose the **Production Iso** button from the **Iso Creation** panel in the **Isos** tab; the **Create Production Iso** dialog box will be displayed. The options in this dialog box are same as in the **Create Quick Iso** dialog box.

In the **Create Production Iso** dialog box, select the lines from the **Line Numbers** list box in the **Display lines** area. Next, select the isometric style from the **Iso Style** drop-down list in the **Output settings** area. You can also select the **Create DWF** check box, if you want to create a Design Web Format (DWF) file from an isometric drawing. If you want to override the existing file with a new file, select the **Override if existing** check box. You can also specify advanced settings by choosing the **Advanced** button. After specifying all the settings, choose the **Create** button; the Production isometric drawing will be created and a message will be displayed at the status bar, refer to Figure 8-3.

Viewing Isometric Results

To view isometric results, double-click on the isometric icon located in the status bar; the **Isometric Creation Results** dialog box will be displayed, as shown in Figure 8-4. To open an isometric file, click on the file path displayed in the dialog box, refer to Figure 8-4; the isometric file will be opened.

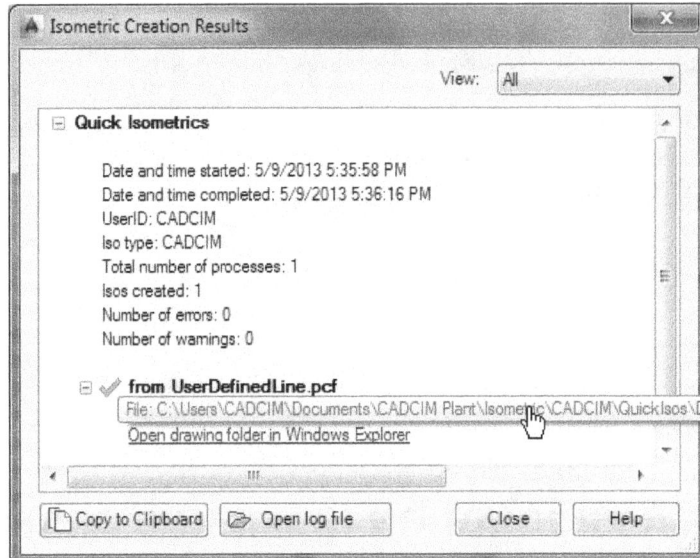

*Figure 8-4 The **Isometric Creation Results** dialog box*

PLACING ISO MESSAGES AND ANNOTATIONS

In AutoCAD Plant 3D, you can place additional information (Iso message) in a 3D model using the **Iso message** tool. The information will be displayed in the isometric drawing. Also, a sphere symbol will be displayed in the 3D model. To create an Iso message, choose the **Iso message** button from the **Iso Annotation** panel in the **Iso** tab; the **Create Iso Message** dialog box will be displayed, as shown in Figure 8-5. In this dialog box, select the message enclosure from the **Enclose message in** drop-down list; the selected message enclosure will be displayed in the preview area. Next, enter the text in the **Message** box. You can select the **Draw dimension to message** check box if you want to place a dimension for locating the message. Next, choose the **OK** button; the dialog box will be closed and you will be prompted to select the insertion point on a pipe or a fitting. After selecting the insertion point, a sphere will be added inside the pipe, as shown in Figure 8-6. You need to switch to the wireframe view, to view this sphere.

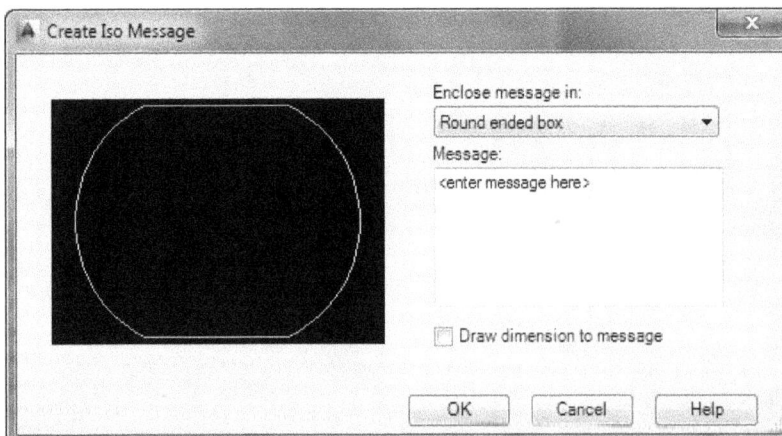

*Figure 8-5 The **Create Iso Message** dialog box*

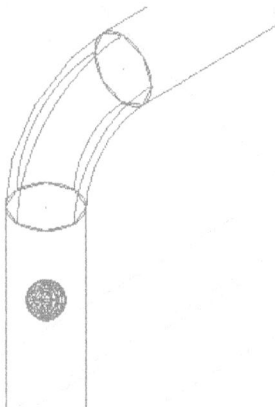

Figure 8-6 *The sphere displayed inside the pipe*

You can place annotation items such as **Floor Symbol**, **Flow Arrow**, **Insulation Symbol**, **Location Point**, and **Break Point** in a 3D model. These items will be displayed in an isometric drawing. To place an annotation item, select it from the **Iso Annotation** panel in the **Isos** tab; you will be prompted to specify the insertion point on a pipe or a fitting. After specifying the insertion point, a sphere will be displayed inside the pipe. However, if you are placing a **Flow Arrow** symbol, you need to specify the insertion point and then choose the **Accept** or **Reverse** option from the Command prompt.

EXPORTING A PIPE COMPONENT FILE

A Pipe Component File(PCF) is the primary input to create an isometric drawing. It is a text file containing information about pipe components and routing. A typical PCF contains information of pipe components (flanges, valve, and so on), coordinate values, and sizes of the components, and shapes to be used to represent the components in an isometric. There will be a separate PCF created for each selected line. Later, you can use this PCF to create an isometric file. To create a PCF, choose the **PCF Export** tool from the **Export** panel in the **Isos** tab; the **Export PCF** dialog box will be displayed, as shown in Figure 8-7. In this dialog box, select the line numbers to be exported from the **Line Numbers** list box and then choose the **Create** button; a separate PCF will be created for each line selected.

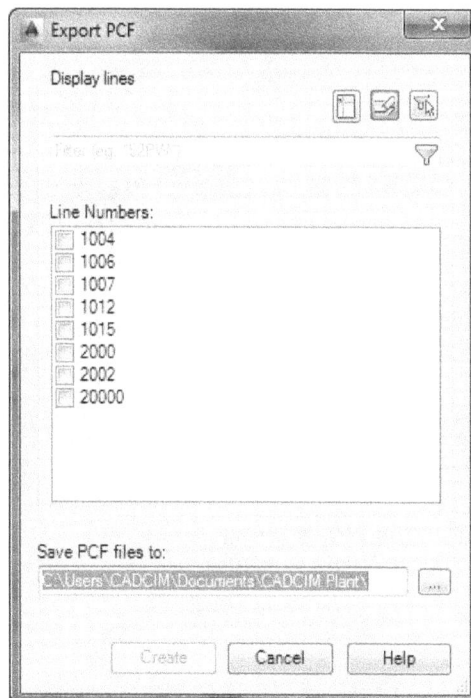

Figure 8-7 *The **Export PCF** dialog box*

Creating an Iso drawing from a Pipe Component File

To create an isometric drawing from a pipe component file, choose the **PCF to Iso** tool from the **Iso Creation** panel in the **Isos** tab; the **Create Iso from PCF** dialog box will be displayed, as shown in Figure 8-8. In this dialog box, choose the **Add** button from the **Display lines** area; the **Pick PCF file** dialog box will be displayed. In this dialog box, browse to the file location and double-click on it; the file name will be displayed in the **PCF files** list box. You can also add more files to the list. After adding the required files, choose the **Create** button; the isometric files will be created.

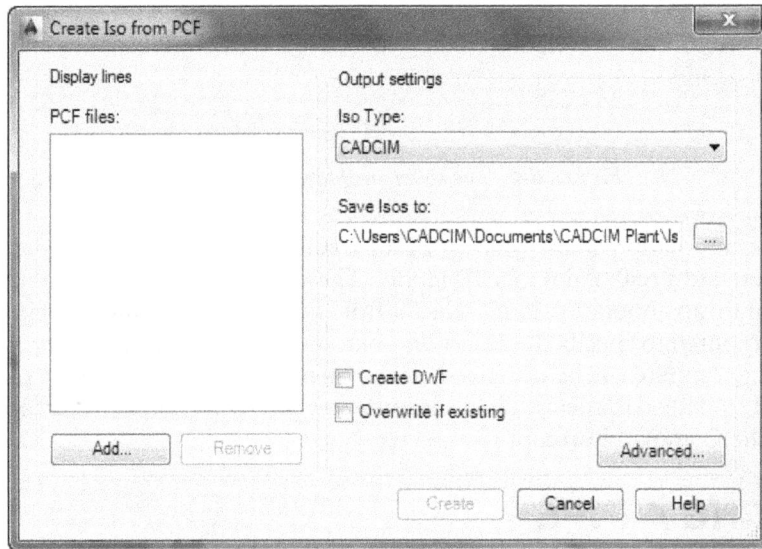

*Figure 8-8 The **Create Iso from PCF** dialog box*

LOCKING A LINE NUMBER

To lock a line, choose the **Isometric DWG** tab in the **Project Manager** and then expand the **Isometric drawing** node; different folders containing isometric drawings are displayed. Expand the required folder to display the line numbers present in the drawing. Next, right-click on the line that you want to lock; a shortcut menu will be displayed. Select the **Lock Line and Issue** option from the shortcut menu; the line will be locked.

CONFIGURING ISOMETRIC DRAWING SETTINGS

By default, four types of Iso styles are available in AutoCAD Plant 3D: Check, Stress, Final, and Spool styles. If you want more styles, you need to change the old styles or create new ones using the **Project Setup** dialog box. To change an existing Iso style, you need to modify any of the following settings:

a. Iso Style
b. Annotations
c. Dimensions
d. Sloped and Offset Piping
e. Title Block and Display

Configuring Iso Style Settings

To configure Iso style settings, choose the **Project Setup** button from the **Project** drop-down in the **Project** panel of the **Home** tab; the **Project Setup** dialog box will be displayed. Next, select **Iso Style Setup** from the **Isometric DWG Settings** node in the **Project Setup** dialog box; the **Iso Style Setup** page will be displayed, as shown in Figure 8-9. Using this page, you can create a new Iso style and configure the settings such as drawing type, file naming convention, field weld control, field fit weld makeup, table overflow settings, spool settings, and iso style paths.

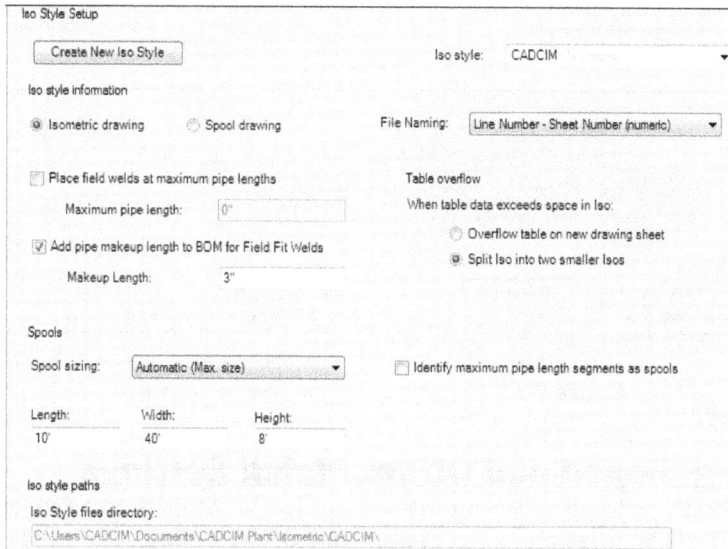

Figure 8-9 *Partial view of the **Iso Style Setup** page*

Configuring Annotation Settings

To configure the annotation settings, select **Annotations** from the **Isometric DWG Settings** node; the **Annotations** page will be displayed, as shown in Figure 8-10. In this page, you can configure the display of annotations related to bill of materials, spool, weld, valves, cut piece, and end connections. You can also set the date format in this page.

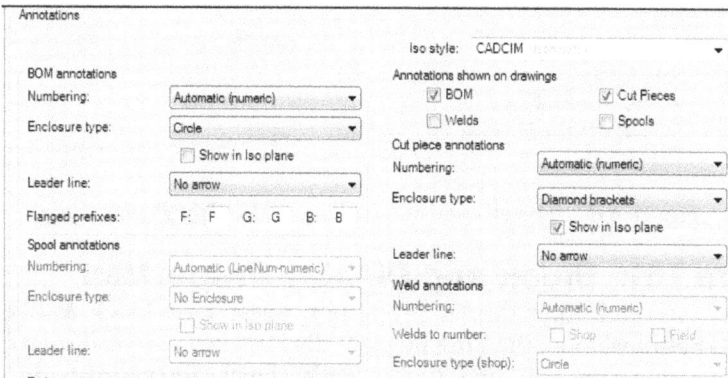

Figure 8-10 *Partial view of the **Annotations** page*

Configuring Dimensional Settings

To configure the dimensional settings, select **Dimensions** from the **Isometric DWG Settings** node; the **Dimensions** page will be displayed, as shown in Figure 8-11. In this page, you can select the required Iso theme and control the display of dimensioning types such as end to end, string, and locating type. Also, you can control the dimensioning behavior and set the other dimensioning options.

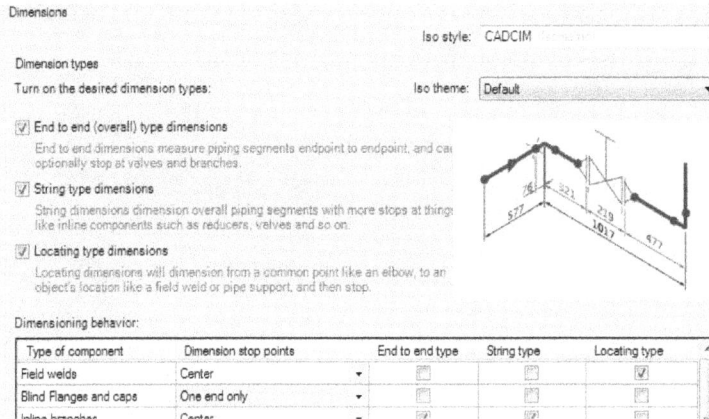

*Figure 8-11 Partial view of the **Dimensions** page*

Configuring Sloped and Offset Piping Settings

To configure the sloped and offset piping settings, select **Sloped and Offset Piping** from the **Isometric DWG Settings** node; the **Sloped and Offset Piping** page will be displayed, as shown in Figure 8-12. In this page, you can set the representation of falls, offset piping, and annotations.

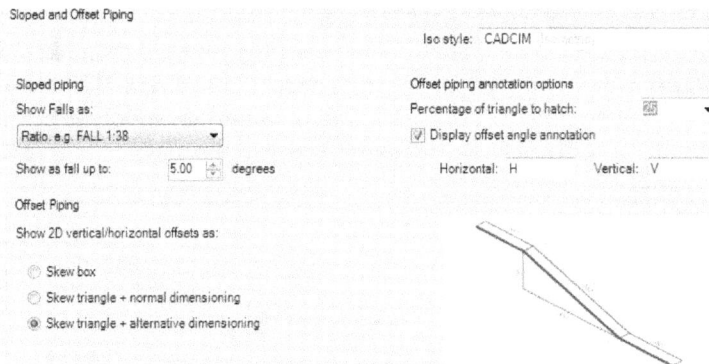

*Figure 8-12 Partial view of the **Slope and Offset Piping** page*

Setting the Title Block and Display Properties

To set the title block and other display properties, select **Title block & display** from the **Isometric DWG Settings** node; the **Title block & display** page will be displayed, as shown in Figure 8-13. Using this page, you can setup a new title block. You can also modify the display of isometric symbols, bends, and elbows using the options in this page. To setup a new title block, choose the **Setup Title Block** button; the title block editor will be displayed. Also, the **Title Block Setup** contextual tab will be displayed in the Ribbon, as shown in Figure 8-14. You can use this tab to setup the title block.

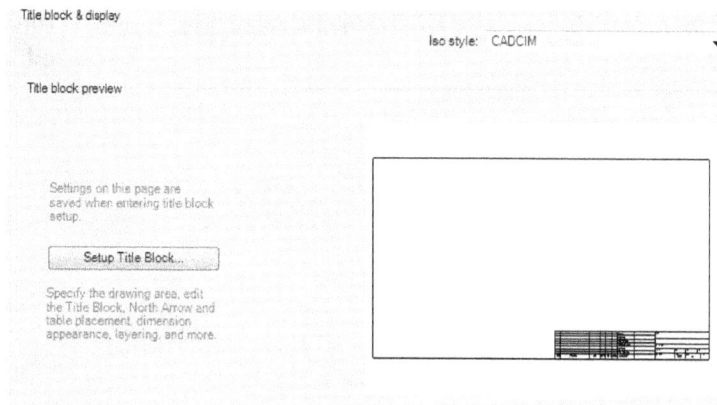

*Figure 8-13 Partial view of the **Title block & display** page*

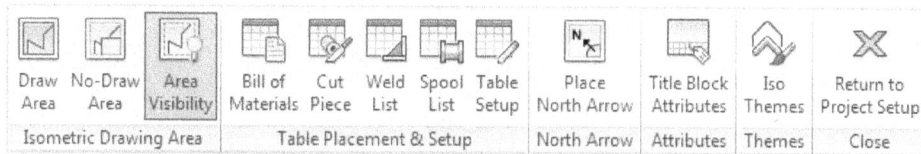

*Figure 8-14 The **Title Block Setup** contextual tab*

Configuring Bill of materials in the Title block

Bill of materials is one of the important parts of an isometric drawing. The default style of bill of materials is available in the title block. You can customize the bill of materials by choosing the **Table Setup** button from the **Table Placement & Setup** panel in the **Title Block Setup** contextual tab. On choosing this button, the **Table Setup** dialog box will be displayed, as shown in Figure 8-15. You can use this dialog box to customize the bill of materials. You can also modify other tables such as Cut Piece list, Weld list, and Spool list by selecting them from the **Table type** drop-down list in this dialog box.

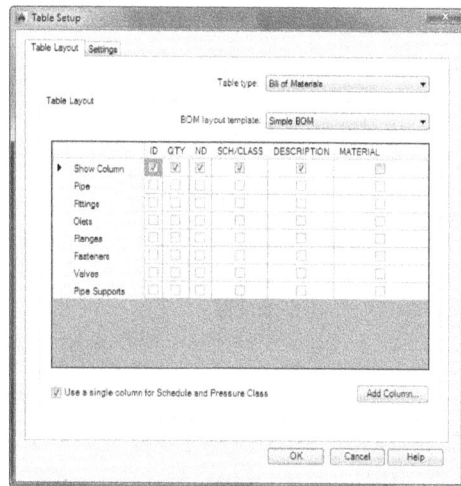

*Figure 8-15 The **Table Setup** dialog box*

TUTORIALS

Tutorial 1

In this tutorial, you will create an isometric drawing of line number 1004, as shown in Figure 8-16.   **(Expected time: 30 min)**

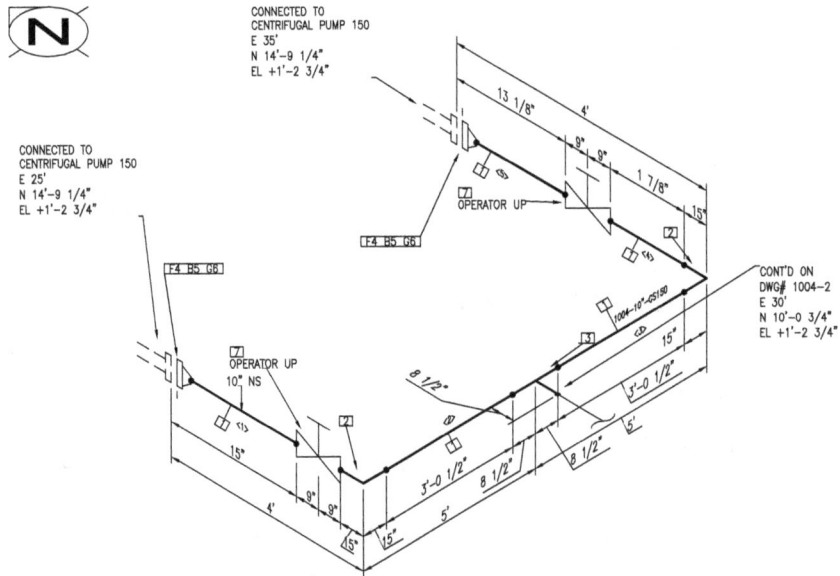

Figure 8-16 Isometric drawing for Tutorial 1

The following steps are required to complete this tutorial:

a.   Open the 3D model file.
b.   Check the line number of the pipe for which the isometric drawing is to be created.
c.   Create the isometric drawing.
d.   Lock the pipe line.

Opening the File

1.   Start AutoCAD Plant 3D and open the **Piping Model.dwg** file from the **Project Manager**.

Checking the line number

Next, you need to create the isometric drawing of the pipe connecting the pumps. To do so, you need to check the line number of the pipe.

1.   Zoom in to the pumps connected to the vertical vessel and hover the cursor on the pipe connecting the inlet nozzles of the pumps, refer to Figure 8-17; the line number of the pipe is displayed.

Creating the isometric drawing

1. Choose the **Production Iso** tool from the **Iso creation** panel in the **Isos** tab; the **Create Production Iso** dialog box is displayed.

2. In this dialog box, select the check box next to the line number 1004 in the **Line Numbers** list box.

3. Select the **Final_ANSI-B** option from the **Iso style** drop-down list in the **Output Settings** area and choose the **Create** button; the isometric drawing creation starts. Next, a balloon is displayed at the right end of the status bar, as shown in Figure 8-18.

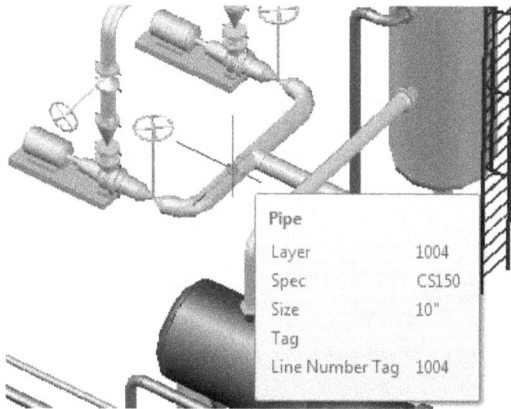

Figure 8-17 Checking the line number of the pipe

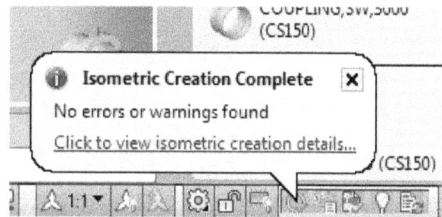

Figure 8-18 Balloon displayed at the status bar

4. Click on the **Click to view isometric creation details** link displayed on the balloon; the **Isometric creation results** dialog box is displayed.

5. In this dialog box, click on the file path of the isometric drawing, refer to Figure 8-19; the isometric drawing is opened.

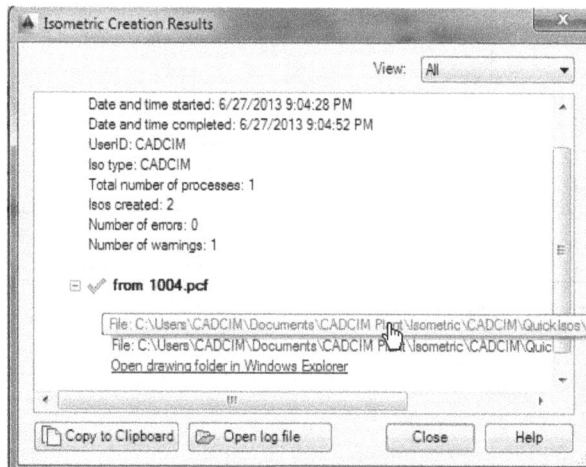

Figure 8-19 Selecting the file path of the isometric drawing

The isometric drawing is displayed, as shown in Figure 8-20 along with the bill of materials and cut piece list, refer to Figures 8-21 and 8-22. Notice that the weld symbols are displayed as points.

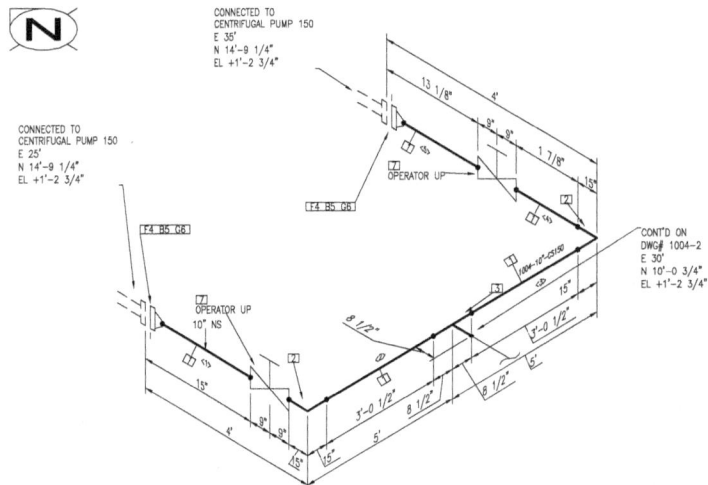

Figure 8-20 *The isometric drawing of line number 1004*

BILL OF MATERIALS				
ID	QTY	ND	SCH/ CLASS	DESCRIPTION
1	7'−11"	10"	40	PIPE, SEAMLESS, PE, ASME B36.10, ASTM A106 GR B SMLS, SCH 40
2	2	10"	40	ELL 90 LR, BW, ASME B16.9, ASTM A234 GR WPB SMLS, SCH 40
3	1	10"	40	TEE, BW, ASME B16.9, ASTM A234 GR WPB SMLS, SCH 40
4	2	10"	150	FLANGE WN, 150 LB, RF, ASME B16.5, ASTM A234 GR WPB
5	24	7/8"X4 1/2"	150	BOLT SET, RF, 150 LB, STUD BOLT
6	2	10"	150	GASKET, SWG, 1/8" THK, RF, 150 LB, ASME B16.20, CS/PTFE
7	2	10"	150	GATE VALVE, DOUBLE DISC, 150 LB, BW, ASME B16.10, ASTM A216 GR WPB, HAND WHEEL

Figure 8-21 *The bill of materials of the selected pipe segment*

CUT PIECE LIST				
ID	LENGTH	ND	END1	END2
1	10 15/16"	10"	BEVEL	BEVEL
2	3'-0 1/2"	10"	BEVEL	BEVEL
3	3'-0 1/2"	10"	BEVEL	BEVEL
4	1 7/8"	10"	BEVEL	BEVEL
5	9 1/16"	10"	BEVEL	BEVEL

Figure 8-22 Cut list of the selected pipe segment

Locking the Pipe Line

1. Choose the **Isometric DWG** tab from the **Project Manager**; the **Isometric Drawings** node is displayed.

2. Expand the **Isometric Drawings** node and then the **Final ANSI-B** node; the line numbers are displayed.

3. Right-click on the 1004 line number and choose the **Lock Line and Issue** option from the shortcut menu displayed, as shown in Figure 8-23; the line is locked, as shown in Figure 8-24 and you cannot make changes to the isometric drawing.

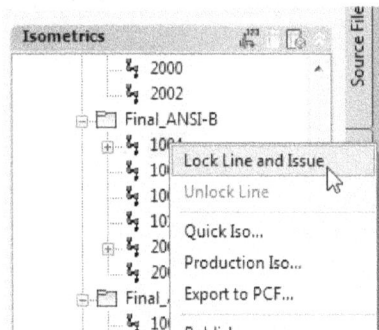

Figure 8-23 Locking the line

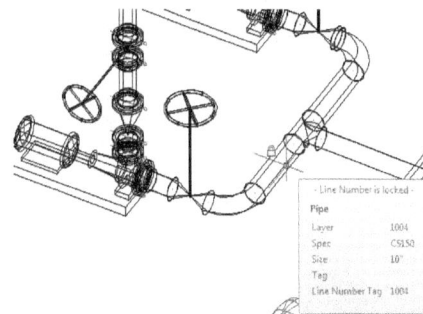

Figure 8-24 Checking the status of the locked line

4. Save and close the 3D model and isometric drawing files.

Tutorial 2

In this tutorial, you will create an isometric drawing of line number 1007. Later, you will modify the pipe attributes and view changes in the isometric drawing.

(Expected Time: 30 min)

The following steps are required to complete this tutorial:

a.   Open the 3D model file.
b.   Check the line number of the pipe for which the isometric drawing is to be created.
c.   Create the isometric drawing.
d.   Change the weld type of the pipe in the 3D model.
e.   Add insulation symbol to the pipe.
f.   Generate the isometric drawing.
g.   Lock the pipe.

Creating the Isometric Drawing

1.   Open the **Piping Model.dwg** file from the **Project Manager**, if it is not already opened.

     Next, you need to create the isometric drawing of the pipe connecting the pumps.

2.   Choose the **Isometric Files** tab from the **Project Manager**.

3.   Select the line number 1007 from **Isometric Drawings > Check_ANSI-B** node.

4.   Right-click on the selected line number and choose the **Quick Iso** option from the shortcut menu displayed; the **Create Quick Iso** dialog box is displayed.

5.   Choose the **Create** button from the **Create Quick Iso** dialog box; the isometric drawing creation starts. Next, a balloon is displayed at the right end of the status bar.

6.   Open the isometric drawing created. Figure 8-25 shows the isometric drawing of the selected line.

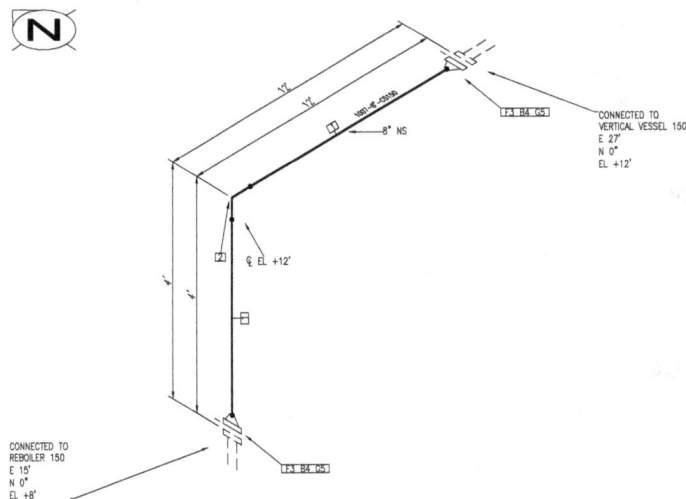

Figure 8-25 The isometric drawing of line number 1007

Changing the Weld Type in the 3D Model

Next, you need to change the weld type in the 3D model and see the representation in the isometric drawing.

1. Open the 3D model, if not already open and then set the view style to Wireframe.

2. Turn on the weld display by choosing the **Toggle Weld Display** button from the **Visibility** panel in the **Home** tab; the pipe connectors are displayed as dots.

3. Select the pipe connector at the elbow by using a selection window, as shown in Figure 8-26; the grips are displayed on it, as shown in Figure 8-27.

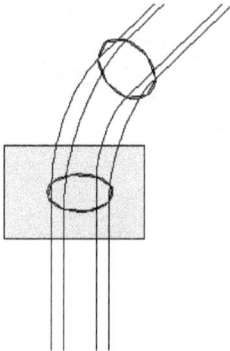

Figure 8-26 Selecting the pipe connector

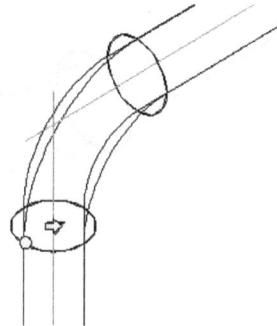

Figure 8-27 Grips displayed on the pipe connector

4. Right-click on any one of the grips and choose the **Properties** option from the shortcut menu; the **Properties** palette is displayed.

Next, you need to change the welding type in the **Properties** palette.

5. In this palette, scroll down to the **General** sub-rollout under the **Plant 3D** rollout and select the **FIELD** option from the **Shop/Field** drop-down list, refer to Figure 8-28; the weld type is changed to field weld.

6. Next, press ESC to exit the **Properties** palette of the connector.

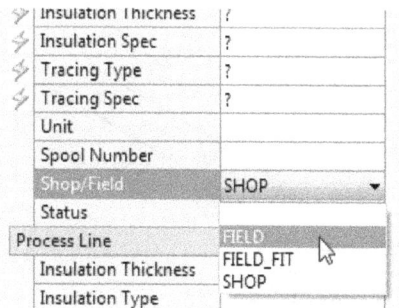

Figure 8-28 Changing the weld type

Adding an Insulation Symbol to the Isometric Drawing

1. Select the pipe connecting the reboiler and the vertical vessel and then right-click on it; a shortcut menu is displayed.

2. Choose **Add to Selection > Entire Line Number** from the shortcut menu; the entire line number is selected, refer to Figure 8-29.

Figure 8-29 Choosing the Entire Line Number option

3. Invoke the **Properties** palette of the pipe and select **0.5-1/2"** from the **Insulation Thickness** drop-down list in the **Process Line** sub-rollout under the **Plant 3D** rollout.

4. Select **HC - Hot Cold** option from the **Insulation Type** drop-down list, as shown in Figure 8-30.

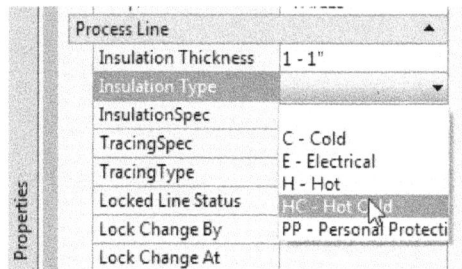

Figure 8-30 Selecting the Insulation Type from the Properties palette

Next, you need to mark the insulation symbol on the line.

5. Choose the **Insulation Symbol** button from the **Iso Annotation** panel of the **Isos** tab and select the insertion point on the pipe, as shown in Figure 8-31. Notice that a sphere is displayed inside the pipe, as shown in Figure 8-32.

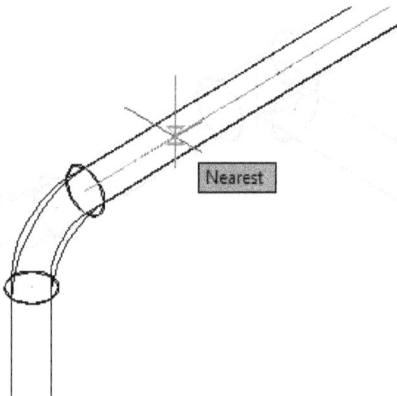

Figure 8-31 Specifying the insertion point of the insulation Symbol

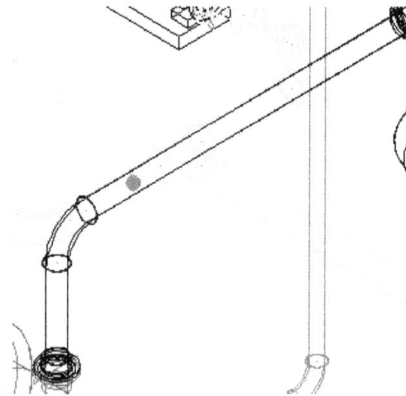

Figure 8-32 Sphere displayed at the selected point

6. Choose the **Save** button from the **Quick Access Toolbar** to save the model.

7. Invoke the **Create Production Iso** dialog box and select the line number 1007. Next, select the **Final_ANSI-B** option from the **Iso style** drop-down list and choose the **Create** button; the isometric drawing is created, as shown in Figure 8-33.

   Notice that the weld representation is changed in the isometric drawing. Also, an insulation symbol is displayed, refer to Figure 8-33.

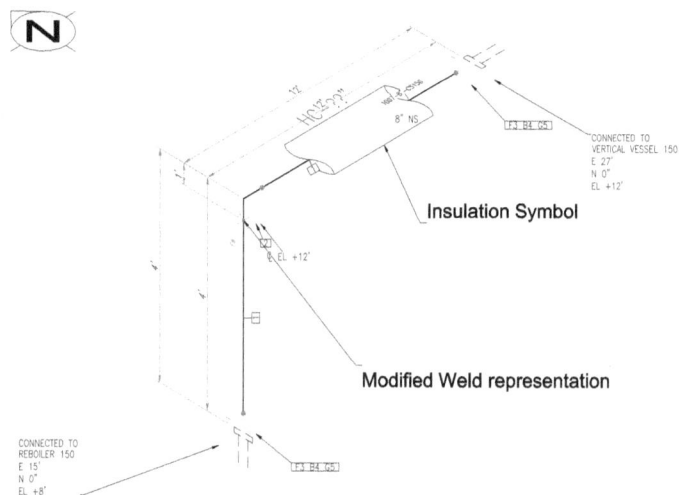

Figure 8-33 The isometric drawing with the modified weld representation and the insulation symbol

8. Lock the line number.

9. Save and close the 3D model and isometric drawing files.

Tutorial 3

In this tutorial, you will create a new Iso style based on the **Final_ANSI-B** style. Next, you will customize this style and create an isometric drawing, as shown in Figure 8-34.

(Expected Time: 45 min)

The following steps are required to complete this tutorial:

a. Create a new Iso style and create an Isometric drawing.
b. Modify the annotation and dimensional settings.
c. Create an isometric drawing and notice the changes.
d. Open the iso.dwt file of the new Iso style.
e. Download and insert the new title block into the iso.dwt file.

f.   Save and close the iso.dwt file.
g.   Invoke the **Title Block Editor** of the newly created iso style.
h.   Define the Draw area, No-Draw area, BOM list and Cut Piece list area.
i.   Save and close the iso.dwt file.
j.   Create an isometric drawing with the customized iso style.

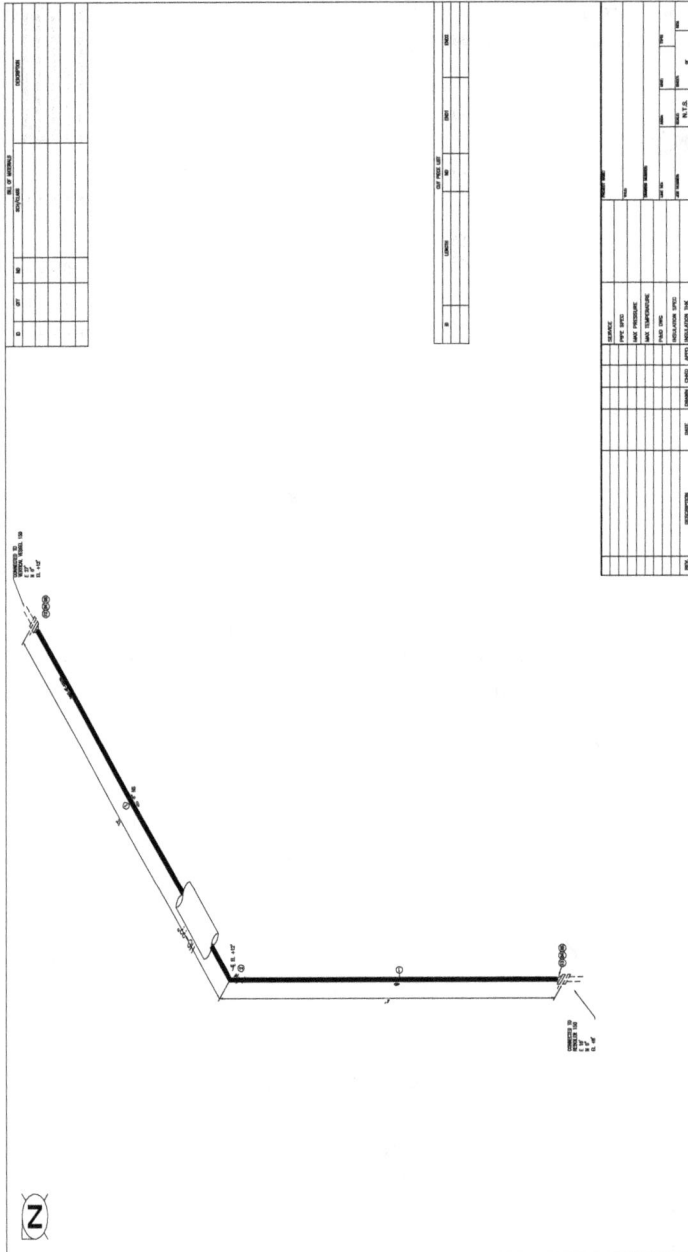

Figure 8-34 Isometric drawing for Tutorial 3

Opening the Project and Creating a New Iso Style

1. Start AutoCAD Plant 3D and open the CADCIM project from the **Project Manager**.

2. Choose the **Project Setup** tool from the **Project** drop-down in the **Project** panel; the **Project Setup** dialog box is invoked.

3. In this dialog box, expand the **Isometric DWG settings** node and select **Iso Style Setup**; the **Iso Style Setup** area is displayed.

4. Choose the **Create New Iso Style** button from the **Iso Style Setup** area; the **Create Iso Style** dialog box is displayed.

5. Enter **CADCIM** in the **Name** edit box and select the **Final ANSI-B** option from the **Base on existing style** drop-down list.

6. Choose the **Create** button from the **Create Iso Style** dialog box; a new iso style is created and listed in the **Iso style** drop-down list of the **Project Setup** dialog box.

7. Choose the **OK** button from the **Project Setup** dialog box.

8. Create an isometric drawing of the line number 1007 using the newly created Iso style. Note that the isometric drawing will be same as the one created in Tutorial 2, as shown in Figure 8-35.

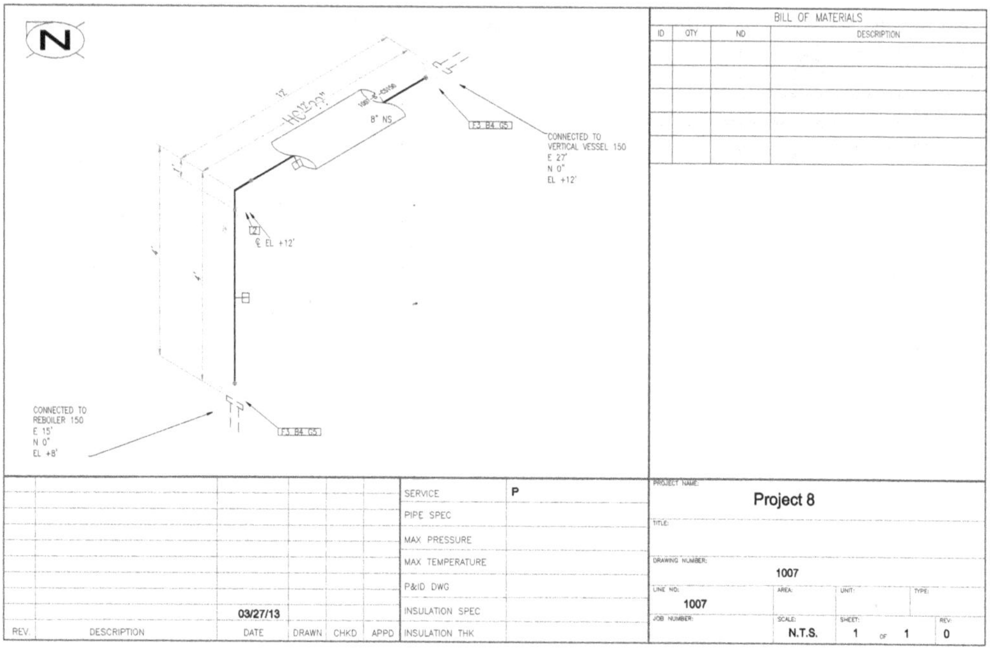

Figure 8-35 Isometric drawing of the line number 1007 created with default settings

Customizing the Iso Style

1. Invoke the **Project Setup** dialog box and select **Annotations** from the **Isometric DWG Settings** node; the **Annotations** page is displayed. Make sure that **CADCIM** is selected in the **Iso style** drop-down list.

2. Select the **Circle** option from the **Enclosure type** drop-down list in the **BOM annotations** area, refer to Figure 8-36.

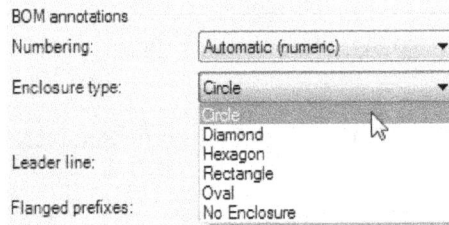

Figure 8-36 *Selecting the enclosure type of BOM annotations*

Next, you need to modify the dimensional settings of the Iso style.

3. Select **Dimensions** from the **Isometric DWG settings** node; the **Dimensions** area is displayed.

4. Clear the **String type dimension** and **Locating type dimension** check boxes and choose the **OK** button.

5. Invoke the **Create Production Iso** dialog box and select **CADCIM** from the **Iso style** drop-down list.

6. Select the **1007** check box from the **Line Numbers** list and choose the **Create** button.

7. Open the isometric drawing. Notice that the BOM annotations are enclosed in circles, as shown in Figure 8-37. Also, the string dimensions and locating dimensions are not displayed in the drawing, refer to Figure 8-37.

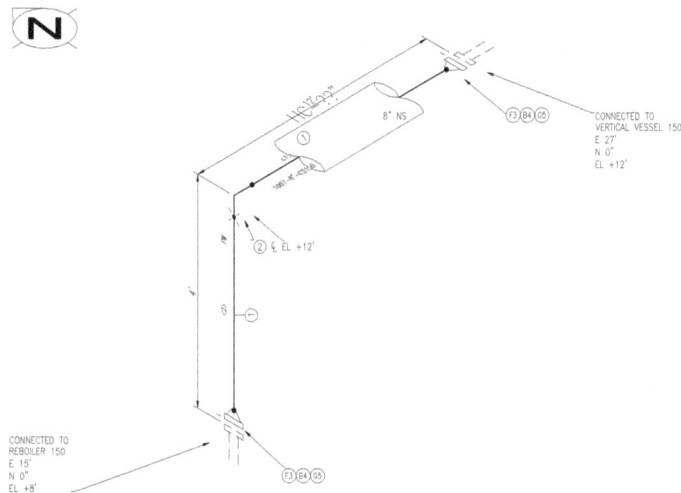

Figure 8-37 *Isometric drawing after modifying the annotation and dimensional settings*

Creating a New Iso Template

In this section, you will create a new template for the isometric drawings.

1. Choose the **Open** button from the **Quick Access Toolbar**; the **Select File** dialog box is displayed.

2. In this dialog box, select the **Drawing Template (.dwt)** option from the **Files of type** drop-down list.

3. Browse to the location *C:\Users\Documents\CADCIM\Isometric\CADCIM* and double-click on the **iso.dwt** file; the template file is opened.

4. Select the entire geometry from the drawing area and then press DELETE to delete it.

5. Enter **PURGE** at the command prompt; the **Purge** dialog box is displayed.

6. In this dialog box, expand the **Blocks** node and select **Title Block**.

7. Choose the **Purge** button; the **Purge-Confirm Purge** message box is displayed.

8. Choose **Purge this item** from the message box; the title block is purged.

9. Close the **Purge** dialog box.

    Next, you need to insert a new title block into the drawing area.

10. Download the **CADCIM_Iso.dwg** file from *http://www.cadcim.com*. The path of the file is as follows:

    Textbooks>CAD/CAM>AutoCAD Plant 3D>AutoCAD Plant 3D for Designers 2014

11. Save the file at the location *C:\Users\Documents\CADCIM Plant\Isometric\CADCIM*.

12. Choose the **Insert** button from the **Block** panel in the **Insert** tab; the **Insert** dialog box is displayed.

13. Choose the **Browse** button and open the *CADCIM_Iso.dwg* file.

14. Accept the default settings in the **Insert** dialog box and choose the **OK** button; the title block is attached to the cursor and you are prompted to specify the angle of rotation.

15. Specify **0** as the angle of rotation; the title block is placed in the drawing area, as shown in Figure 8-38.

16. Enter **RENAME** at the Command prompt; the **Rename** dialog box is displayed.

17. Select **Blocks** from the **Named Objects** list box and **CADCIM_Iso** from the **Items** list box.

18. Enter **Title Block** in the **Rename To** edit box and choose the **Rename To** button.

19. Choose the **OK** button from the dialog box to close it.

    Next, you need to set the limits for the drawing area.

20. Enter **LIMITS** at the Command prompt.

21. Press **ENTER** to accept the origin as the first corner.

22. Select the top-right corner of the title block to define the second corner, refer to Figure 8-39.

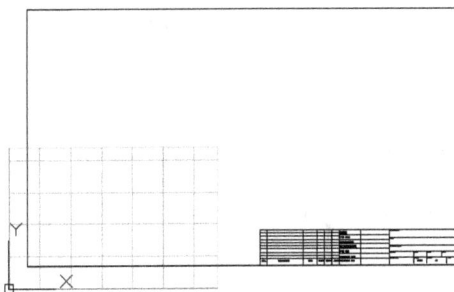

Figure 8-38 The **iso.dwt** file after placing the title block

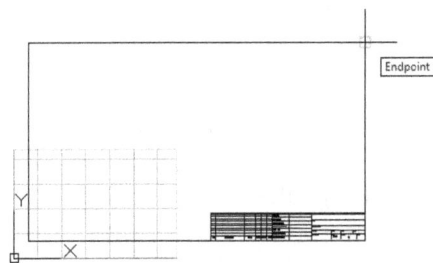

Figure 8-39 Specifying the second corner of the drawing sheet

23. Choose the **Save** button from the **Quick Access Toolbar** and close the file.

Customizing the Title Block

In this section, you will customize the drawing and BOM area of the title block.

1. Invoke the **Project Setup** dialog box and select **Title Block and Display** from the **Isometric DWG Settings** node; the **Title block & display** page is displayed.

2. Make sure that **CADCIM** is selected in the **Iso Style** drop-down list.

3. Choose the **Setup Title Block** button from the **Title Block & Display** area; the dialog box is closed and the **Title Block Editor** is displayed. Also, the **Title Block Setup** contextual tab is displayed.

4. Choose the **Draw Area** button from the **Isometric Drawing area** panel of the **Title block Setup** contextual tab; you are prompted to specify the first corner point of the drawing area.

5. Select the top-left corner of the title block, as shown in Figure 8-40; you are prompted to select the second corner.

6. Select the second corner of the drawing area, as shown in Figure 8-41; the drawing area is defined.

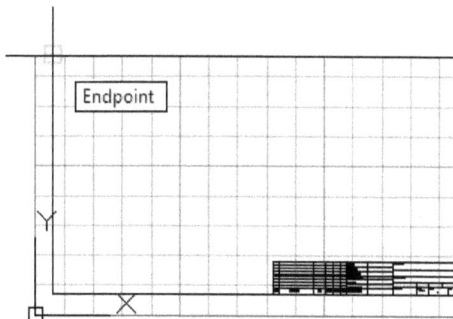

Figure 8-40 *Selecting the first corner of the drawing area*

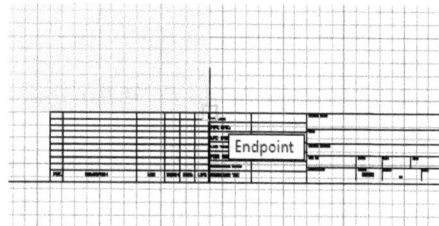

Figure 8-41 *Selecting the second corner of the drawing area*

Next, you need to place the north arrow.

7. Choose the **Place North Arrow** button from the **North Arrow** panel; you are prompted to specify the default north arrow direction.

8. Choose the **upper left** option from the Command prompt and place the north arrow symbol at the top-left corner, as shown in Figure 8-42.

9. Choose the **No-Draw Area** button from the **Isometric Drawing** area panel and define the **No-draw** area, as shown in Figure 8-43.

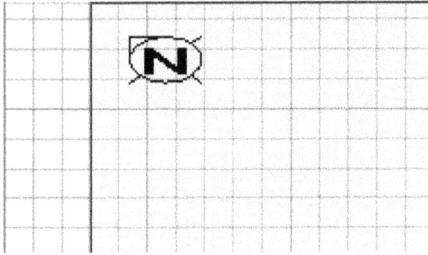

Figure 8-42 Position of the North Arrow

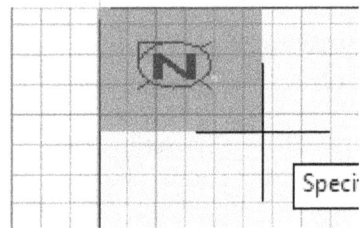

Figure 8-43 Defining the No-draw area

Next, you need to define the BOM list and Cut Piece list area.

10. Choose the **Bill of Materials** button and define the **BOM list** area, as shown in Figure 8-44.

11. Choose the **Cut Piece** button and define the **Cut Piece** list area, as shown in Figure 8-45.

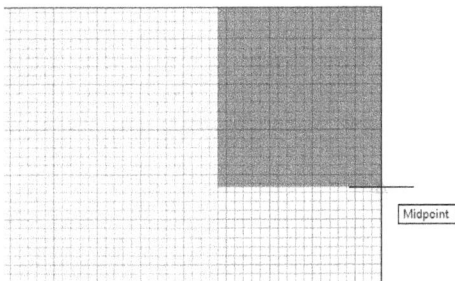

Figure 8-44 Defining the **BOM list** area

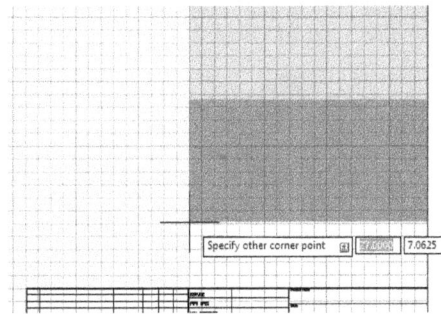

Figure 8-45 Defining the **Cut Piece list** area

The drawing sheet after configuring the settings is shown in Figure 8-46.

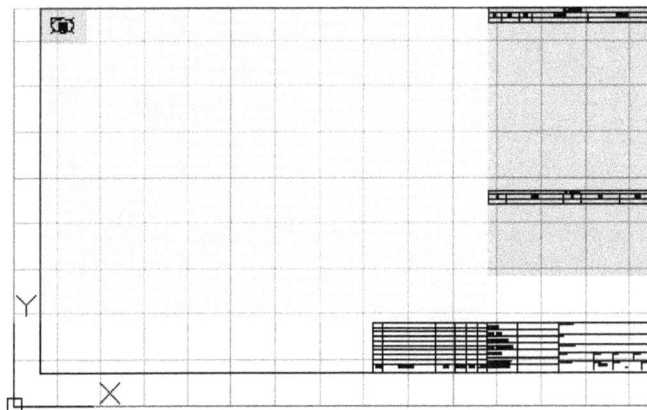

Figure 8-46 The drawing sheet after configuring all the settings

12. Choose the **Return to Project Setup** button from the **Close** panel in the **Title Block Setup** contextual tab; the **Block-Change Not Saved** message box is displayed.

13. Choose the **Save Changes to "iso.dwt"** button from the message box; the changes made to the **iso.dwt** file are saved. Also, the **Project Setup** dialog box is displayed.

14. Choose the **OK** button from the **Project Setup** dialog box to close it.

15. Create the Production Iso of the line number 1007 and open it; the isometric drawing is displayed, as shown in Figure 8-47.

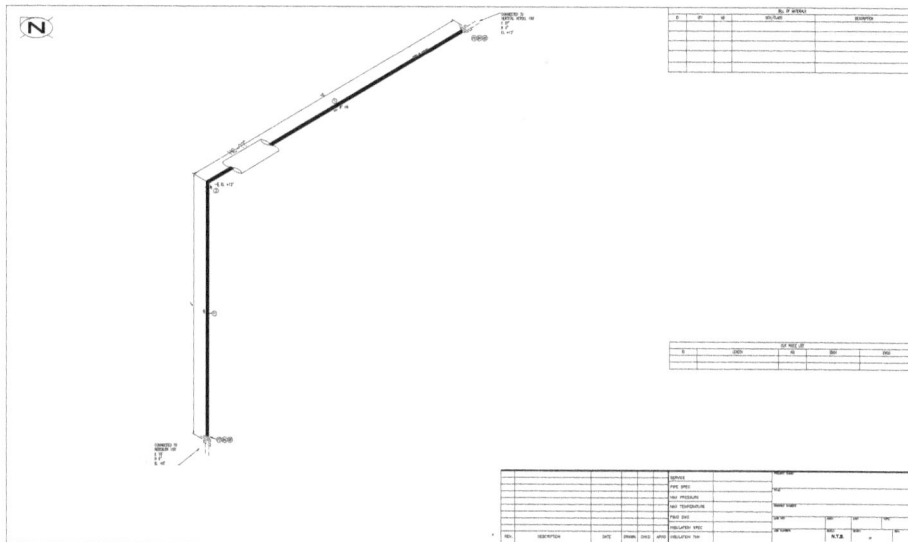

Figure 8-47 *The isometric drawing created after modifying the Iso style*

16. Save and close all the files.

Self-Evaluation Test

Answer the following questions and then compare them to those given at the end of this chapter:

1. A _____ drawing is created to ensure that the final isometric drawing will be created without errors.

2. The _____ tool is used to define the area in which the isometric drawing will be created.

3. You can set the date format of an isometric drawing in the _____ page of the **Project Setup** dialog box.

4. The _____ drawing will present the geometric data which is precise and relevant to the stress analysis of pipes.

5. The _____ tool is used to create an isometric file from a PCF (Pipe Component File).

6. To create a new iso style, choose the _____ button from the **Iso style Setup** area in the **Project Setup** dialog box.

7. By default, there are _____ types of Iso styles available in AutoCAD Plant 3D.

8. A balloon will be displayed at the _____ after the isometric drawing is created.

9. You can control the display of dimensions in the _____ page of the **Project Setup** dialog box.

10. A _____ is the final isometric drawing separated in individual sections.

Review Questions

Answer the following questions:

1. The _____ drawing will have a bill of material (BOM) and it is used for carrying out the fabrication and construction process.

2. You can create _____ by selecting pipes either from the drawing area or from the line list.

3. After creating an Iso message, a _____ will be added inside the pipe.

4. You can select predefined isometric styles from the _____ drop-down list in the **Create**

Answers to Self-Evaluation Test
1. Check Isometric, **2. Draw Area, 3. Annotation, 4.** Stress isometric, **5. PCF to Iso, 6. Create New Iso Style, 7.** four, **8.** status bar, **9. Dimensions, 10.** Spool drawing.

Chapter 9

Creating Orthographic Drawings

Learning Objectives

After completing this chapter, you will be able to:
- *Create Orthographic views*
- *Add annotations and dimensions to views*
- *Update Views*

INTRODUCTION

After creating a 3D model, you need to create their orthographic views. Orthographic views are the two-dimensional (2D) representations of a 3D model. In this chapter, you will learn to create orthographic and sectional views of the 3D model and then update the views as per the changes made in the 3D model. You will also learn to add dimensions and annotations to the 2D views.

CREATING AN ORTHOGRAPHIC DRAWING

You can create an orthographic drawing in AutoCAD Plant 3D. To do so, choose the **Orthographic DWG** tab from the **Project Manager**; the **Orthos** tree view will be displayed. In this tree view, right-click on the **Orthographic Drawings** node and choose the **New Drawing** option from the shortcut menu displayed, refer to Figure 9-1; the **New DWG** dialog box will be displayed. Next, enter the file name and author name in their respective fields and then choose the **OK** button to create a new orthographic drawing.

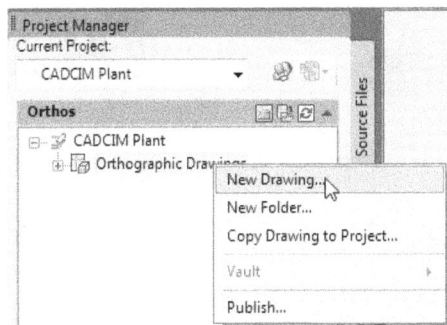

Figure 9-1 Selecting the New Drawing option

The orthographic drawing contains a drawing sheet and the **Ortho View** tab. The **Ortho View** tab is shown in Figure 9-2. The tools in this tab are used to perform operations such as editing, updating, annotating the drawing view, and so on.

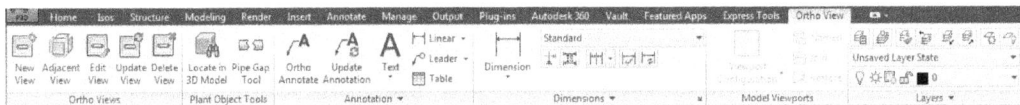

Figure 9-2 The Ortho View tab

Generating the First View

To generate the first view of the drawing, choose the **New View** button from the **Ortho Views** panel in the **Ortho View** tab; the **Select Reference Models** dialog box will be displayed, as shown in Figure 9-3. In this dialog box, select the models to be included from the **Project models** list box. You need to select the check boxes adjacent to the drawing files, to include them in the orthographic view to be created. Next, choose the **OK** button from the dialog box; the **Orthographic View Selection** window will be displayed. Also, the **Ortho Editor** contextual tab will be displayed, as shown in Figure 9-4. The options in the tab help you to create the orthographic view. Also, the OrthoCube is displayed in the drawing area, as shown in Figure 9-5, which will help you to specify the view boundary.

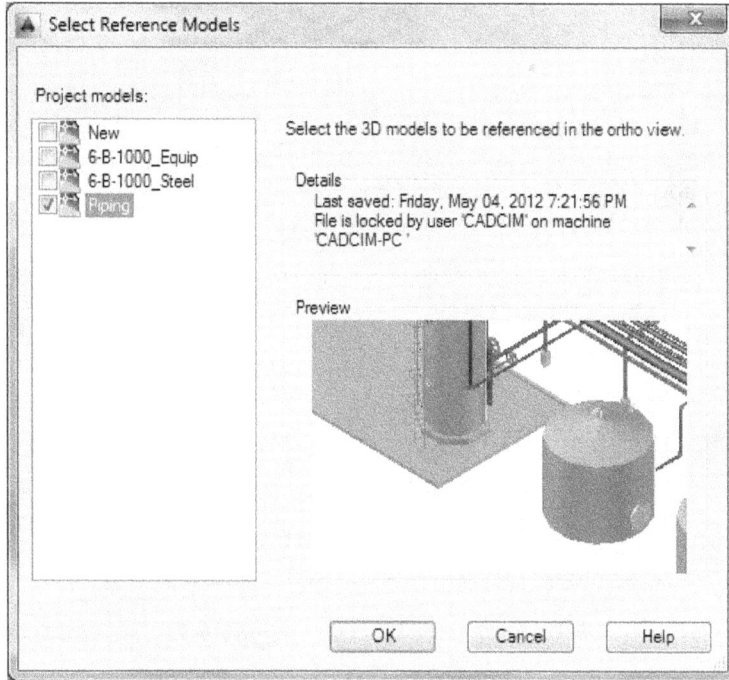

Figure 9-3 The **Select Reference Models** *dialog box*

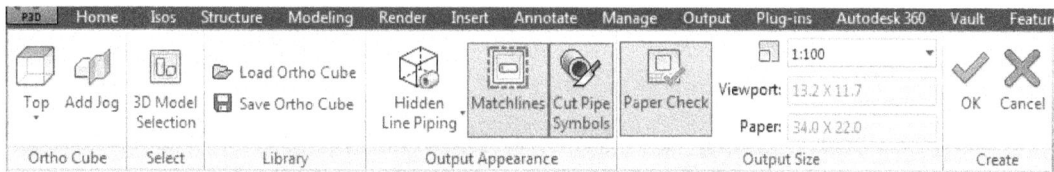

Figure 9-4 The **Ortho Editor** *tab*

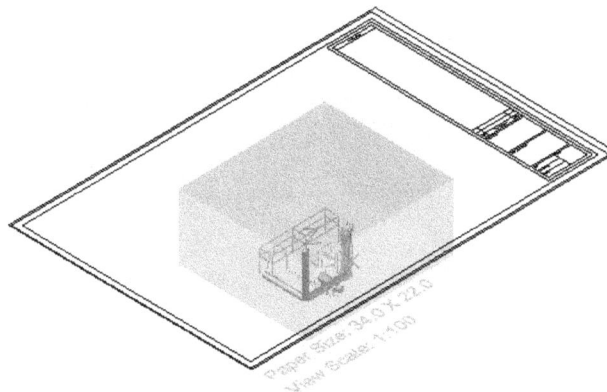

Figure 9-5 The *OrthoCube*

The options in the **Ortho Editor** tab are discussed next.

Ortho Cube Panel

In this panel, the View drop-down is used to select the view to be created. You can select the required orthographic view or isometrics view from this drop-down. The **Add Jog** tool is used to add a jog to the OrthoCube. A jog is used to exclude a portion of the model from the view to be created. To add a jog, choose the **Add Jog** tool from the **Ortho Cube** panel and select an edge the OrthoCube, refer to Figure 9-6; the jog will be added, as shown in Figure 9-7.

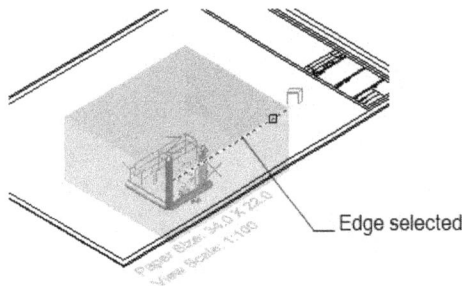

Figure 9-6 Selecting the edge of the OrthoCube

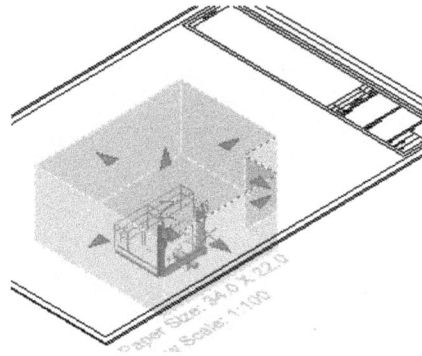

Figure 9-7 Jog added to the OrthoCube

Select Panel

In this panel, the **3D Model Selection** tool is used to invoke the **Select Reference Models** dialog box. Using this dialog box, you can select piping models for creating orthographic views.

Library Panel

Using the tools in this panel, you can save the settings made in the **Ortho View Selection** window. Also, you can load the previously saved settings.

Output Appearance Panel

The tools in this panel are used to the set the display of the view outputs. The **View Style** drop-down in this panel is used to set the display style of the orthographic views. The **Hidden Line Piping** option in this drop-down is used to display the pipes which are hidden, refer to Figure 9-8. The **All Hidden Lines** option is used to display all the hidden objects, refer to Figure 9-9. If you choose the **No Hidden Line** option, the hidden objects will not be displayed, refer to Figure 9-10.

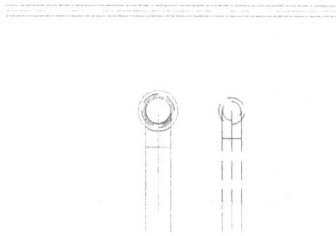

Figure 9-8 *A view with **Hidden Line Piping** option selected*

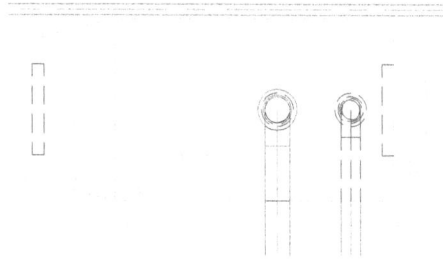

Figure 9-9 *A view with **All Hidden Lines** option selected*

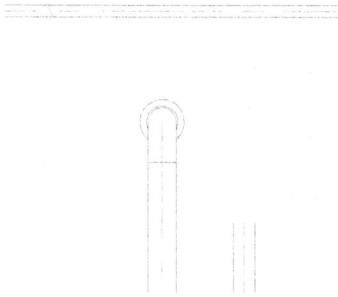

Figure 9-10 *A view with **No Hidden Lines** option selected*

The **Matchlines** tool is used to display the matchlines on the boundary of an orthographic view, as shown in Figure 9-11.

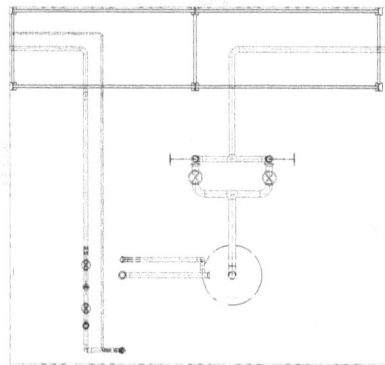

Figure 9-11 *A view with matchlines*

The **Cut Pipe Symbols** tool is used to display a cut pipe symbol on the pipe when it is sectioned and only a portion of pipe is displayed in the view.

Output Size Panel

The **Papercheck** tool is used to display the drawing sheet in the graphics window, refer to Figure 9-5. You can adjust the OrthoCube and scale the view according to the paper size. The **Scale** drop-down is used to set the scaling factor of the view. The **Viewport** and **Paper Size** display boxes display the viewport and the paper size, respectively in the drawing. Note that these values are read only.

Create panel

The **OK** button in this panel is used to accept the settings and generate the orthographic views. You can choose the **Cancel** button to cancel the view generation.

After specifying the required settings in the **Orthographic View Selection** window, choose the **OK** button from the **Create** panel; the viewport will be attached to the cursor and you need to specify its insertion point in the drawing sheet. At this stage, you can specify the scale value of the viewport by choosing the **Scale** option from the Command Prompt. Click to specify the insertion point of the viewport; the **Ortho Generation** message box indicating the Ortho generation progress will be displayed, refer to Figure 9-12. Once the view generation is completed, the orthographic view will be placed in the drawing sheet.

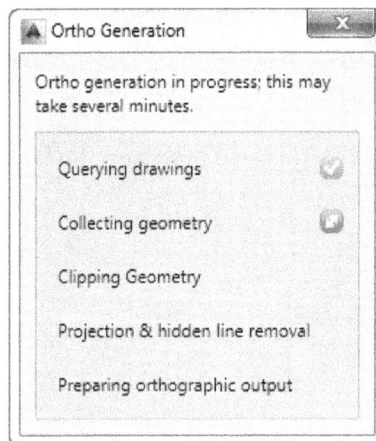

Figure 9-12 *The **Ortho Generation***
message box

Creating the Adjacent View

To create an adjacent view, first you need to open the Orthographic drawing by using the **Project Manager**. To do so, choose the **Orthographic DWG** tab and then expand the tree view. Next, double-click on the required orthographic drawing to open it.

In the orthographic drawing, choose the **Adjacent View** button from the **Ortho Views** panel in the **Ortho View** tab; you will be prompted to select an orthographic view to create an adjacent view. Select an existing orthographic view from the drawing sheet; the **Create an Adjacent View** dialog box will be displayed, as shown in Figure 9-13. Next, select the required orthographic view from this dialog box. Choose the **OK** button; the selected view will be attached to the cursor and you will be prompted to specify its insertion point. At this stage, you can scale or rotate the view by choosing the **Scale** or **Rotate** option from the Command Prompt. Next, place the view depending upon its orientation with the existing view.

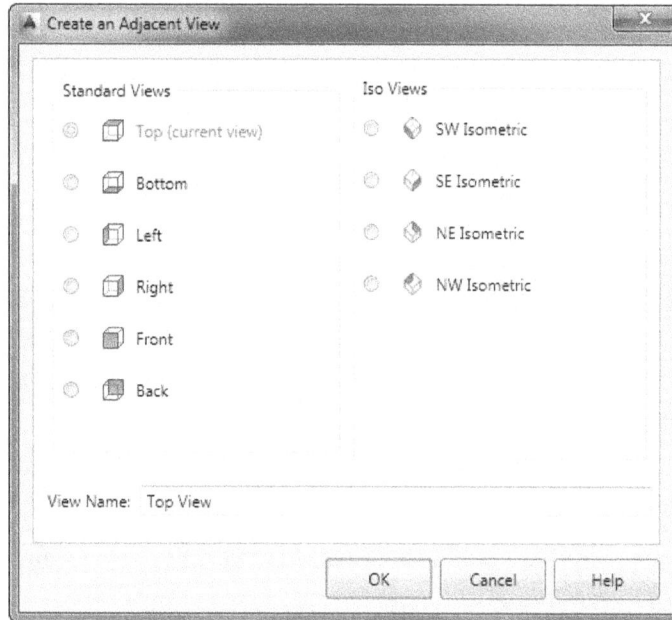

Figure 9-13 The *Create an Adjacent View* dialog box

Adding Annotations to the Drawing

To add annotations to the drawing, choose the **Ortho Annotate** tool from the **Annotation** drop-down in the **Annotation** panel; you will be prompted to select the component to annotate. Select the component to be annotated from any of the orthographic views in the drawing sheet; you will be prompted to specify the annotation style. Press ENTER, if you want to select the annotation displayed by default. Otherwise, you can specify any annotation style at the Command Prompt. If you want to know the available annotation styles, enter **?** at the Command Prompt and then press ENTER; a list of styles will be displayed at the Command Prompt. After specifying the annotation style, press ENTER; you will be prompted to specify the position of the annotation. Position the annotation anywhere near the component.

You can also add annotation to the drawing by right-clicking in the drawing area and choosing the required annotation from the **Ortho Annotate** sub menu, refer to Figure 9-14. The **Ortho Annotate** sub menu contains the options to annotate pipes, steel structures, equipment, instruments and so on. You can select the required annotation from this sub menu and select the component to be annotated from the orthographic view. For example, to place the coordinate position of a pipe, choose the **Piping Coordinate [Position]** option from the **Ortho Annotate** sub menu and select the pipe to be annotated from the orthographic view.

Ortho Annotate	Elevations and Coordinates
Plant3D	Bottom of Pipe [BOP]
Clipboard	Center of Pipe [COP]
Isolate	Face of Flange [FOF]
Undo Plantorthomenuannotate	Piping Coordinate [Position]
Redo Ctrl+Y	Top of Pipe [TOP]
Pan	Bottom of Steel [BOS]
Zoom	Center of Steel Elevation [COS Z]
SteeringWheels	Center of Steel X (Easting) [COS X]
Action Recorder	Center of Steel Y (Northing) [COS Y]
Subobject Selection Filter	Column Coordinate [COS]
Convert MS to PS	Top of Steel [TOS]
Quick Select...	Other Annotations
QuickCalc	Equipment
Find...	Instrumentation
Options...	Piping
	Structural

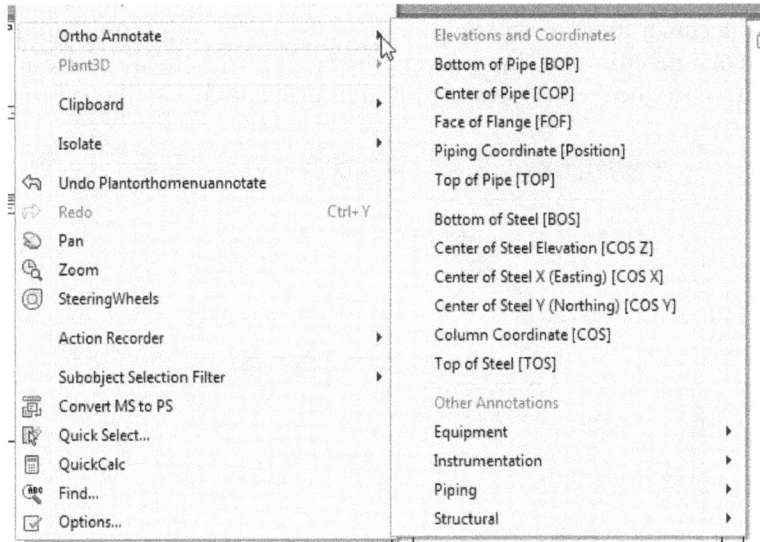

*Figure 9-14 The **Ortho Annotate** sub menu*

Note
*You can update an annotation as per the changes made in the 3D model. To do so, choose the **Update Annotation** tool from the **Annotation** drop-down in the **Annotation** panel; the annotation gets updated.*

Adding Dimensions to the Drawing

Linear   You can add dimensions to a drawing. For example, to add a linear dimension to the drawing, choose the **Linear** tool from the Dimension drop-down of the **Annotation** panel in the **Ortho View** tab; you will be prompted to specify the first extension line origin. Specify the origin point of the dimension; you will be prompted to specify the second extension line origin. After specifying the second extension line origin, the dimension will be attached to the cursor and you will be prompted to specify the location of the dimension line. Move the cursor and click to specify the location of the dimension.

Locating a Component in the 3D Model

Locate in   To locate a component in the 3D model, choose the **Locate in 3D model** tool from
3D Model   the **3D models** panel; you will be prompted to select a component to locate from the 3D model. Select the component from the orthographic view; the 3D model will be opened and it will zoom to the selected component.

Editing a Drawing View

Edit   To edit the drawing view, choose the **Edit View** tool from the **Ortho Views** panel; you
View   will be prompted to select a viewport. On selecting the required viewport from the drawing sheet, the **Ortho View Selection** window will be displayed. Use the options in the **Ortho Editor** tab in this window to edit the view settings. You can also use the

OrthoCube to redefine the boundaries of the view. After modifying all the settings, choose the **Update Ortho View** button from the **Confirm/Cancel** panel in the **Ortho Editor** tab; the selected view will be updated.

Updating a View

During piping design, a lot of changes are made in a 3D model during a project. You need to update the drawing views when changes have been made in the 3D model. To do so, choose the **Update View** button from the **Ortho Views** panel in the **Ortho View** contextual tab, and then select the view to be updated; the view will be updated. Note that the dimensions will not get updated along with the view. You need to manually update the dimensions.

Adding Gaps to Pipes

You can add gaps to a pipe to display the pipe hidden behind it. To add a gap, choose **Pipe Gap Tool** from the **Plant Objects Tools** panel and select the pipe; a gap will be added to the pipe. Next, choose the **Update View** button and update the view; the hidden pipe will be visible.

TUTORIAL

Tutorial 1

In this tutorial, you will create orthographic views of a 3D model and add dimensions and annotations to the views. Also, you will modify the 3D model and update the orthographic views.   **(Expected time: 30 min)**

The following steps are required to complete this tutorial

a.   Open the **Piping Model.dwg** file from the **Project Manager**.
b.   Create the Top view.
c.   Create the adjacent views
d.   Add Dimensions and annotations to the views
e.   Save the drawing file.
f.   Locate a pipe in the 3D Model and modify it.
g.   Validate the views and then update them.
h.   Manually update the dimension.

Creating the Orthographic Drawing and Configure the Settings

1.   Open the **Piping Model.dwg** file.

2.   Choose the **Create Ortho View** tool from the **Ortho Views** panel in the **Home** tab; the **Select Orthographic drawing** dialog box is displayed.

3.   Choose the **Create New** button from this dialog box; the **New DWG** dialog box is displayed.

4. In this dialog box, enter **OrthoViews** in the **File name** edit box and then choose the **OK** button; the **Ortho Editor** contextual tab is displayed along with the OrthoCube in the drawing area.

5. Select the Orthocube to display grips on it. Next, use the grips and modify the Orthocube, as shown in Figure 9-15. Make sure that all the components are covered under the Orthocube.

    Next, you need to specify the orientation of the first view.

6. Select the **Top** option from the **View** drop-down list in the **Ortho Cube** panel, if not selected.

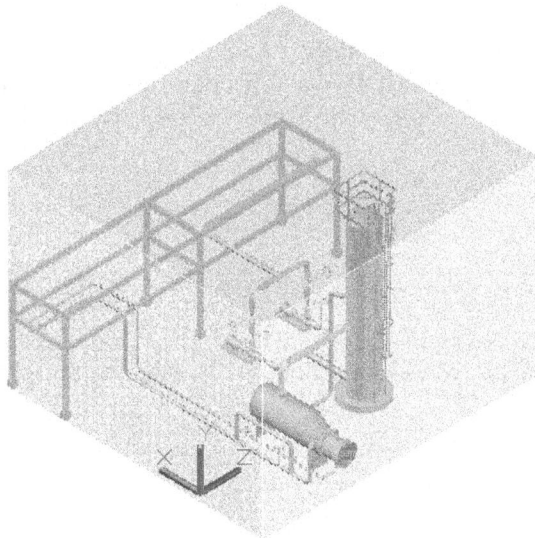

Figure 9-15 *The modified OrthoCube*

7. Select **1:100** from the **Scale** drop-down list.

    Next, you need to save the configured settings.

8. Choose the **Save Ortho Cube** tool from the **Library** panel; the **Save View** dialog box is displayed.

9. In this dialog box, enter **Top View** in the **View Name** edit box and choose the **OK** button; the view settings are saved. Now, these settings can be used later.

10. Choose the **OK** button from the **Create** panel in the **Ortho Editor** contextual tab; the view is attached to the cursor and you are prompted to specify the insertion of the view.

11. Place the view at the top right of the layout. Figure 9-16 shows the top view of the model.

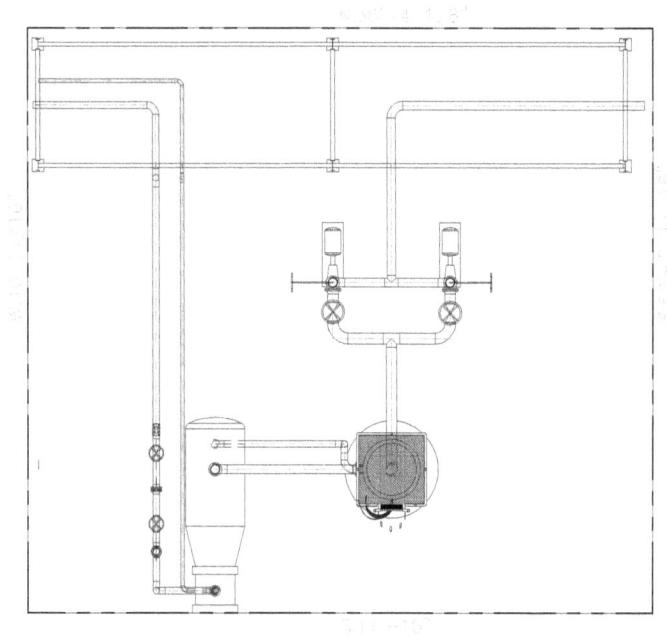

Figure 9-16 *Top View of the 3D model*

Creating the Adjacent Views

In this section, you will create front and right views.

1. Choose the **Adjacent Views** tool from the **Ortho Views** panel; you are prompted to select an orthographic view to create an adjacent view.

2. Select the view frame of the Top view; the **Create Adjacent View** dialog box is displayed.

3. In this dialog box, select the **Front** radio button and then choose the **OK** button; the front view is attached to the cursor and you are prompted to specify the insertion point of the view.

4. Place the view below the top view, refer to Figure 9-17. Note that you need to use object tracking to position the front view precisely with respect to the top view.

5. Next, invoke the **Adjacent Views** tool and create the right view of the model, refer to Figure 9-17. Note that you need to select the front to create this view.

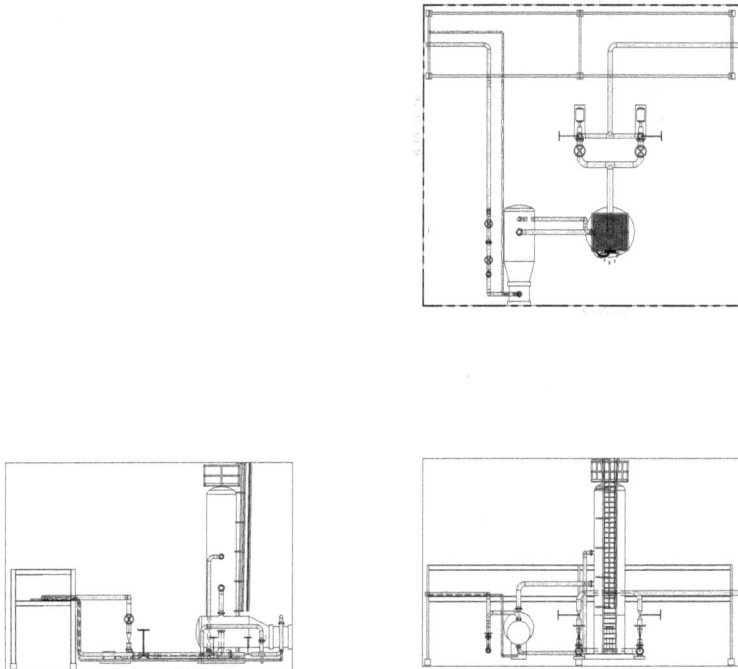

Figure 9-17 *Orthographic views of the 3D model*

Adding Annotations and Dimensions

In this section, you will add annotations and dimensions.

1. Choose the **Ortho Annotate** tool from the **Annotation** panel; you are prompted to select a component to annotate.

2. Zoom in to the **Top View** and then select the reboiler from the view; you are prompted to specify the annotation style.

3. Choose the **Equipment Annotation Tag** option from the Command Prompt; you are prompted to specify the annotation position.

4. Place the annotation below the reboiler, as shown in Figure 9-18.

5. Select the thin pipe line connecting the reboiler and enter **Full Line Number Callout**; the annotation is attached to the cursor. Place the annotation near the pipe line, as shown in Figure 9-19.

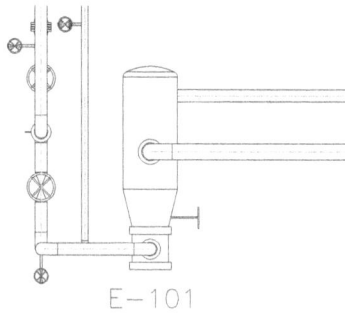

Figure 9-18 *The annotation placed below the reboiler*

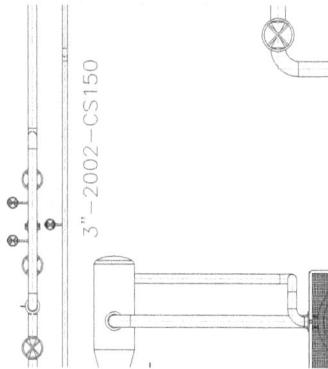

Figure 9-19 *The pipe being annotated*

Next, you need to create dimensions.

6. Choose the **Linear** tool from the **Dimension** drop-down in the **Dimension** panel; you are prompted to specify the first extension line origin.

7. Zoom to the reboiler in the **Front View** and select the end point of the pipe connecting the bottom nozzle of the reboiler, as shown in Figure 9-20.

8. Next, specify the second extension line origin, refer to Figure 9-20; the dimension line is attached to the cursor and you are prompted to specify its location.

9. Place the dimension line, as shown in Figure 9-21. Note that this dimension will be used later in the tutorial to explain the **Update** tool.

10. Choose the **Save** button from the **Quick Access toolbar**; the drawing file is saved.

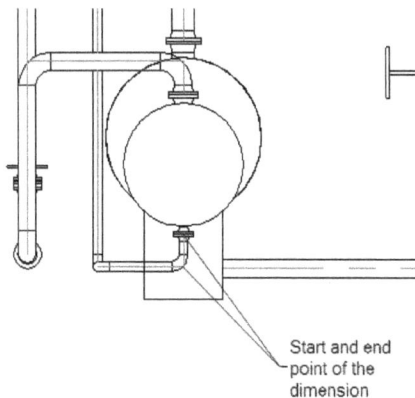

Start and end point of the dimension

Figure 9-20 *The origin points of the dimension extension lines*

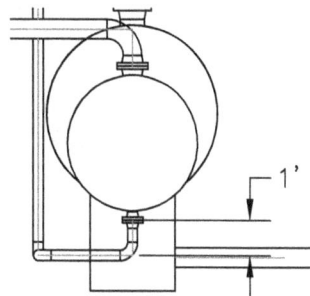

Figure 9-21 *Location of the dimension*

Updating the 3D Model

In this section, you will update the 3D model.

1. Choose the **Locate in 3D Model** tool from the **3D Model** panel; you are prompted to select a component to be located in the 3D model.

    Locate in
    3D Model

2. Select the horizontal pipe, as shown in Figure 9-22; it is located in the 3D model.

3. Select the highlighted pipe in the 3D model; the Move Part grip is displayed, as shown in Figure 9-23.

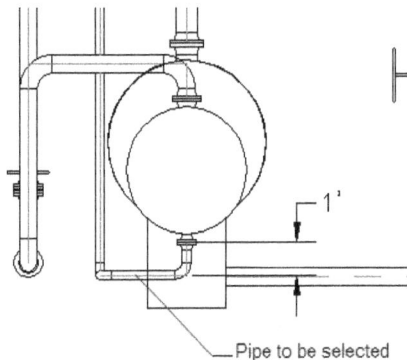

Figure 9-22 The pipe to be located in the 3D model

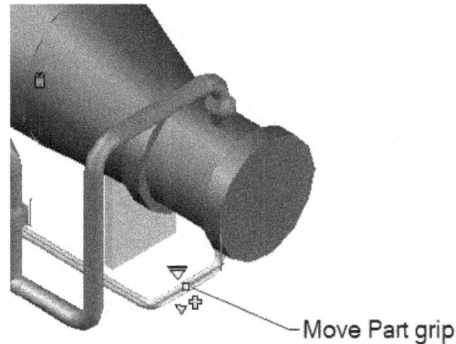

Figure 9-23 The Move Part grip displayed on the pipe

4. Select the Move Part grip and move the cursor downward and enter 6" at the Command prompt; the pipe is modified, as shown in Figure 9-24.

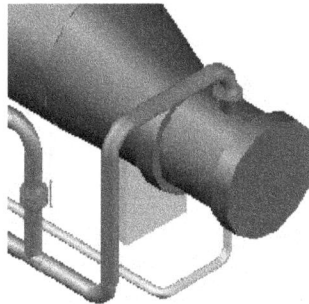

Figure 9-24 The pipe after modification

5. Choose the **Save** button on the Quick Access Toolbar.

Validating the Views and then Updating them

1. Right-click on **OrthoViews** under the **Orthographic Drawings** node in the **Orthographic**

DWG tab of the **Project Manager** and choose the **Validate Views** option, refer to Figure 9-25; all the views turn red and a message is displayed showing that the referenced models have changed.

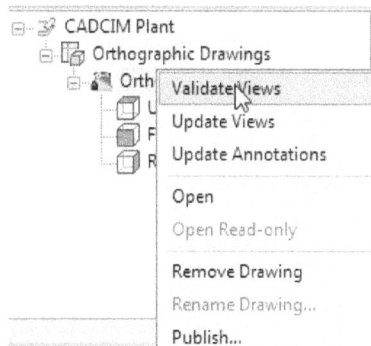

Figure 9-25 *Choosing the Validate Views option*

2. Right-click on **OrthoViews** in the **Project Manager** and choose the **Update Views** option; the views are updated.

Note that the dimension is not updated, refer to Figure 9-26. You need to manually update the dimension.

3. Select the origin point of the second extension line of the dimension and snap to endpoint of the pipe, as shown in Figure 9-27; the dimension value is modified.

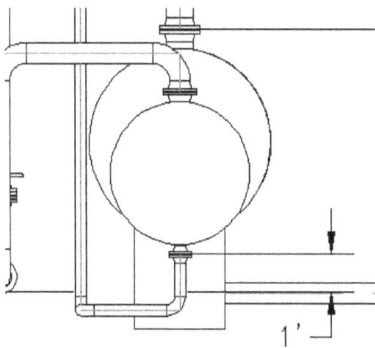

Figure 9-26 *The dimension not updated in the drawing view*

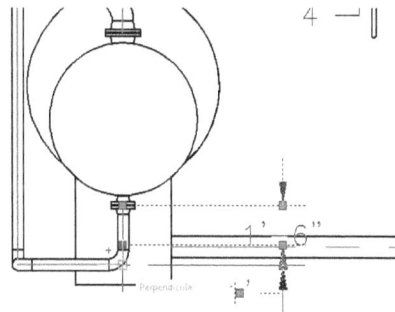

Figure 9-27 *Modifying the dimension*

4. Choose **Close > All Drawings** from the **Application** menu; the **AutoCAD** message box is displayed.

5. Choose the **Yes** button in this message box to omit the changes made to the files.

Self-Evaluation Test

Answer the following questions and then compare them to those given at the end of this chapter:

1. In the _____window, you need to use the _____to specify the boundaries of the view.

2. You can save the orthographic view settings by choosing the _____button.

3. The tools in the _____tab are used to perform operations such as editing, updating, annotating the drawing view, and so on.

4. You can annotate an equipment or a part by using the _____ button.

Review Questions

Answer the following questions:

1. You can use the _____ tool to locate a component in the 3D model.

2. On updating a drawing view, the dimensions will also get updated. (T/F)

3. You can also include all the 3D model files present in the current project, while creating the orthographic view. (T/F)

4. You can also scale the drawing view at the time of placing it. (T/F)

Answers to Self-Evaluation Test
1.Orthographic View Selection, OrthoCube, **2. Save Ortho Cube, 3. Ortho View, 4. Ortho Annotate.**

Chapter 10

Managing Data and Creating Reports

Learning Objectives

After completing this chapter, you will be able to:
- *Understand the concept of data available in the Data Manager*
- *View project and drawing data*
- *Create reports*
- *Export project data*
- *Import project data*

INTRODUCTION

In this chapter, you will learn to use the **Data Manager** to view the information in P&IDs and Plant 3D drawings, and export and import the data. In addition to that, you will be able to filter the data in the **Data Manager** so as to view a particular information.

Also, you will learn to generate, export, and import reports using the **Data Manager**.

DATA MANAGER

The **Data Manager** is used to view, modify, export, and import the project data. In addition to that, you can arrange the information within the **Data Manager** into different forms and then generate reports. To display the **Data Manager**, choose the **Data Manager** button from the **Project** panel in the **Home** tab. Figure 10-1 shows a typical **Data Manager** and its components.

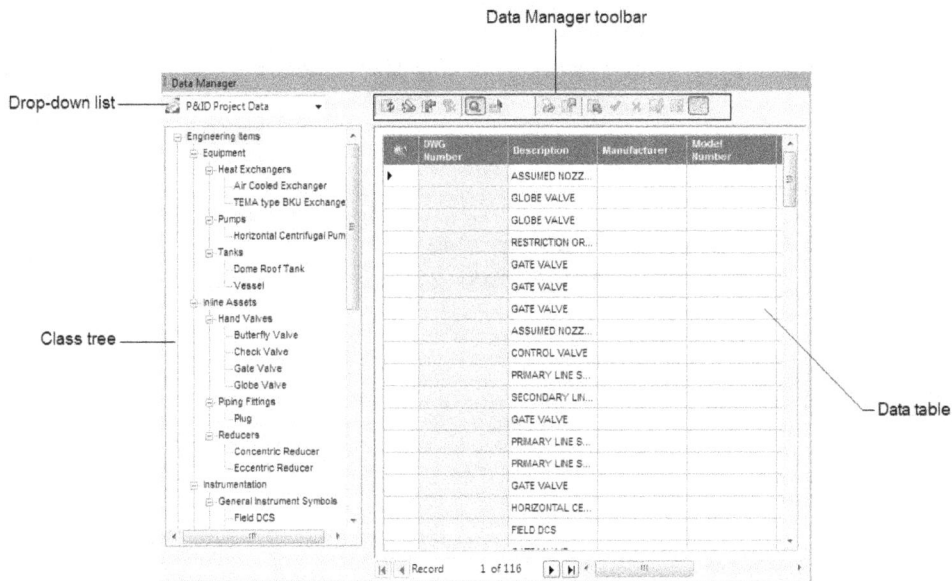

Figure 10-1 The Data Manager

The components of the **Data Manager** are discussed next.

Drop-down list

It is used to display different data types such as **Current Drawing data**, **Project data**, and **Project Reports**.

Class Tree

Using the Class tree, you can display the data related to a particular component of the drawing.

Data Manager toolbar

Using the Data Manager toolbar, you can perform tasks such as to refresh data, export and import data, print, and so on.

Data table

It is the place where the project data is arranged in rows and columns.

VIEWING DATA IN THE DATA MANAGER

In the **Data Manager**, you can view the project data related to Plant 3D and P&ID drawings. You can also view the project reports. To view Plant 3D data, first you need to open a Plant 3D model. Next, invoke the **Data Manager** and select the **Plant 3D Project Data** option from the drop-down list available at the top left; the Plant 3D project data will be displayed, as shown in Figure 10-2. Next, select the required class from the Class tree present on the left side of the **Data Manager**; the data related to the selected component will be displayed in the Data table.

To view the P&ID data, first you need to open a P&ID file. Next, invoke the **Data Manager** and select the **P&ID Project Data** option from the drop-down list in the **Data Manager**; the P&ID Project Data will be displayed in the **Data Manager**, as shown in Figure 10-3.

Figure 10-2 The Pant 3D Project data   *Figure 10-3* The P&ID Project data

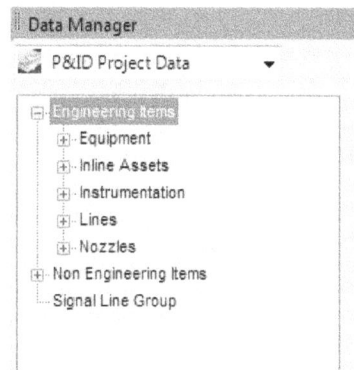

To view the data of the current drawing file, select the **Current Drawing Data** option from the drop-down list available in the **Data Manager**. Next, select the required class from the Class tree; the data related to the current drawing file is displayed.

MODIFYING THE DISPLAY OF DATA

You can modify the display of data by using the options available in the **Data Manager**. These options are discussed next.

Displaying the Data by Object Type and Area

You can display the data based on object type or area. To do so, select the **Order by Object type** or **Order by Area** option from the drop-down list located at the bottom of the Class tree, refer to

Figure 10-4; the data displayed will be modified according to the option selected. Figures 10-5 and 10-6 show the data displayed based on the object type and area, respectively.

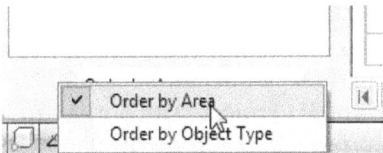

Figure 10-4   *The drop-down list located at the bottom of the Class tree*

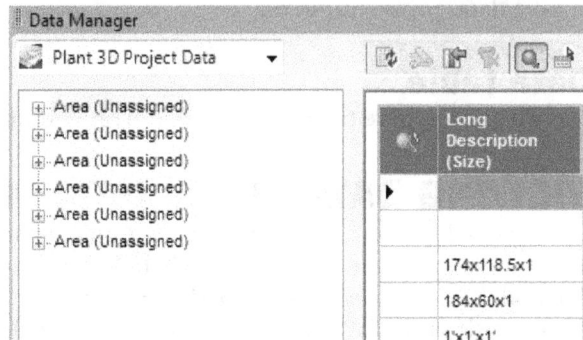

Figure 10-5   *The project data displayed based on area*

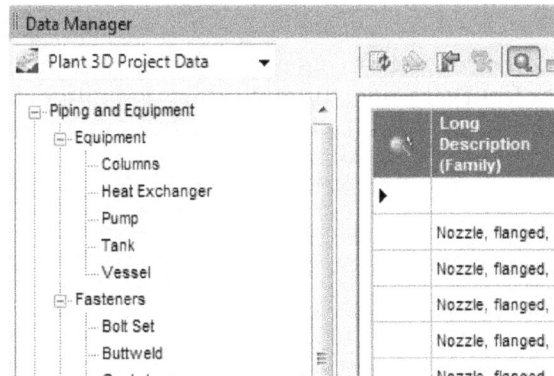

Figure 10-6   *The project data displayed based on the object type*

> **Tip.** *You can modify the display of nodes in the Class tree. To do so, right-click in the Class tree and choose the* **Show All Nodes** *or* **Show Only Nodes with Content** *option; the Class tree will be modified depending upon the option chosen.*

ZOOMING TO PLANT OBJECTS

You can zoom to an object in the drawing area by using the **Data Manager**. Similarly, you can locate the object in the **Data Manager** by selecting it from the drawing area. The various methods to zoom and scroll through objects in the **Data Manager** and locate them in the drawing area are discussed next.

Locating an Object in the Drawing Area

You can locate an object in the drawing area by using the **Data Manager**. To do so, select the **Current Project data** option from the drop-down list located at the top left of the **Data Manager**. Next, select the required component from the tree; the related data will be displayed in the spreadsheet. In the spreadsheet, click on the header row of the object, refer to Figure 10-7; the object will be located in the drawing area.

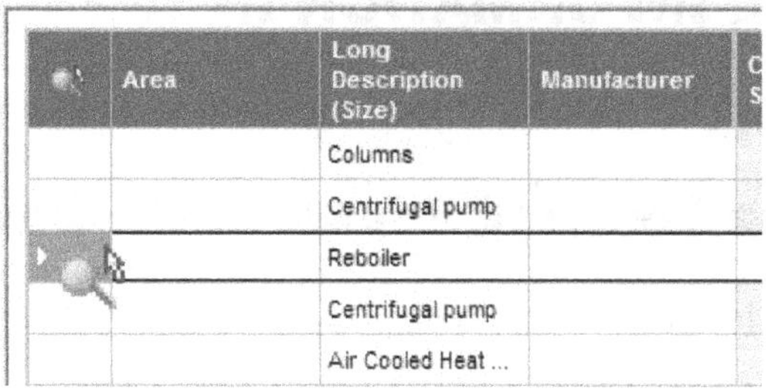

Figure 10-7 *Clicking in the header row*

Scrolling through the Data in the Data Manager

You can scroll through the objects in the **Data Manager** and locate them in the drawing area using the arrow buttons located at the bottom of the Data table, refer to Figure 10-8. On choosing the right arrow button, the next record will be highlighted in the Data table. Also, the object will be located in the drawing area. On choosing the left arrow button, the previous record will be highlighted in the Data table.

Figure 10-8 *Arrow buttons located at the bottom of the Data table*

Tip. *You can highlight the first or last object of the Data table by using the arrow buttons with a vertical line.*

EDITING THE DATA IN THE DATA MANAGER

You can edit the data in the **Data Manager** by manually changing the data or copying it from one cell to another. Note that you cannot change the value of cells displayed in shaded background.

PLACING ANNOTATIONS IN A P&ID USING THE DATA MANAGER

You can place annotations in a P&ID using the **Data Manager**. To do so, open the P&ID and then display the **Data Manager**. In the **Data Manager**, select the **Current Drawing Data** option from the drop-down list located at the top of the Class tree; the data related to the current drawing file will be displayed. Next, select the required class from the Class tree; the data related to the selected class will be displayed in the Data table. Click in the required cell in the Data table and drag the cursor into the drawing area; the object corresponding to the selected cell will be highlighted in the P&ID and you will be prompted to specify the location of the annotation. Specify the location of the annotation near the highlighted object.

FILTERING THE INFORMATION IN THE DATA TABLE

You can filter the data in the data table so that you can view only the required data and export it. The various methods to filter the information in the Data table are discussed next.

Viewing Only the Selected Record in the Data Table

To view the only the selected item in the data table, right-click on it and choose the **Filter by Selection** option from the shortcut menu displayed; all the records except the selected one will be hidden.

Viewing All the Records in the Data Table Except the Selected One

To view all the items in the data table except the selected one, right-click on it and choose the **Filter Excluding Selection** option from the shortcut menu displayed; all the records except the selected one will be displayed.

Viewing the Data of the Objects Selected in the Drawing Area

You can view the data of the objects selected in the drawing area. To do so, select the object from the drawing area, as shown in Figure 10-9. Next, choose the **Show Selected Items** button from the Data Manager, as shown in Figure 10-10; the object will be placed in the Data table, refer to Figure 10-11.

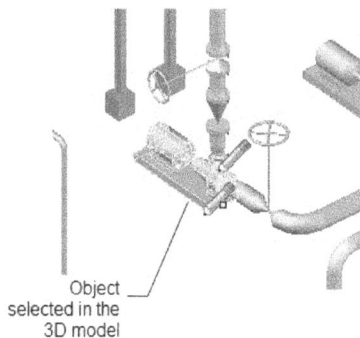

Figure 10-9 Object selected in the 3D model

*Figure 10-10 Choosing the **Show Selected Items** button*

*Figure 10-11 Object placed in the **Data Manager***

To remove all the filters, choose the **Remove Filter** button from the Data Manager toolbar; all the filters will be removed from the Data table.

EXPORTING DATA FROM THE DATA MANAGER

To export the data from the **Data Manager**, follow the steps given next:

1. Select a node from the Class tree and choose the **Export** button from the **Data Manager** toolbar; the **Export Data** dialog box will be displayed, as shown in Figure 10-12.

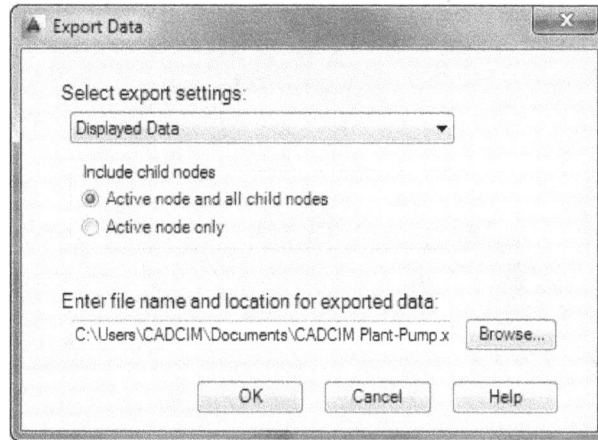

*Figure 10-12 The **Export Data** dialog box*

2. In this dialog box, select an option from the **Select export settings** drop-down list. Next, select the required option from the **Include child nodes** area. If you select the **Active node and all child nodes** radio button, the node selected from the Class tree and all the sub-nodes under it will be included while exporting the data. For example, if you select the **Equipment** node from the Class tree, all the sub-nodes under it will be included and separate worksheets will be created from each node. If you select the **Active node only** radio button, only the selected node will be included while exporting the data.

3. Choose the **Browse** button from the dialog box; the **Export To** dialog box will be displayed.

4. In this dialog box, specify the location of the exported file, enter its name and select the file type from the **File type** drop-down list.

5. Choose the **Save** button and then the **OK** button; the data will be exported to the file format selected in the **File type** drop-down list.

IMPORTING DATA TO THE DATA MANAGER

To import data to the **Data Manager**, follow the steps given next.

1. Select the node from the Class tree to which the data is to be imported and choose the **Import** button from the Data Manager toolbar; the **AutoCAD Plant** message box will be displayed showing that the log file for the accepted or rejected changes will be created in the same folder of the imported file.

2. Choose the **OK** button; the **Import From** dialog box will be displayed.

3. In this dialog box, select the required option from the **File type** drop-down list and browse to the location of the file from which you want to import the data.

4. Select the file and choose the **Open** button; the **Import Data** dialog box will be displayed, as shown in Figure 10-13.

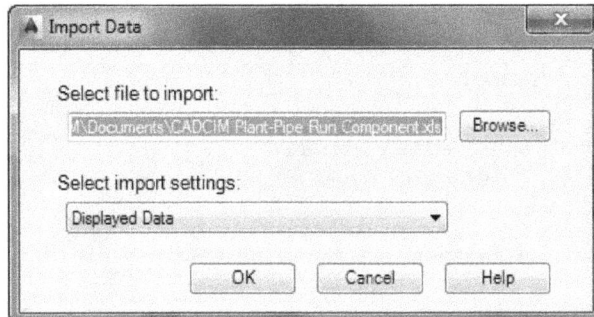

Figure 10-13 The **Import Data** *dialog box*

5. Choose the **OK** button; the data will be imported to the data table.

Accepting or Rejecting Changes in the Imported Data

The imported data may undergo some modifications. After importing the data you need to accept or reject the changes. To do so, first select the node from the Class tree to which the data is imported. You will notice that the changes made in the data are highlighted in yellow background. Choose the **Accept** button to accept the changes one by one or choose the **Accept All** button to accept all the changes at a time. Similarly, you can reject the changes by choosing the **Reject** or **Reject All** button.

VIEWING REPORTS IN THE DATA MANAGER

You can view, export, and import project peports using the **Data Manager**. To view Project Reports, click on the **Reports** button in the **Project Manager**; a flyout will be displayed. Choose the **Reports** option from the flyout; the project report data will be displayed in the **Data Manager**, as shown in Figure 10-14. In this section, you will learn how to import and export the project reports.

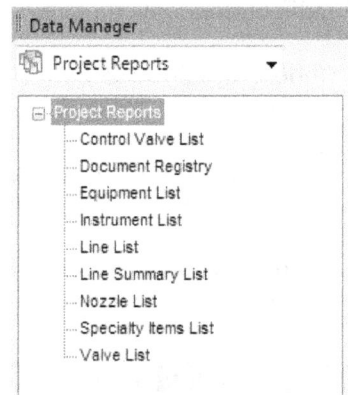

Figure 10-14 The project report data

Exporting the Project Reports

To export project reports, first make sure that the project reports data is displayed in the **Data Manager**. Next, choose the **Export** button from the **Data Manager**; the **Export Report Data** dialog box will be displayed, as shown in Figure 10-15. In this dialog box, select the check boxes from the **Reports** list box. Next, choose the **Browse** button; the **Export To** dialog box will be displayed. Select the file format from the **File type** drop-down list and specify the location of the file. Next, choose the **Save** button and then the **Export** button; the project report will be exported.

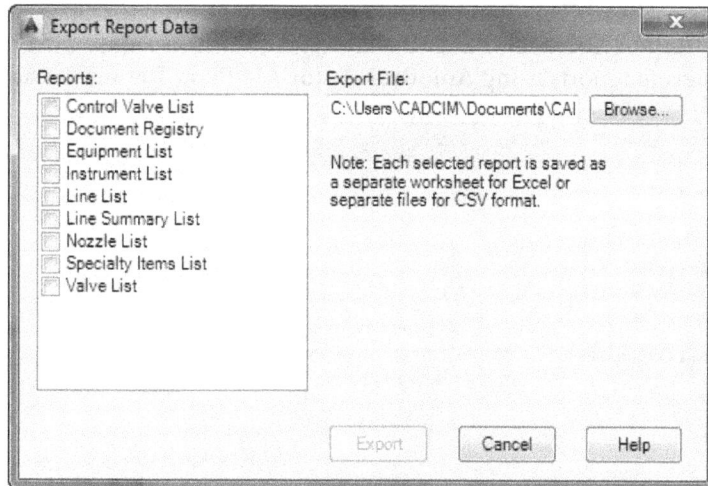

Figure 10-15 The Export Report Data dialog box

Importing the Project Reports

To import project reports, select the **Project Reports** node in the Class tree and choose the **Import** button; the **AutoCAD Plant** message box will be displayed showing that the log file for the accepted or rejected changes will be created in the same folder of the imported file. Choose the **OK** button; the **Import From** dialog box will be displayed. Select the type of file from which you want to import the data (*.xls,.xlsx,.csv*) and browse to the file location. Select the file and choose the **Open** button; the **Project Report Selection** dialog box will be displayed, as shown in Figure 10-16. In this dialog box, select the report to be imported from the **Select project report to import** drop-down list and choose the **OK** button; the selected report will be imported.

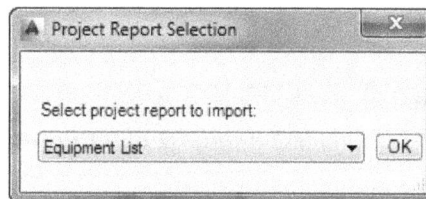

Figure 10-16 The Project Report Selection dialog box

WORKING WITH THE REPORT CREATOR

AutoCAD Plant 3D Report Creator is an additional application which comes along with **AutoCAD Plant 3D**. It is used to generate reports without starting the **AutoCAD Plant 3D** application. To start this application, choose **All Programs > Autodesk > AutoCAD Plant 3D 2014 > Autodesk AutoCAD Plant Report Creator 2014** from the **Start** menu; **Autodesk AutoCAD Plant Report Creator** will be displayed, as shown in Figure 10-17. Now, you will learn how to generate reports using **Autodesk AutoCAD Plant Report Creator**.

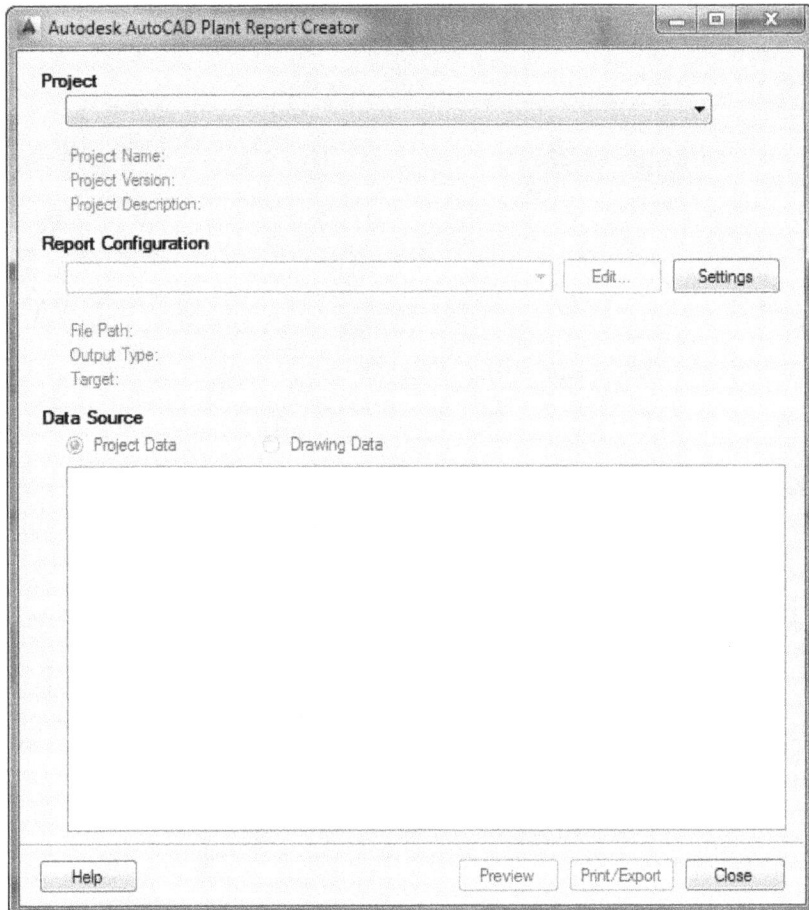

Figure 10-17 Autodesk AutoCAD Plant Report Creator

Generating Reports Using the AutoCAD Plant 3D Report Creator

To generate reports, invoke **Autodesk AutoCAD Plant Report Creator** and select the required project from the **Project** drop-down list. If it is not available in this drop-down list, select the **Open** option from this drop-down list; the **Open** dialog box will be displayed. Browse and select **Project.xml** file. Next, choose the **Open** button; the data configurations related to the selected project will be displayed in **Autodesk AutoCAD Plant Report Creator**. Select the required configuration from the **Report Configuration** drop-down list, refer to Figure 10-18. Select the **Project Data** or **Drawing Data** radio button from the **Data Source** area. If you select the **Drawing Data** radio button, you need to select a drawing or multiple drawings from the **Data Source** list box. Next, choose the **Preview** button; the **Preview** window will be displayed, as shown in Figure 10-19. Close the **Preview** window and choose the **Print/Export** button; the **Export Results** window will be displayed, as shown in Figure 10-20. Choose the **OK** button; the report will be exported to a PDF file and placed in the **Reports** folder of the respective project.

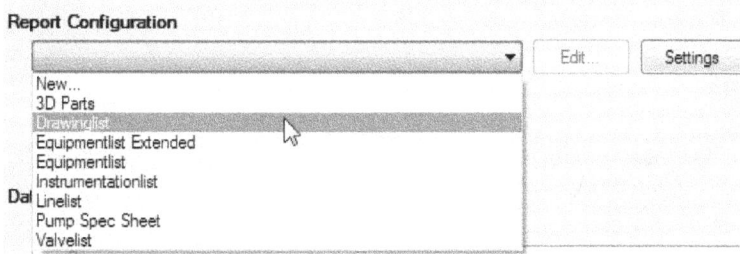

Figure 10-18 The Report Configuration drop-down list

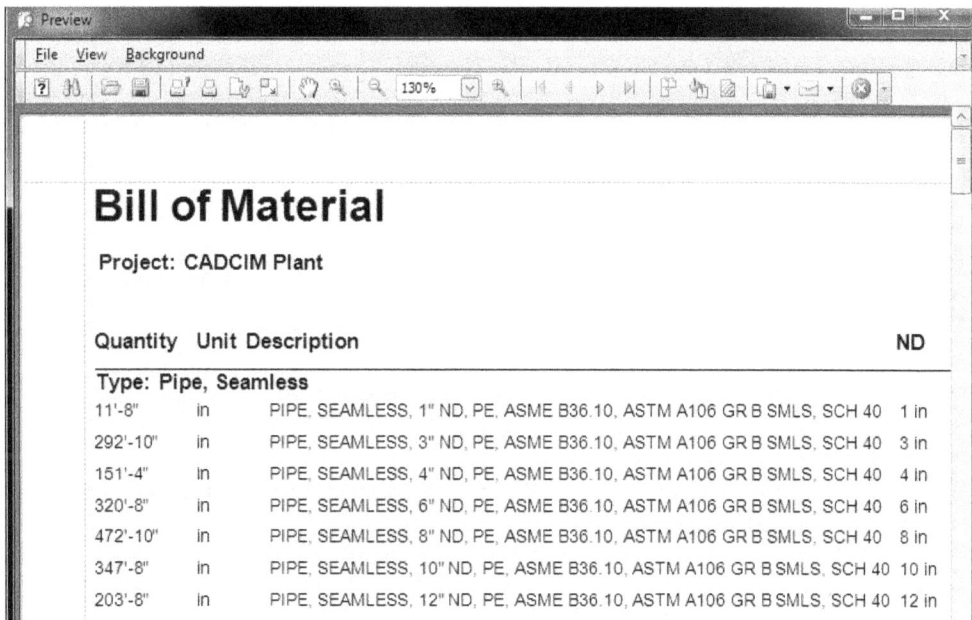

Figure 10-19 The Preview window

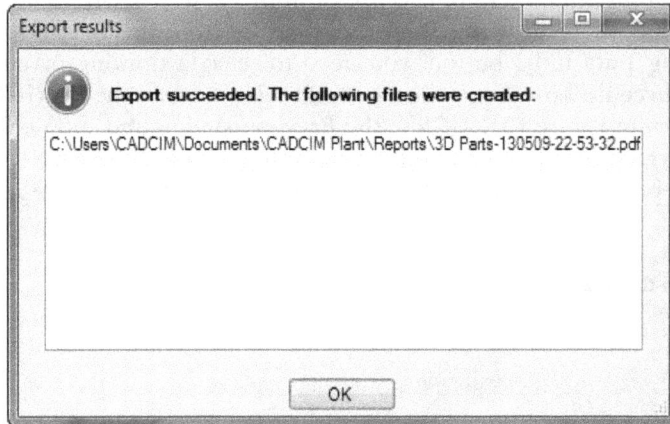

*Figure 10-20 The **Export results** window*

TUTORIALS

Tutorial 1

In this tutorial, you will filter the data and export it from the **Data Manager**. Next, you will modify the data, import it, and then apply the changes.   **(Expected time: 20 min)**

The following steps are required to complete this tutorial

a. Open the 3D model file.
b. Invoke the **Data Manager**.
c. Filter the data and export it.
d. Open the exported file and modify the data.
e. Import the modified data into the **Data Manager**.

Starting AutoCAD Plant 3D and Opening the Project

1. Start AutoCAD Plant 3D from the Start menu.

2. Invoke the **Project Manager** and select the **Open** option from the **Project** drop-down list; the **Open** dialog box is displayed.

3. In this dialog box, browse to the location *C:\Documents\CADCIM* and select the **Project.xml** file.

4. Choose the **Open** button; the **CADCIM** project is opened in the **Project Manager**.

5. Open the **Piping_Model.dwg** file from the **Plant 3D Drawings** node in the **Project Manager**.

Invoking the Data Manager and Viewing the Data

1. Choose the **Data Manager** button from the **Project** panel in the **Home** tab; the **Data Manager** will be displayed.

2. In the **Data Manager**, select the **Plant 3D Project Data** option from the drop-down list located at the top-left refer to Figure 10-21; the plant 3D project data is displayed.

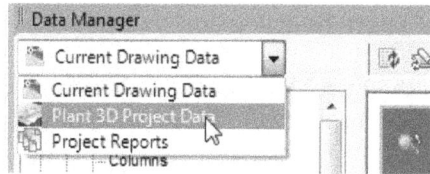

Figure 10-21 Selecting the **Plant 3D Project Data** option

3. Select the **Equipment** node from the Class tree; the data related to equipment is displayed in the Data table.

Filtering and Exporting the Data

1. Right-click in the **Centrifugal Pump** column in the Data table and choose the **Filter by Selection** option from the shortcut menu displayed, refer to Figure 10-22; only the Centrifugal Pump data is displayed in the Data table.

Figure 10-22 Choosing the **Filter By Selection** button

Next, you need to export the data.

2. Choose the **Export** button from the Data Manager toolbar, refer to Figure 10-23; the **Export Data** dialog box is displayed.

Figure 10-23 *Choosing the* **Export**
button

3.   Select the **Active node only** radio button from the **Include child nodes** area and choose the **Browse** button; the **Export To** dialog box is displayed.

4.   In this dialog box, select the **Excel Workbook (*.xlsx)** from the **Files of type** drop-down list, refer to Figure 10-24, and specify the file location as *C:\Users\Documents\CADCIM*.

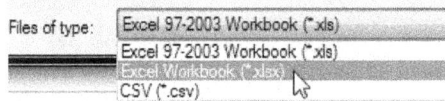

Figure 10-24 *Selecting the* **Excel Workbook**
option

5.   Choose the **Save** button; the **Export To** dialog box is closed.

6.   Choose the **OK** button from the **Export Data** dialog box; the data is exported to the excel file.

Modifying the Exported Data

1.   Browse to the location *C:\Users\Documents\CADCIM* and open the **CADCIM-Equipment.xlsx** file; the spreadsheet is displayed.

2.   In the spreadsheet, scroll to the **Spec** column and enter **CS150** in the first cell, refer to Figure 10-25.

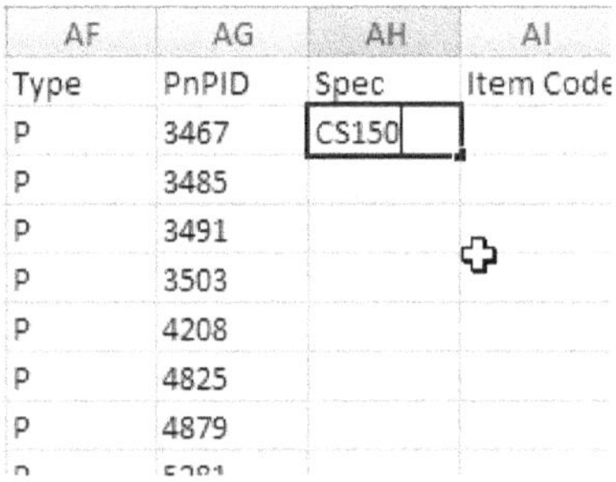

Figure 10-25 *Entering a value in the*
first cell of the **Spec** *column*

3.   Similarly, enter **CS150** in all cells of the **Spec** column.

4.   Save and close the excel file.

Importing the Data

1.   Switch to AutoCAD Plant 3D and open the **Data Manager**, if closed.

2.   In the **Data Manager**, make sure that **Plant 3D Project Data** is selected in the drop-down list located at the top and Equipment data is displayed in the Data table.

3.   Choose the **Import** button from the Data Manager toolbar, refer to Figure 10-26; the **AutoCAD Plant 3D** message box is displayed.

Figure 10-26 *Choosing the* **Import** *button*

4.   Choose the **OK** button; the **Import From** dialog box is displayed.

5.   Browse to the location *C:\Users\Documents\CADCIM* and double-click on the **CADCIM-Equipment.xlsx** file; the **Import Data** dialog box is displayed.

6.   Choose the **OK** button from this dialog box; the data in the excel file is imported into the **Data Manager**, refer to Figure 10-27. Also, the **Spec** column is highlighted with yellow background.

	Type	PnPID	Spec	Item Code
	P	3467	CS150	
	P	3485	CS150	
	P	3491	CS150	
	P	3503	CS150	
	P	4208	CS150	
	P	4825	CS150	
	P	4879	CS150	
	P	5281	CS150	

Figure 10-27 *The Data table after importing the data*

7. Choose the **Accept All** button from the Data Manager toolbar, to accept the changes made in the data, refer to Figure 10-28.

Figure 10-28 Choosing the Accept All button

8. Choose **Close > All Drawings** from the **Application menu** to close the drawing.

Tutorial 2

In this tutorial, you will use **Autodesk AutoCAD Plant Report Creator** to create reports.
(Expected time: 10 min)

The following steps are required to complete this tutorial:

a. Start **AutoCAD Plant 3D Report Creator**.
b. Select the required configuration.
c. Select the Data Source.
d. Export the data.

Starting AutoCAD Plant 3D Report Creator and Generating Reports

1. Choose **All Programs > Autodesk > AutoCAD Plant 3D 2014 > Autodesk AutoCAD Plant Report Creator 2014**; the **Autodesk AutoCAD Plant Report Creator** dialog box is displayed.

2. In this dialog box, select the **Open** option from the **Project** drop-down list; the **Open** dialog box is displayed.

3. Browse to the location *C:\Users\Documents\CADCIM* and double-click on the **Project.xml** file; the data sources related to the CADCIM project are displayed in the **Autodesk AutoCAD Plant Report Creator** dialog box.

4. Select the **Valvelist** option from the **Report Configuration** drop-down list.

5. Select the **Project Data** radio button from the **Data Source** area.

6. Choose the **Print/Export** button from the **Autodesk AutoCAD Plant Report Creator** dialog box; the **PDF Export Options** dialog box is displayed.

7. Enter the values in this dialog box, as shown in Figure 10-29, and choose the **OK** button; the report is exported to a PDF file and the **Export results** window is displayed.

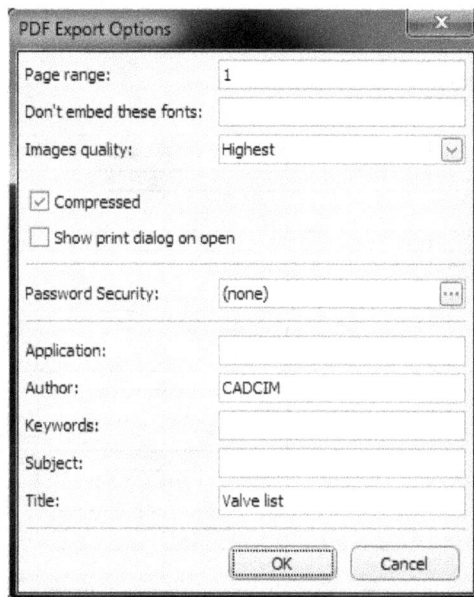

Figure 10-29 *The **PDF Export Options** dialog box*

8. In this window, double-click on the file path, refer to Figure 10-30; the PDF file containing the valve report is displayed, as shown in Figure 10-31.

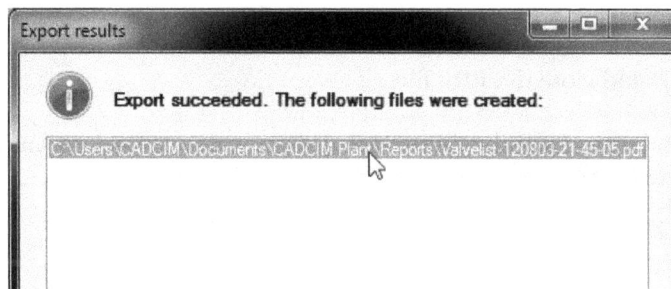

Figure 10-30 *Clicking on the file path in the **Export results** window*

Valvelist   CADCIM

Project: CADCIM Plant

Tag	Size	Spec	Manufacturer	Model No.	Supplier	Description
HA-101	8"	CS150				GATE VALVE
HA-102	8"	CS150				GATE VALVE
HA-103	8"	CS150				CHECK VALVE
HA-104	8"	CS150				CHECK VALVE
HA-105	8"	CS150				GATE VALVE
HA-106	8"	CS150				GATE VALVE
HA-107?	4"	CS150				GATE VALVE
HA-108?	4"	CS150				GATE VALVE
HA-110	8"	CS150				GATE VALVE
HA-111	8"	CS150				GATE VALVE
HA-112	4"	CS150				GLOBE VALVE
HA-113	4"	CS150				GLOBE VALVE
HA-114	12"	CS150				GATE VALVE
HA-115	12"	CS150				GATE VALVE
HA-116	10"	CS150				BUTTERFLY VALVE
HA-117	6"	CS150				GATE VALVE
HA-118	6"	CS150				GATE VALVE
HA-120	1"	CS150				BUTTERFLY VALVE
HA-135	1"	CS150				GATE VALVE
HA-136	1"	CS150				GATE VALVE
HA-142	10"	CS150				GATE VALVE
HA-143	10"	CS150				GATE VALVE
HA-144	8"	CS150				CHECK VALVE
HA-145	8"	CS150				CHECK VALVE
HA-146	8"	CS150				GATE VALVE
HA-147	8"	CS150				GATE VALVE
HA-148	6"	CS150				GATE VALVE
HA-149	1"	CS150				GATE VALVE
HA-150	1"	CS150				GATE VALVE
HA-151	4"	CS150				GATE VALVE

Figure 10-31 Partial view of the valve report in the PDF file

9. View the report and close the PDF file.

10. Choose the **OK** button from the **Export results** window and close the **Autodesk AutoCAD Plant Report Creator** dialog box.

Self-Evaluation Test

Answer the following questions and then compare them to those given at the end of this chapter:

1. The _____ is the location where all the project data is concentrated.

2. You can select the _____ option from the drop-down list located at the bottom of the Class tree to display the data based on area.

3. You can right-click on a selected item in the data table and choose the _____ option from the shortcut menu displayed to hide all the other items.

4. You can right-click on a selected item in the data table and choose the _____ option from the shortcut menu displayed to hide it from the data table.

5. You can choose the _____ button from the Data Manager to view the data related to the objects selected in the drawing area.

Review Questions

Answer the following questions:

1. If you select the _____ radio button, the node selected from the Class tree and all the sub-nodes under it will be included while exporting the data.

2. You can export the data from the Data Manager into an _____ or _____ file.

3. Changes made in the data are highlighted in yellow background. (T/F)

4. Report Creator is used to generate reports without starting AutoCAD Plant 3D application. (T/F)

5. You can view P&ID data in the Data Manager without opening a P&ID file. (T/F)

Answers to Self-Evaluation Test
1. Data Manager, 2. Order by Area, 3. Filter by Selection, 4. Filter Excluding Selection,
5. Show Selected Items

Index

Other Publications by CADCIM Technologies

The following is the list of some of the publications by CADCIM Technologies. Please visit www.cadcim.com for the complete listing.

Autodesk Inventor Textbooks
- Autodesk Inventor 2013 for Designers
- Autodesk Inventor 2012 for Designers
- Autodesk Inventor 2011 for Designers
- Autodesk Inventor 11 for Designers

Solid Edge Textbooks
- Solid Edge ST5 for Designers
- Solid Edge ST4 for Designers
- Solid Edge ST3 for Designers
- Solid Edge ST2 for Designers

NX Textbooks
- NX 8.5 for Designers
- NX 8 for Designers
- NX 7 for Designers

SolidWorks Textbooks
- SolidWorks 2013 for Designers
- SolidWorks 2012: A Tutorial Approach
- Learning SolidWorks 2011: A Project based Approach
- SolidWorks 2012 for Designers

EdgeCAM Textbooks
- EdgeCAM 11.0 for Manufacturers
- EdgeCAM 10.0 for Manufacturers

CATIA Textbooks
- CATIA V5-6R2012 for Designers
- CATIA V5R21 for Designers
- CATIA V5R20 for Designers

Creo Parametric and Pro/ENGINEER Textbooks
- Creo Parametric 2.0 for Designers
- Creo Parametric 1.0 for Designers
- Pro/ENGINEER Wildfire 5.0 for Designers

Autodesk Alias Textbooks
• Learning Autodesk Alias Design 2012
• Learning Autodesk Alias Design 2010

ANSYS Textbooks
• ANSYS Workbench 14.0: A Tutorial Approach
• ANSYS 11.0 for Designers

Customizing AutoCAD Textbooks
• Customizing AutoCAD 2013

AutoCAD LT Textbooks
• AutoCAD LT 2013 for Designers
• AutoCAD LT 2012 for Designers

AutoCAD Electrical Textbooks
• AutoCAD Electrical 2013 for Electrical Control Designers
• AutoCAD Electrical 2012 for Electrical Control Designers
• AutoCAD Electrical 2011 for Electrical Control Designers

Autodesk Revit Architecture Textbooks
• Autodesk Revit Architecture 2013 for Architects and Designers
• Autodesk Revit Architecture 2012 for Architects and Designers
• Autodesk Revit Architecture 2011 for Architects and Designers

Autodesk Revit Structure Textbooks
• Exploring Autodesk Revit Structure 2014
• Exploring Autodesk Revit Structure 2013

AutoCAD Civil 3D Textbooks
• Exploring AutoCAD Civil 3D 2013
• Exploring AutoCAD Civil 3D 2012
• AutoCAD Civil 3D 2009 for Engineers

AutoCAD Map 3D Textbooks
• Exploring AutoCAD Map 3D 2013
• Exploring AutoCAD Map 3D 2012
• Exploring AutoCAD Map 3D 2011

AutoCAD MEP Textbooks
• AutoCAD MEP 2014 for Designers

3ds Max Design Textbooks
- Autodesk 3ds Max Design 2013: A Tutorial Approach
- Autodesk 3ds Max Design 2012: A Tutorial Approach
- Autodesk 3ds Max Design 2011: A Tutorial Approach

3ds Max Textbooks
- Autodesk 3ds Max 2013: A Comprehensive Guide
- Autodesk 3ds Max 2012: A Comprehensive Guide
- Autodesk 3ds Max 2011: A Comprehensive Guide

Maya Textbooks
- Autodesk Maya 2013: A Comprehensive Guide
- Autodesk Maya 2012: A Comprehensive Guide
- Autodesk Maya 2011: A Comprehensive Guide

Fusion Textbook
- The eyeon Fusion 6.3: A Tutorial Approach

Flash Textbook
- Adobe Flash Professional CS6: A Tutorial Approach

Premiere Textbooks
- Adobe Premiere Pro CS6: A Tutorial Approach
- Adobe Premiere Pro CS5.5: A Tutorial Approach

Computer Programming Textbooks
- Learning Oracle 11g: A PL/SQL Approach
- Learning ASP.NET AJAX
- Learning Java Programming
- Learning Visual Basic.NET 2008
- Learning C++ Programming Concepts
- Learning VB.NET Programming Concepts

Paper Craft Book
- Constructing 3-Dimensional Models: A Paper-Craft Workbook

AutoCAD Textbooks Authored by Prof. Sham Tickoo and Published by Autodesk Press
- AutoCAD: A Problem-Solving Approach: 2013 and Beyond
- AutoCAD 2012: A Problem-Solving Approach
- AutoCAD 2011: A Problem-Solving Approach
- AutoCAD 2010: A Problem-Solving Approach
- Customizing AutoCAD 2010

Textbooks Authored by CADCIM Technologies and Published by Other Publishers

3D Studio MAX and VIZ Textbooks
• Learning 3ds max5: A Tutorial Approach
• Learning 3ds Max: A Tutorial Approach, Release 4
  Goodheart-Wilcox Publishers (USA)
• Learning 3D Studio VIZ: A Tutorial Approach
  Goodheart-Wilcox Publishers (USA)
• Learning 3D Studio R4: A Tutorial Approach
  Goodheart-Wilcox Publishers (USA)

Coming Soon from CADCIM Technologies
• AutoCAD 2014: A Problem-Solving Approach
• Autodesk Inventor 2014 for Designers
• Creo Direct 2.0 and Beyond for Designers
• Autodesk Simulation Mechanical 2014 for Designers
• NX Nastran 8.5 for Designers
• The Foundry NukeX 7 for Compositors
• Pixologic ZBrush 4R5: A Comprehensive Guide
• Autodesk 3ds Max Design 2014: A Tutorial Approach
• Autodesk 3ds Max 2014: A Comprehensive Guide
• Autodesk Softimage 2014: A Tutorial Approach
• Exploring Autodesk Revit Architecture 2014 for Architects and Designers
• Autodesk Maya 2014: A Comprehensive Guide
• Exploring Autodesk Navisworks 2014
• Exploring AutoCAD Civil 3D 2014

Online Training Program Offered by CADCIM Technologies
CADCIM Technologies provides effective and affordable virtual online training on various software packages including computer programming languages, Computer Aided Design, Manufacturing, and Engineering (CAD/CAM/CAE), animation, architecture, and GIS. The training will be delivered 'live' via Internet at any time, any place, and at any pace to individuals as well as the students of colleges, universities, and CAD/CAM training centers. For more information, please visit the following link: *http://www.cadcim.com*

www.ingramcontent.com/pod-product-compliance
Lightning Source LLC
Chambersburg PA
CBHW082138210326
41599CB00031B/6020